Fundamentals of
Enhanced Oil Recovery

Fundamentals of Enhanced Oil Recovery

H. K. van Poollen and Associates, Inc.

PennWell Books
Division of
PennWell Publishing Company
Tulsa, Oklahoma

Copyright © 1980 by
PennWell Publishing Company
1421 South Sheridan Road/P. O. Box 1260
Tulsa, Oklahoma 74101

Library of Congress cataloging in publication data

van Poolen (H. K.) and Associates.
Fundamentals of enhanced oil recovery.

Includes bibliographies.
1. Secondary recovery of oils.   I.   Title.
TN871.V29   1981   622'.3382   80-21106
ISBN 0-87814-144-8

All rights reserved. No part of this book may be
reproduced, stored in a retrieval system, or
transcribed in any form or by any means, electronic
or mechanical, including photocopying and recording,
without the prior written permission of the publisher.

Printed in the United States of America

2 3 4 5   84 83 82 81

Preface

For the last two decades, scientists have been searching for techniques to recover more oil from "depleted" reservoirs, which still contain as much as 50% of the original oil in place. More than 100 billion Bbls of oil may still remain in these depleted reservoirs in the U.S. alone.

During the next decade, world production capabilities by conventional means will not meet energy demands. Therefore, oil prices will continue to soar. Some speculate that the soaring oil prices could make enhanced oil recovery very economically attractive, and that could be the beginning of an era when unconventional petroleum sources become economic. These potential oil resources include oil shale, tar sands, heavy oils, and enhanced recovery from known reservoirs.

Viewed from this perspective, the future of the petroleum industry is indeed bright, even in the face of dwindling new discoveries. Although rising oil and gas prices could improve the economic climate for enhanced oil recovery, they could also pave the way for massive development of nonfossil energy resources — solar, nuclear, and geothermal — especially since operating costs for enhanced oil recovery increase with rising oil prices.

Although petroleum is a very convenient and adaptable energy resource, it is nonrenewable. However, through enhanced oil recovery techniques, it would be possible to increase or maintain the present levels of production for many years to come. Enhanced oil recovery techniques must, therefore, be developed to their full potential in order to supply the needed energy demands. This can only be accomplished with a free exchange of theoretical, experimental, and field application information pertaining to each enhanced oil recovery process.

Fundamentals of Enhanced Oil Recovery brings together published information on this vast subject. The technical aspects of each recognized process are thoroughly reviewed and general criteria for process selection are presented.

All of the methods discussed in this book are aimed at producing oil economically by these processes or, better yet, at a monetary profit. That means short payout times. EOR will make unmoveable oil moveable, but for all except a few processes great monetary gains may be doubtful.

This book is a result of the concerted efforts of a number of individuals. From H. K. van Poollen and Associates Inc., Dr. M. A. Sabet wrote the chapters on thermal process and miscible hydrocarbon methods; J. C. Baptist wrote the chapters on carbon dioxide and inert gas injection. From EXOIL Services, H. Surkalo wrote the chapters on chemical processes.

H. K. van Poollen

Table of Contents

Preface
Table of Contents
Executive Summary x
Introduction xvii

Part I — Thermal Processes

   Steam Stimulation 3
      Introduction 3
      Technical Discussion 3
         Model Studies 3
         Experimental Studies 7
      Case Histories 8
         Tia Juana Field, Venezuela 8
         Duri Field, Indonesia 10
         Huntington Beach Field, California (USA) 10
      Bibliography 12
   Steam Flooding (including Hot Water Injection) 14
      Introduction 14
      Technical Discussion 15
         Properties of Steam 15
         Model Studies 16
         Experimental Studies 23
         Wellbore Heat Loss 24
         Surface Heat Loss 29
         Oil Recovery Estimates 30
      Conclusions 31
      Case Histories 31
         South Belridge Field, California (USA) 32
         Tia Juana Field, Venezuela 33
         Schoonebeek Field, The Netherlands 34
         Cat Canyon Field, California (USA) 36
      Bibliography 38
   In Situ Combustion 41
      Introduction 42
      Technical Discussion 43
         Ignition 43
         Air Requirements 45
         Experimental Studies 47
         Model Studies 48
         Economic Evaluation 49
         Oil Recovery 49
      Conclusions 49
      Case Histories 50
         Sloss Field, Nebraska (USA) 50
         Miga Field, Venezuela 50
         Heidelberg Field, Mississippi (USA) 51
      Bibliography 52

Part II — Chemical Processes

   Surfactant-Polymer Injection 58

Introduction	58
Technical Discussion	59
Surfactant Slug	59
External Phase	60
Mobility Control	60
Surfactant Adsorption	61
Petroleum Sulfonates	63
Phase Diagrams	65
Laboratory Design	66
Description of Various Surfactant-Polymer Processes	69
Field Handling Systems	71
Water Treating System	71
Surfactant Slug Blending System	71
Polymer Blending System	71
Fluid Injection System	73
Case Histories	73
Robinson 119-R Project, Illinois (USA)	74
Robinson 219-R Project, Illinois (USA)	74
Robinson M-1 Project, Illinois (USA)	74
North Burbank Project, Oklahoma (USA)	75
Bibliography	75
Polymer Flooding	83
Introduction	83
Technical Discussion	83
Reservoir Heterogeneity	83
Mobility Ratios	83
Polymer Chemistry	84
Displacement Mechanisms	85
Reservoir Selection	93
Laboratory Design	94
Feasibility Analysis	94
Field Handling Systems	97
Polymer Blending System	97
Fluid Injection System	98
Case Histories	98
East Coalinga Field, California (USA)	98
North Stanley Field, Oklahoma (USA)	98
Bibliography	99
Caustic Flooding	104
Introduction	104
Technical Discussion	104
Rock Properties	104
Displacement Mechanisms	105
Chemicals	105
Laboratory Design	106
Laboratory Core Flooding	106
Field Operations	107
Case Histories	107
Singleton Field, Nebraska (USA)	107
North Ward-Estes Field, Texas (USA)	107
Midway-Sunset Field, California (USA)	107
Bibliography	107

Part III — Miscible Displacement Processes

Miscible Hydrocarbon Displacement	114
Introduction	114
Technical Discussion	118

Experimental Studies	118
Model Studies	121
Conclusions	121
Case Histories	121
Midway-Sunset Field, California (USA)	121
Pembina Field, Canada	122
Levelland Field, Texas (USA)	125
Ante Creek Field, Canada	127
Block 31 Field, Texas (USA)	128
Swanson River Field, Alaska (USA)	129
Hassi-Messaoud Field, Algeria	129
Bibliography	130
Carbon Dioxide Injection	132
Introduction	132
Technical Discussion	132
Criteria for Application of Carbon Dioxide Injection	132
Phase Behavior and Miscibility	133
Displacement Mechanisms	134
Project Design	135
Sources of Carbon Dioxide	136
Conclusions	136
Case Histories	137
Kelly-Snyder Field, Texas (USA)	138
Crossett Field, Texas (USA)	139
Mead-Strawn Field, Texas (USA)	140
Wasson Field, Texas (USA)	141
Bibliography	141
Inert Gas Injection	146
Introduction	146
Technical Discussion	146
Criteria for Application	146
Phase Behavior and Miscible Displacement	146
Conclusions	147
Advantages of Inert Gas Injection Process	147
Disadvantages of Inert Gas Injection Process	148
Case Histories	148
Block 31 Field, Texas (USA)	148
Hawkins Field, Texas (USA)	148
Bibliography	149

Part IV — Conclusions

State of the Art	154

Executive Summary

The *Enhanced Oil Recovery Digest* contains a review of recognized enhanced recovery methods reported in the literature. Recovery processes have been subdivided into three major categories and presented as follows:

Part I   THERMAL PROCESSES
              Steam Stimulation
              Steam Flooding (Including Hot Water Injection)
              In Situ Combustion

Part II   CHEMICAL PROCESSES
              Surfactant-Polymer Injection
              Polymer Flooding
              Caustic Flooding

Part III   MISCIBLE DISPLACEMENT PROCESSES
              Miscible Hydrocarbon Displacement
              Carbon Dioxide Injection
              Inert Gas Injection

A technical review and case history description are provided for each enhanced oil recovery method. In addition, a bibliography is presented. A brief summary of each enhanced recovery technique follows.

ENHANCED RECOVERY METHODS

Steam Stimulation

Steam stimulation is also known as: cyclic steam injection, steam soak or huff and puff. In this process steam (± 1000 Bbls/day) is injected into a producing well for a specified period of time (normally 2-3 weeks). Following this, the well is shut-in for a few days (to allow sufficient heat dissipation) and then placed on production. Heat from the injected steam increases the reservoir temperature, resulting in a pronounced increase in mobility of heavy oils and a corresponding improvement in producing rates. Other positive benefits that may contribute to production stimulation include: 1) thermal expansion of fluids; 2) compression of solution gas; 3) reduced residual oil saturation; 4) wellbore cleanup effects. The technique has gained wide acceptance because quick payouts result from successful applications, but many project failures have also been reported (due principally to improper design).

Many initial applications result in production increases considerably greater than predicted by model studies. This is mainly due to well cleanup and permeability improvement in the vicinity of the wellbore. Many oil wells are given successive treatments; however, in almost all reported cases, production response to a second stimulation cycle is usually less than first cycle, and the period of each cycle is usually shorter.

This enhanced oil recovery technique is primarily a stimulation method. Many operators, after the application of several steaming cycles, will convert to a steam flooding project.

Steam Flooding (Including Hot Water Injection)

Steam flooding is a process similar to waterflooding. A suitable well pattern is chosen and steam is injected into a number of wells while the oil is produced from adjacent wells. Ideally, the steam forms a saturated zone around the injection well. The temperature of this zone is nearly equal to that of the injected steam. As the steam moves away from the well, its temperature drops as it continues to expand in response to pressure drop. At some distance from the well, the steam condenses and forms a hot water bank. In the steam zone, oil is displaced by steam distillation and gas (steam) drive. In the hot water zone, physical changes in the characteristics of the oil and reservoir rock take place and result in oil recovery. These changes are thermal expansion of the oil, reduction of viscosity and residual saturation and changes in relative permeability.

Most steamfloods are developed on a ± 5 acre spacing but other physical parameters vary widely from project to project. Initial investment requirements are substantial, so care must be taken in the design and implementation of a project. Excessive heat losses (surface and reservoir) and mechanical failures must be avoided.

Steam flooding is currently the principal enhanced oil recovery method. The number of steam injection wells in California alone has risen from 140 in 1966 to 1630 in 1975. More impressive yet is the daily oil recovery by steam injection which amounts to 405 MBPD worldwide (U.S. 240, Venezuela 148, Canada 9 and others 8). This is compared with 55 MBPD by steam stimulation plus in situ combustion and 221 MBPD by all other enhanced recovery methods combined.

An advantage of steam injection over other enhanced oil recovery methods is that steam can be applied to a wide variety of reservoirs. Two limiting factors are: depth (less than 5000 ft) and reservoir thickness (greater than 10 ft). The depth limitation is imposed by the critical pressure of steam (3202 psia); the reservoir thickness is determined by the rate of heat loss to base and cap rock.

Design of steamflood projects requires a clear understanding of steam properties and the physical mechanisms involved in oil displacement by both steam and water. It is also necessary to be able to estimate heat losses in order to properly calculate the capacities of needed steam generating equipment.

Hot water injection has been attempted in reservoirs containing oils in the viscosity range of 100-1000 cp. Because of the excessive heat losses on the surface, in the wellbore and in the reservoir rock, steam injection is generally preferred. Furthermore, field tests of hot water flooding have frequently been unsuccessful because of viscous fingering and low volumetric sweep efficiency.

In Situ Combustion

There are two fundamentally different processes of in situ combustion: 1) forward combustion and 2) reverse combustion. In forward combustion, the reservoir is ignited in the vicinity of an air injection well, and the combustion front propagates away from the well. Continued injection of air drives the combustion zone through the reservoir to nearby producing wells.

The reverse combustion process is started in the same manner as forward combustion, but (after burning out a short distance from the ignition wells) air injection is switched to adjacent wells. This drives the oil towards the previously ignited well, while the combustion front travels in the opposite direction toward the adjacent wells. The process was developed as a method for improving recovery in reservoirs containing extremely heavy crudes.

In situ combustion has a very broad application, having been successfully conducted in reservoirs with a wide range of physical characteristics. Several projects have been technically successful but unprofitable. Project economics must be accurately determined prior to committing funds for in situ combustion projects.

There are many variations of the forward in situ combustion process. Wet and partially quenched combustion are two of these variations. Although these methods have not progressed beyond laboratory experimentations and pilot field studies, the consensus is that they offer significantly higher potential than dry combustion. In dry combustion, considerable heat is left in the burned zone to dissipate through the reservoir cap and base rock. However, when water in moderate amounts is injected simultaneously with the air, it flashes into superheated steam a short distance from the injection well. As the superheated steam (mixed with air) reaches the combustion front, only the oxygen is utilized in the burning process. Upon crossing the combustion front, the superheated steam is mixed with nitrogen from the air and flue gas consisting mainly of CO and CO_2. This mixture of gases displaces the oil in front of the combustion zone. Laboratory results showed that optimal wet combustion requires only about one-third as much air as that needed for dry combustion.

The main factors which govern the volume of air required for in situ combustion are: amount of fuel content (coke) of the oil being burned and the efficiency of oxygen utilization. The combustion zone can only move as fast as it depletes the coke. If the amount of coke deposited by the crude oil is excessive, the rate of advance of the combustion zone would be slow and air requirement would be high. However, if sufficient fuel is not available the combustion process cannot be sustained.

Proper design of an in situ combustion project depends on the correct evaluation of a host of parameters. Laboratory studies for fuel content and air requirements will establish the feasibility of initiating a field test.

Surfactant-Polymer Injection

This process is conducted in two steps: 1) injection of a surfactant slug and 2) injection of a polymer mobility buffer. The surfactant slug has also been referred to as micellar solution, microemulsion, soluble oil and swollen micelle. The primary purpose of surfactants is to lower interfacial tension and displace oil that cannot be displaced by water alone. The purpose of the polymer is to provide mobility control for more effective piston-like displacement.

In practice, surfactant-polymer systems have been difficult to control and predict in actual field applications. Because of continual chemical reaction and change that occurs with slug movement through the reservoir, laboratory simulations often do not represent total field conditions. The surfactant-polymer process has excellent potential but more field testing is required.

There are basically two different surfactant processes: high concentration surfactant slugs and low concentration surfactant slugs. In addition to the surfactant concentration, these slugs contain various amounts of cosurfactants (alcohols), water, hydrocarbons and salts. These variations in concentrations of ingredients result in an infinite number of slug formulations.

Adsorption of surfactants on the reservoir rock is an important consideration in slug design. It is one of the major causes of slug breakdown. The higher equivalent weight sulfonates, which are primarily responsible for the lowering of interfacial tension, are readily adsorbed. As this occurs the slugs ability to mobilize oil deteriorates.

Some authors have suggested that injection of a preflush ahead of the surfactant slug will reduce adsorption. Laboratory core flooding has demonstrated that this technique is a viable one.

Most of the work on surfactant-polymer design for field application considers visual techniques to determine the solubilization of reservoir oil and brine in the surfactant slug. Interfacial tension measurements are also determined and used to screen various slugs. All companies use oil recovery as a measure of the effectiveness of a surfactant-polymer system. Although Berea rock is used in testing, the final design work is usually done in the reservoir rock.

Although much laboratory work has been done on this process, no field project has as yet been reported as economic. This process has the potential to recover a significant portion of the residual oil-in-place. Continued research and field application could result in routine application of this process in the future.

Polymer Flooding

The polymer flooding technique involves the addition of a thickening agent (polymer) to the injected water to increase its viscosity. Although ultimate residual oil saturation is essentially unaffected, the addition of a polymer yields two benefits: 1) reduces the total volume of water required to reach the ultimate residual oil saturation and 2) increases sweep efficiency due to improved mobility ratio. The economic success of reported projects has varied considerably. Successful projects have yielded increased recoveries of 5-15%. Many unprofitable projects were the result of inadequate reservoir description prior to startup.

Polymers now used in oil recovery are divided into two classes, polyacrylamides and polysaccharrides.

The polyacrylamides come in a variety of forms, ranging in molecular weight from a few hundred thousand to many millions. They may be purchased in a dry, emulsion or gel state. Product costs range from $1.50 to $2.25 per pound and are generally used at concentrations of 50-1000 ppm. Polyacrylamides decrease the mobility of the injected fluid by decreasing the permeability of the reservoir rock. Generally, the permeability reduction is related directly to the molecular weight of the polymer. Because of the large variations in pore geometries that exist in sandstone reservoirs, the extent of permeability reduction cannot be determined by the permeability value of the rock alone. A polyacrylamide that is ideal for a particular 100 md reservoir may completely plug another 100 md reservoir. For this reason laboratory data should first be obtained to determine the suitability of a specific polymer.

A polysaccharride reduces the mobility of the injected fluid by increasing the viscosity of the fluid with very low levels of permeability reduction occurring in the reservoir rock. Depending on the permeability of the reservoir, filtration may also be necessary prior to the injection of the biopolymer.

Considerable work has been done in the laboratory on the flow of polymers through porous media. Polyacrylamide solutions are classified as non-Newtonian fluids because their flow behavior is too complex to be characterized by the single parameter, viscosity. Despite the complex flow behavior, the apparent viscosities of these fluids are significantly higher than the viscosity of water. Other factors affecting polymer character are: 1) solvent and salt concentration; 2) the molecular weight; and 3) the extent of hydrolysis.

Flow through porous media should also consider the retention, adsorption, inaccessible pore volume and entrapment caused by the polymer.

The flexible long chain polyacrylamides are susceptible to shear degradation. This degradation involves breakage of the polymer chain into several shorter molecular chains. When handling these polymers in the field, care should be taken to minimize any shearing effects.

Caustic Flooding

Caustic or alkaline flooding is a process in which the pH of the injected water is controlled at a value of 12-13 to improve recovery beyond that of ordinary waterflooding.

Existing waterfloods can be rather easily converted to a caustic flood by the addition of 1-5% by weight NaOH. Only a relatively few pilot tests have been reported and results are not encouraging. A considerable amount of further study is needed to better understand the improved recovery mechanisms and permit a proper choice of reservoirs for application of this technique.

Although the benefits were recognized early, the mechanism actually responsible for improving oil recovery was not well understood. At the present time, at least six methods are postulated as follows:

1. Lowering of interfacial tension
2. Wettability change from oil-wet to water-wet
3. Wettability change from water-wet to oil-wet
4. Emulsification and entrapment
5. Emulsification and entrainment
6. Solubilizing the rigid films at the oil-water interface.

With such a list of mechanisms that can benefit the ultimate recovery by caustic addition to the injection water, it appears that some logical laboratory procedure can be formulated which could evaluate a reservoir's potential. The ultimate goal should be to determine the benefit to be obtained and not necessarily to determine the dominant mechanism.

The economics of caustic flooding are appealing because investment for the caustic chemicals is relatively low when compared to most other enhanced oil recovery processes. Although the ultimate oil production benefits may not be as great, the economic return may be substantial.

Miscible Hydrocarbon Displacement

The miscible displacement process involves introduction of a fluid (solvent) that will completely dissolve the reservoir oil, eliminating the forces that cause oil retention in the rock matrix, and sweep the solvent-oil mixture to the producing well. This solvent can be alcohol, refined hydrocarbons, condensed hydrocarbon gases, carbon dioxide, liquefied petroleum gases or exhaust gas.

First a slug of solvent (miscible with the reservoir oil) is injected. This is followed by injection of a liquid or gas to force the solvent-oil mixture to the producing wells. Unfortunately the solvent (miscible slug) becomes concentrated with oil as it moves through the reservoir, changing its composition and diminishing its ability to dissolve oil; as a result, theoretical recoveries are never achieved. There are a number of successful miscible hydrocarbon displacement projects active today. Reservoir parameters vary widely.

There are three different miscible hydrocarbon displacement processes for improved oil recovery. The first method, known as the miscible slug process, consists of injecting a slug of liquid hydrocarbon equivalent to 50% PV followed by natural gas, or gas and water, in order to drive the slug through the reservoir. The second technique, known as the enriched gas process, consists of injecting a slug of enriched natural gas followed by lean gas or lean gas and water. The size of the slug usually ranges between 10-20% PV, and it is enriched with ethane through hexane (C_2-C_6). The third method, known as the high pressure lean gas process, consists of injecting lean gas at a high pressure in order to cause retrograde evaporation of the crude oil and formation of a miscible phase, consisting of C_2-C_6, between the gas and oil. Thus, the main difference between enriched gas and high pressure lean gas processes is that in the former, C_2-C_6 components are transferred from the gas to the oil, whereas in the latter, C_2-C_6 components are transferred from the oil to the gas.

Design of a miscible hydrocarbon displacement process requires laboratory model studies followed by field pilot tests. The reservoir parameters involved in the design are: depth and thickness of reservoir rock, viscosity of oil, reservoir temperature and pressure, permeabilities and dip angle. The studies should yield optimum slug size, rate of flooding and recovery efficiency.

Carbon Dioxide Injection

The mechanism for achieving miscibility between CO_2 and oil is similar to that of the high-pressure gas mechanism for hydrocarbon miscible displacement with lean natural gas. Although CO_2 is not miscible with reservoir oil upon initial contact, it can create a miscible front with the right conditions of pressure, temperature and oil composition. Under the right conditions, the gas will extract heavier hydrocarbons from oil and concentrate them at the displacement front. Thus, the carbon dioxide and oil become miscible, form a single liquid phase and efficiently sweep the reservoir oil to the producing well.

Miscibility can be attained at pressures as low as 1500 psi at moderate reservoir temperatures. Impurities in the carbon dioxide (such as methane or nitrogen) increase the pressure of miscibility, while hydrogen sulfide and propane will lower miscibility pressure.

The main advantages of miscible CO_2 injection are summarized as follows:

1. Swells oil and reduces viscosity.
2. Develops miscibility with oil by extraction, vaporization and chromatographic transport.
3. Even if complete miscibility is not achieved, CO_2 will function as a solution gas drive.
4. The miscible front, if lost, will regenerate itself as it does with the lean gas process.
5. Unlike LPG, CO_2 will become miscible with oils that have been depleted of the C_2-C_4 fractions.
6. Carbon dioxide is soluble in both water and oil and causes the fluids to swell. Swelling of the oil will result in a lower residual oil saturation.
7. Miscibility can be attained at relatively low pressure in many reservoirs.
8. Carbon dioxide is a non-hazardous, non-explosive gas that causes no environmental concern if large quantities are lost to the atmosphere.
9. May be available as a waste gas (gas-processing plants or industrial plants) or may be produced from reservoirs containing CO_2.

Some of the disadvantages of CO_2 injection as an enhanced oil recovery process are summarized as follows:

1. Solubility of CO_2 in water may increase volume needed for oil miscibility, but this disadvantage may be partly or wholly overcome by the increased volume of the CO_2-saturated water.
2. The low viscosity of any free CO_2 gas at low reservoir pressure will promote early breakthrough to the producing well, reducing sweep efficiency. Production of large volumes of diluted gas requires expensive gas processing and recycling facilities.
3. After miscibility is attained the oil is less viscous than reservoir oil, causing fingering and premature breakthrough.
4. Injection of slugs of water is often necessary to reduce fingering.
5. Carbon dioxide with water forms highly corrosive carbonic acid. Special metal alloys and coatings for facilities are needed. Corrosion mitigation can be a considerable part of the cost of the process.
6. The alternate injection of slugs of CO_2 and water requires a dual injection system, adding to the cost and complexity of the project.
7. Large volumes of CO_2 are needed. It may take 5-10 MSCF of gas to produce one barrel of stock tank oil.
8. Carbon dioxide is usually not available locally, requiring long-distance pipelines. Experience has shown that CO_2 pipelines are more subject to breakdown than natural gas pipelines, thus causing expensive delays that may interfere with the technical success of the project.

As yet, carbon dioxide flooding has not progressed far enough to fully determine the economic potential of the process. Various studies have suggested that as much as 40% of the total oil to be recovered by all enhanced oil recovery methods could be recovered by carbon dioxide miscible flooding. The full potential of this process will not be known until additional field projects demonstrate its economic viability.

Inert Gas Injection

The phenomena of miscible displacement of reservoir fluids by inert gas occurs only in a narrow range of fluid compositions, pressures and temperatures. The number of reservoirs that are candidates for this enhanced oil recovery method is therefore limited.

A relatively few number of projects are reported and process techniques vary considerably.

Some laboratory studies conclude that N_2 is not a suitable agent for miscible displacement because of the high pressure needed to obtain miscibility. Although such data appear to make a strong case against N_2 injection, these experiments only measured "first contact" miscibility. Under actual reservoir conditions, inert gas will repeatedly contact reservoir fluids. These repeated contacts strip lighter hydrocarbons from the crude and (similar to the lean gas hydrocarbon miscible process) will result in a miscible displacement. Even if miscibility is not attained, vaporization of the condensed liquids will occur as they are contacted by fresh N_2.

An advantage of inert gas over natural gas is that combustion increases the volume of natural gas by a factor of 5-10. Inert gas, when injected at sufficiently high pressures, will miscibly displace oil in a manner similar to lean natural gas.

STATE OF THE ART

Although the potential for enhanced oil recovery is immense, the application of specific techniques is in its infancy. Until the technologies are well understood and the economic return and risks are acceptable, the growth of enhanced oil recovery application will remain slow.

Thermal Processes

Thermal methods account for about 70% of the world's enhanced oil recovery production. Their application to reservoirs having low gravity, high viscosity and high porosity have become almost routine. There is every indication that this segment of enhanced oil technology will continue to grow.

Chemical Processes

The chemical processes for recovering additional oil account for less than 1% of the enhanced oil recovered in the United States. Although these processes have the best chance for recovering oil from reservoirs that have been successfully waterflooded (but still contain considerable oil), development has been slow because of associated high costs, high risk and complicated technology.

Miscible Displacement Processes

Inert gas miscible projects are on the increase in recent years, in contrast to hydrocarbon miscible projects which are declining because of the high cost and limited supply of injected hydrocarbons. Recent reports state that CO_2 miscible flooding could potentially recover 40% of the total projected enhanced reserves in the USA.

Introduction

At the beginning of 1979 world crude oil reserves were estimated at 641.5 billion barrels, and U.S. reserves dropped to 28.5 billion barrels. In the United States some 440 billion barrels of oil-in-place have been found to date, yet the estimated ultimate recovery is only 145 billion barrels, or about one-third of that discovered. Throughout the entire world a resource close to 2.0 trillion barrels of oil can be classified as unrecoverable by existing technology. Continued development of enhanced oil recovery processes can add significantly to the world supply of oil. In the U.S. alone, a resource of some 295 billion barrels is classified as unrecoverable by existing primary and secondary recovery technology. A realization of the magnitude of this resource along with the decline in production rates in the U.S. has prompted a concerted effort to develop novel methods for increasing oil recovery.

These methods or processes have as their objectives: 1) to increase recovery from reservoirs considered depleted by secondary recovery methods of waterflooding or gas injection and 2) to increase recovery from reservoirs which would not respond to conventional waterflooding or gas injection. Processes developed to meet the first objective were appropriately termed tertiary recovery projects, but the name was ill-suited for all other methods. This prompted introduction of the term "enhanced oil recovery" (EOR), which is now the generally accepted designation for all unconventional recovery processes.

Enhanced oil recovery processes can be separated into three major categories and the text is subdivided according to this classification:

Part I   Thermal Processes
Part II   Chemical Processes
Part III Miscible Displacement Processes

Part I

THERMAL PROCESSES

Chapter 1
STEAM STIMULATION

INTRODUCTION

It has long been recognized that heat is needed to improve the productivity of wells producing low API gravity oils. Some operators installed bottom-hole heaters; but they soon discovered that although the generated heat facilitated lifting, it did not improve inflow performance (Dietz, 1975). This is because heat given off by a bottom-hole heater spreads through the reservoir by conduction and does not travel far from the wellbore before it is brought back by the fluids flowing into the well.

Now it is evident that better results can be obtained if production is stopped for a period of time in order to allow the heat to spread into the reservoir. However, this simple method was not utilized until 1959 when steam erupted at the surface during an early trial of steam injection in the Mene Grande Tar Sands, Venezuela. Then it was decided to relieve the pressure by backflowing the steam injection well (deHaan and van Lookeren, 1969). This resulted in impressively high oil production rates, especially since this reservoir could not be produced by primary means.

Steam stimulation (also known as cyclic steam injection, steam soak or huff and puff) has become an established enhanced oil recovery method. In California, oil production had declined to 300 MMBO per year in 1964, but upon application of thermal recovery methods (steam stimulation in particular) annual oil production increased to 360 MMBO by 1967.

The steam stimulation process involves the injection of 5000-15000 Bbls of 80% quality steam into a producing well. The well is then produced for a few months or a year. Some operators shut the well in before allowing it to produce in order to "soak" the reservoir. However, there is a difference of opinion regarding the advantages of this practice. On the one hand, if the well is produced immediately following steam injection, large amounts of heat will be lost with the produced steam. On the other hand, since heat dissipation in base and cap rock is a function of time, large amounts of heat will be lost in heating base and cap rock during the "soak" period. In general, a short soak period of 1 to 5 days is now an accepted practice.

Steam stimulation is, in general, less expensive than steam injection. However, recovery by steam stimulation is usually less than recovery by steam injection. For this reason, some operators first apply steam stimulation and then convert to steam injection following the first or second cycle. When the reservoir rock is so discontinuous that steam injection must be ruled out, steam stimulation would be the most feasible alternative.

Reservoir response to first-cycle steaming usually exceeds that predicted by model studies. This is mainly due to well cleanup and improvement of permeability in the vicinity of the well. Production as high as 20-fold that predicted by model studies has been reported. But in almost all reported cases, reservoir response to a second stimulation cycle is usually less than that of the first cycle, and the duration of each successive cycle is usually shorter.

An estimation of expected oil production rate after steaming (Q_{hot}, Bbls/day) can be made by the following simple relation:

$$Q_{hot} = Q_{cold} \frac{\text{viscosity of oil before steaming}}{\text{viscosity of oil after steaming}}$$

However, predictions of reservoir response and oil-steam ratio require model studies. A simple analytical model given by Boberg and Lantz (1966) usually provides good estimates.

TECHNICAL DISCUSSION

Model Studies

Numerical and analytical steam stimulation models are utilized to calculate the reservoir response to different injection and production strategies, thus enabling the engineer to select optimum operating conditions. At optimum conditions, the incremental oil-steam ratio (defined as the ratio of stimulation less primary oil production to the amount of steam injected expressed as barrels of condensate) should be the highest. The incremental oil-steam ratio heavily influences project economics. Model studies are often used to determine this variable and estimate potential profitability of a proposed steam stimulation project.

In addition to reservoir rock and fluid properties, input variables needed in a steam stimulation model study are: injection rate, injection pressure, length of injection time, length of soak time and number of steam stimulation cycles. The response variables are: incremental oil-steam ratio and water and steam production.

Numerous steam stimulation models that vary in complexity and sophistication are available in the literature. Boberg and Lantz (1966) formulated a model based on the Marx and Langenheim model discussed in Chapter 2. The Boberg and Lantz model accounts for both radial and vertical heat conduction. It is useful in predicting fluid production in reservoirs with some primary production. Martin (1967) formulated an analytical, non-conduction model in which heat is transported only by fluid flow. Kuo et al (1968) studied the problem of gravity drainage that was not previously considered, but did not address the specific performance during each stimulation cycle. Seba and Perry (1969) also addressed gravity drainage, but assumed constant hot zone temperature and radius, and harmonic decline in oil production. Closmann et al (1970) considered the effects of changing fluid saturations and vertical flow across interbedded shale, but ignored gravity drainage. Williamson et al (1976) designed a model specifically for use in a numerical isothermal reservoir simulator which treated the effects of transient relative fluid saturation and changes in viscosity with temperature in individual wells. Jones (1977) formulated a simplified model based on the model by Boberg and Lantz. Jones' model has the advantage that it accounts for gravity drainage, provides for calculating injection pressure and temperature, and can easily be programmed for a small computer.

In the following text, the model by Boberg and Lantz is presented with some of the modifications made by Jones (1977).

Boberg and Lantz Model

This model (Figure 1) is based on the assumption that the oil-bearing strata are uniformly and radially invaded by injected steam. For wells producing from multiple horizons, each horizon is assumed to be invaded to the same radial distance. Energy losses from the wellbore and conduction to adjacent impermeable rock are taken into account when calculating the heated radius r_h. After steam injection is stopped, heat conduction is assumed to continue. This results in cooling the producing horizons within the radius $r < r_h$ while warming the unheated portion of the producing layer and the surrounding impervious layers in the interval $r > r_h$. However, in computing oil production rate, it is assumed

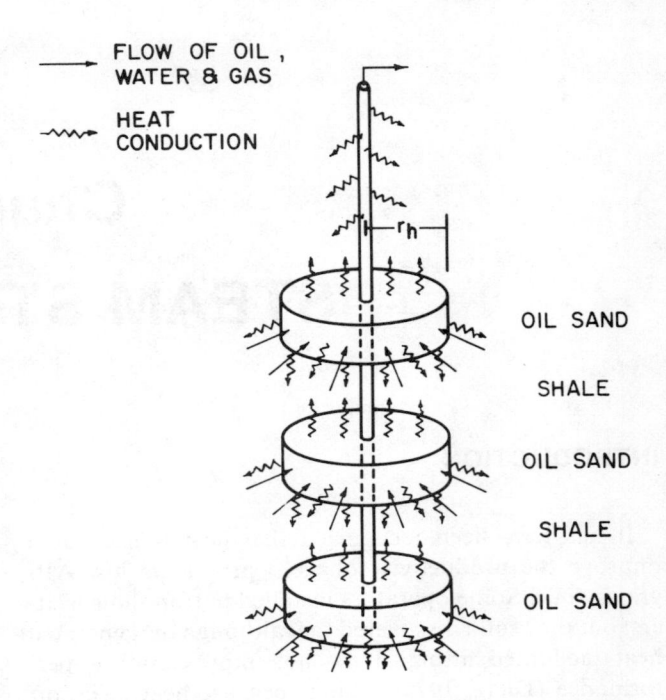

Fig. 1. Schematic representation of heat transfer and fluid flow calculated by mathematical model.
(after Boberg and Lantz, 1966)

that temperature distribution in the reservoir takes the form of a step function. The average temperature T_{avg} extends from the well to the heated radius r_h, and the original reservoir temperature T_r extends through the region $r > r_h$. The oil production rate is calculated by a steady-state radial approximation which accounts for the reduced oil viscosity in the region $r < r_h$.

The necessary calculations proceed according to the following steps. Steps 1 through 3 are related to steam injection, and Steps 4 and 5 are related to production.

Step 1

Calculate the cumulative wellbore heat losses Q_{hl} (Btu):

$$Q_{hl} = 2\pi z \, K \, r_c^2 \left(T_s - T_r + \frac{az}{2} \right) I/D \tag{1}$$

where:

z = reservoir depth, ft
K = overburden thermal conductivity, Btu/ft day °F
r_c = inside radius of injection pipe (tubing or casing), ft
T_s = steam temperature, °F
T_r = reservoir temperature, °F
a = geothermal gradient, °F/ft
D = thermal diffusivity of overburden, ft^2/day

STEAM STIMULATION

I = dimensionless factor read from Figure 2 as a function of τ ($\tau = Dt_i/r_c^2$)
t_i = time of injection (current cycle), days

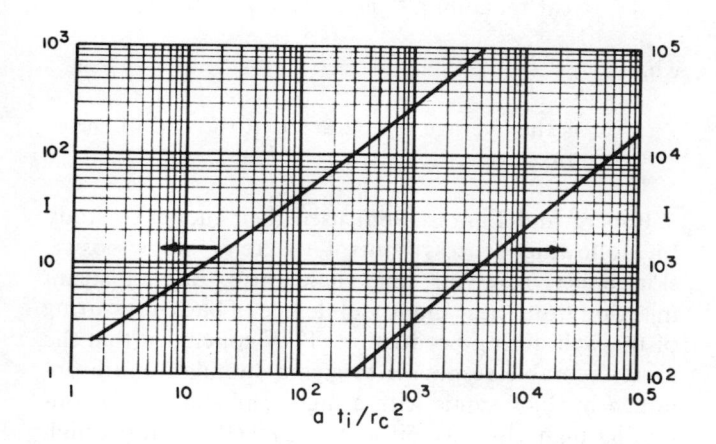

Fig. 2. *I* factor for wellbore heat loss determination. (after Boberg and Lantz, 1966)

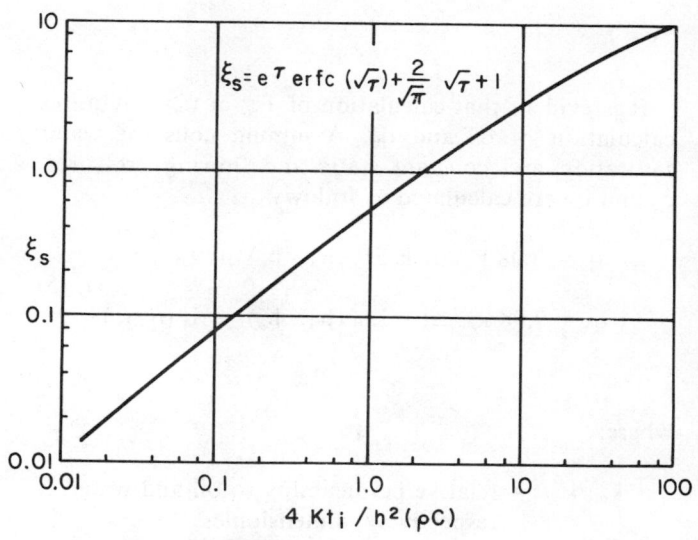

Fig. 3. Dimensionless position of steamed-out region. (after Boberg and Lantz, 1966)

Step 2

Calculate average downhole steam quality $\bar{X}_i$ for the entire steam injection period.

$$\bar{X}_i = X_{surf} - Q_{hl}/350\, t_i q_s \ell \qquad (2)$$

where:

X_{surf} = steam quality at wellhead
q_s = steam injection rate, Bbls/day
ℓ = latent heat of steam, Btu/lb
Q_{hl} = cumulative wellbore heat loss, Btu

Step 3

Calculate the average heated radius $\bar{r}_h$ by using the Marx and Langenheim model. (Equation 6, Chapter 2)

$$\bar{r}_h^2 = \frac{350\, h^2\, q_s\, (\bar{X}_i \ell + h_w - h_{w,r})\, \xi_s}{4K\pi(T_s - T_r)\, \Sigma h_i} \qquad (3)$$

and,

$$\bar{r}_h^2 = \Sigma\, \bar{r}_h^2\, h_i / \Sigma h_i$$

where:

h_w = specific enthalpy of water at T_s, Btu/lb
$h_{w,r}$ = specific enthalpy of water at T_r, Btu/lb
h_i = thickness of producing layer number i, ft
Σh_i = sum of thicknesses of individual producing layers, ft
h = average thickness of sand, ft
ξ_s = $\exp(\tau)\, \text{erfc}(\sqrt{\tau}) + (2/\sqrt{\pi})\sqrt{\tau} - 1$, Figure 3
$\tau = 4Dt_i/(\Sigma h_i)^2$

Step 4

For each time step $\Delta t = t - t_i$ ($t \geq t_i$), the average reservoir temperature T_{avg} (°F) is calculated:

$$T_{avg} = T_r + (T_s - T_r)[V_r\, V_z\, (1 - \delta) - \delta] \qquad (4)$$

(Farouq Ali, 1974)

where:

$V_r = 0.180304 - 0.41269\chi + 0.18217\chi^2 + 0.149516\chi^3 + 0.024183\chi^4$

$\chi = \log_{10}(Dt/\bar{r}_h^2)$

$V_z = 0.474884 - 0.56832y - 0.239719y^2 - 0.035737y^3$

$y = \log_{10}(4\, Dt/\bar{h}^2)$

$\bar{h} = \dfrac{350\, q_s\, t_i(\bar{X}_i \ell + h_w - h_{w,r})}{\pi\, \bar{r}_h^2\, M(T_s - T_r)}$, ft

$M = (1 - \phi)\rho_r C_r + S_w \phi \rho_w C_w + S_o \phi \rho_o C_o$, Btu/ft³-lb (see Chapter 2)

$\delta = [H_{ow}/2\pi\, \Sigma h_i\, \bar{r}_h^2\, M(T_s - T_r)]$, dimensionless

$\delta = 0$ at $t = t_i$

H_{ow} = heat removed with produced fluids, Btu

$H_{ow} = 5.615\, (q_o \rho_o C_o + q_w \rho_w C_w)(T_{avg} - T_r)$, Btu

ρ_o, ρ_w = density of oil and water, respectively, lb/cu ft

C_o, C_w = specific heat of oil and water, respectively, Btu/lb °F

q_o, q_w = oil and water production, respectively, Bbl.

Step 5

It is evident that calculation of T_{avg} at $t > t_i$ requires calculation of q_o and q_w. Assuming constant water saturation and constant static and flowing pressures, q_o and q_w are calculated as follows:

$$q_o = 7.08 \, k_{ro} \, \Delta t \, k \, \Sigma h_i \, (h_{st} - h_w) / \mu_w \, \ln(\bar{r}_h/r_w) \quad (5)$$

$$q_w = 7.08 \, k_{rw} \, \Delta t \, k \, \Sigma h_i \, (h_{st} - h_w) / \mu_o \, \ln(\bar{r}_h/r_w)$$

where:

- k_{ro}, k_{rw} = relative permeability to oil and water, respectively, dimensionless
- k = absolute permeability, darcies
- Δt = $t - t_i$, $t \geq t_i$, days
- h_{st}, h_w = static fluid level and wellbore fluid level, respectively, ft
- μ_o, μ_w = viscosity of oil and water, respectively, cp
- $r_w = \bar{r}_w \, e^{-S}$, ft
- $\bar{r}_w$ = actual wellbore radius, ft
- S = skin factor of well, dimensionless
- $\mu_w = 0.66 \, (T_{avg}/100)^{-1.14}$, cp
- μ_o = ρ_o at T_{avg} times v, the kinematic viscosity in centistokes (for calculation of ρ_o at T_{avg} and for estimation of v, see Figure 4)

For the second cycle, the heat remaining in the reservoir at the end of the first cycle is given by:

$$\text{Heat remaining} = \pi \, \bar{r}_h^2 \, M \, \Sigma h_i \, (T_{avg} - T_r)$$

where:

T_{avg} is the average temperature at the end of the first cycle.

Boberg and Lantz (1966) used their model to study the following process control variables: oil viscosity, skin effect, reservoir sand-shale ratio, rate of steam injection, total heat input and degree of back-pressuring of the well during production. They concluded that the most favorable responses to steam stimulation are obtained in thick sands with a high sand-shale ratio containing high viscosity oil and from wells having a high skin factor prior to stimulation. Furthermore, injection of steam at the highest rate possible reduces wellbore heat losses and the time period during which the well is off production. As the cumulative steam input is increased, the incremental oil-steam ratio increases and attains a maximum before it declines, which suggests that for a given set of operating conditions there seems to be an optimum level of steam input. Back pressuring the steam-stimulated well early in the production phase of the cycle has the effect of reducing heat losses by preventing the produced water from flashing into steam.

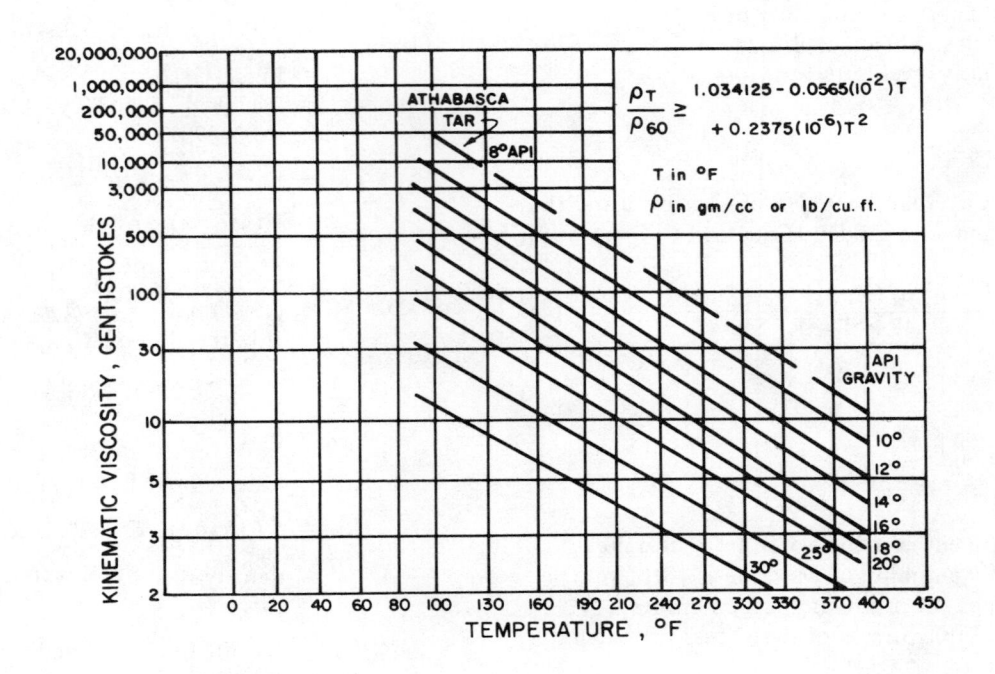

Fig. 4. Viscosities of some crude oils as functions of temperature. (after Farouq Ali, 1974)

Experimental Studies

Although numerical and analytical steam stimulation models are prevalent in the literature, there is a lack of published experimental studies. This is probably due to the fact that the scaling laws of steam stimulation processes require that the compressibility of the model oil be 50 to 100 times as high as the compressibility of the reservoir oil. Since there is no oil available with such a high compressibility, steam stimulation processes cannot be simulated in a sand pack.

Niko and Troost (1970) designed a hydraulic analog to simulate radial compressible fluid flow. The analog consisted of a series of capacitor and resistor tubes arranged as shown in Figure 5. With this model, Niko and Troost studied the effects of the following process variables and formation parameters on the production performance: steam injection rate, soak time, initial reservoir pressure, oil viscosity, layer thickness, steam slug size and cycle length. Their results are in general agreement with the results obtained by Boberg and Lantz. However, their conclusions regarding the effects

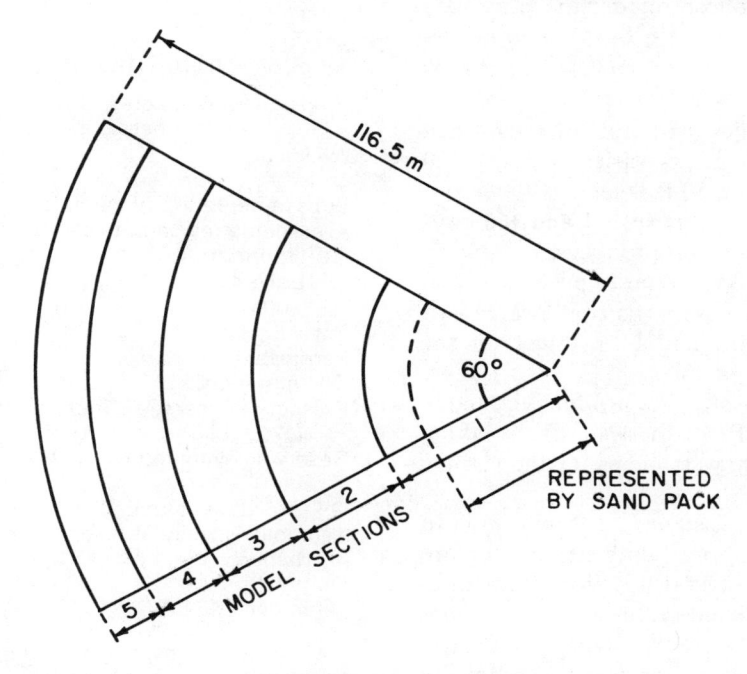

Fig. 5A. Reservoir section.

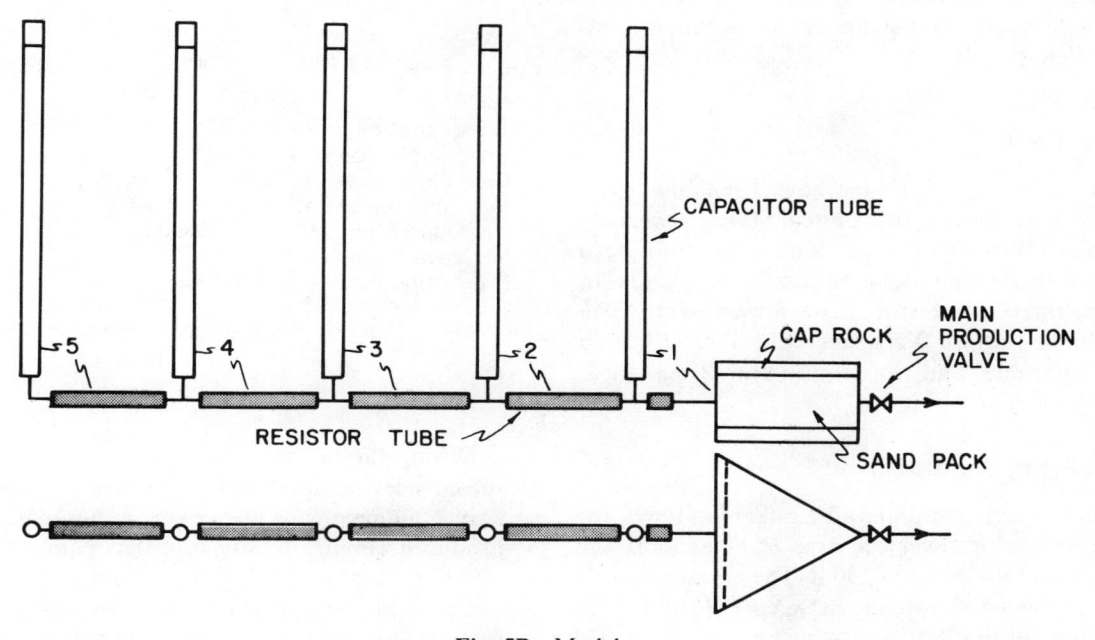

Fig. 5B. Model.
(after Niko and Troost, 1970)

of layer thickness are different from those given by Boberg and Lantz, and provide significant insight in the steam stimulation process. Niko and Troost observed that because of gravity, the injected steam tended to accumulate at the top of the oil sand. This resulted in uneven vertical distribution of heat. Moreover, the average thickness of the steam zone was found independent of the layer thickness. They concluded that if the same amount of steam is injected per unit of thickness, the areal extent of the steam zone would be larger when the reservoir rock is thicker. Since the major part of the viscous resistance is concentrated in the immediate vicinity of the wellbore, the injected heat would be less effective in thick reservoirs. Other conclusions given by Niko and Troost are as follows:

1. Production performance was not affected when the steam injection rate was varied between 120 and 360 Bbls/day into a 30 ft layer.
2. Variation of soak period between 1 and 160 days had a negligible effect on oil production.
3. Oil production was proportional to pressure drawdown. Non-uniform pressure distribution, caused by primary production, had little effect on the process.
4. For a given steam slug size, the productivity index ratio (defined as the PI of the well after soaking to that before soaking) was higher for the viscous oils.
5. Because most of the resistance to flow occurs in the vicinity of the wellbore, small steam slugs are more effective. After several injection-production cycles, the cumulative oil-steam ratio was independent of the steam slug size. During the early cycles, however, oil-steam ratio was higher for the small slug sizes.
6. In a multiple well project, the effect of steam slug size on cumulative oil-steam ratio was found to be insignificant.

CASE HISTORIES

Steam stimulation operations have been most successful in Venezuela and the United States, especially in California. Successful operations have also been reported in Canada, Indonesia and the Netherlands. In this section, three case histories are presented: the Tia Juana Field in Eastern Venezuela; the Duri Field in Sumatra, Indonesia; and the Huntington Beach Field in California, U.S.A.

Tia Juana Field, Venezuela

According to deHaan and van Lookeren (1969), the Tia Juana Reservoir covers an area of 8800 acres and has a net oil sand thickness of 130 ft. The initial reservoir conditions are given in Table I. Between 1936 and 1963, 12.5% of the oil-in-place was recovered through 540 wells, and the reservoir pressure declined from 960 to 550 psi. Table II shows the status of the field in October 1963, before steam injection was begun.

The reservoir consists of many sand bodies. Their origin is believed to be a system of distributary channels that formed in a coastal plain. Well-to-well correlation is difficult, but the reservoir has been divided into upper and lower zones on the basis of oil viscosity and pressure differences (Table II). The upper zone is characterized by lower pressure, higher steam injectivity and lower oil viscosity than the lower zone.

TABLE I
INITIAL RESERVOIR CONDITIONS

Project area (A), acres	850
hectares	343
Well spacing, ft	700
meters	231
Subsea depth top of sand, ft	1500-2000
Average net oil sand thickness, ft	
Upper zone	80
Lower zone	97
Total	177
Porosity ϕ	0.40
Permeability k, darcies	2
Oil saturation S_{oi}	0.80
Pressure at average depth top of sand (1750 ft subsea), psig	780
Reservoir temperature T_f, °C	43
°F	110
Oil formation volume factor	1.04
Tank oil viscosity at reservoir temperature, cp	700-7000
Oil gravity, °API	15-11
Oil in place (STOIIP), Bbl	355.6×10^6

(after de Haan and van Lookeren, 1969)

TABLE II
STATUS AT START OF PROJECT
(OCT. 1, 1963)

Cumulative oil produced, Bbl	62.15×10^6
Producible potential, BOPD	5110
Cumulative GOR, cu ft/bbl	400
Average pressure at top of sand, psig	
Upper Zone (1750 ft subsea)	240
Lower Zone (1950 ft subsea)	400
Production decline rate, %/year	6

(after de Haan and van Lookeren, 1969)

During the primary production period, considerable subsidence was observed in the field. A plot of cumulative volume of subsidence versus cumulative net liquid produced (Figure 6) suggests that compaction has essentially been the principal production mechanism during most of the primary production period. During the early stages, solution gas drive contributed about 5% of the initial oil-in-place.

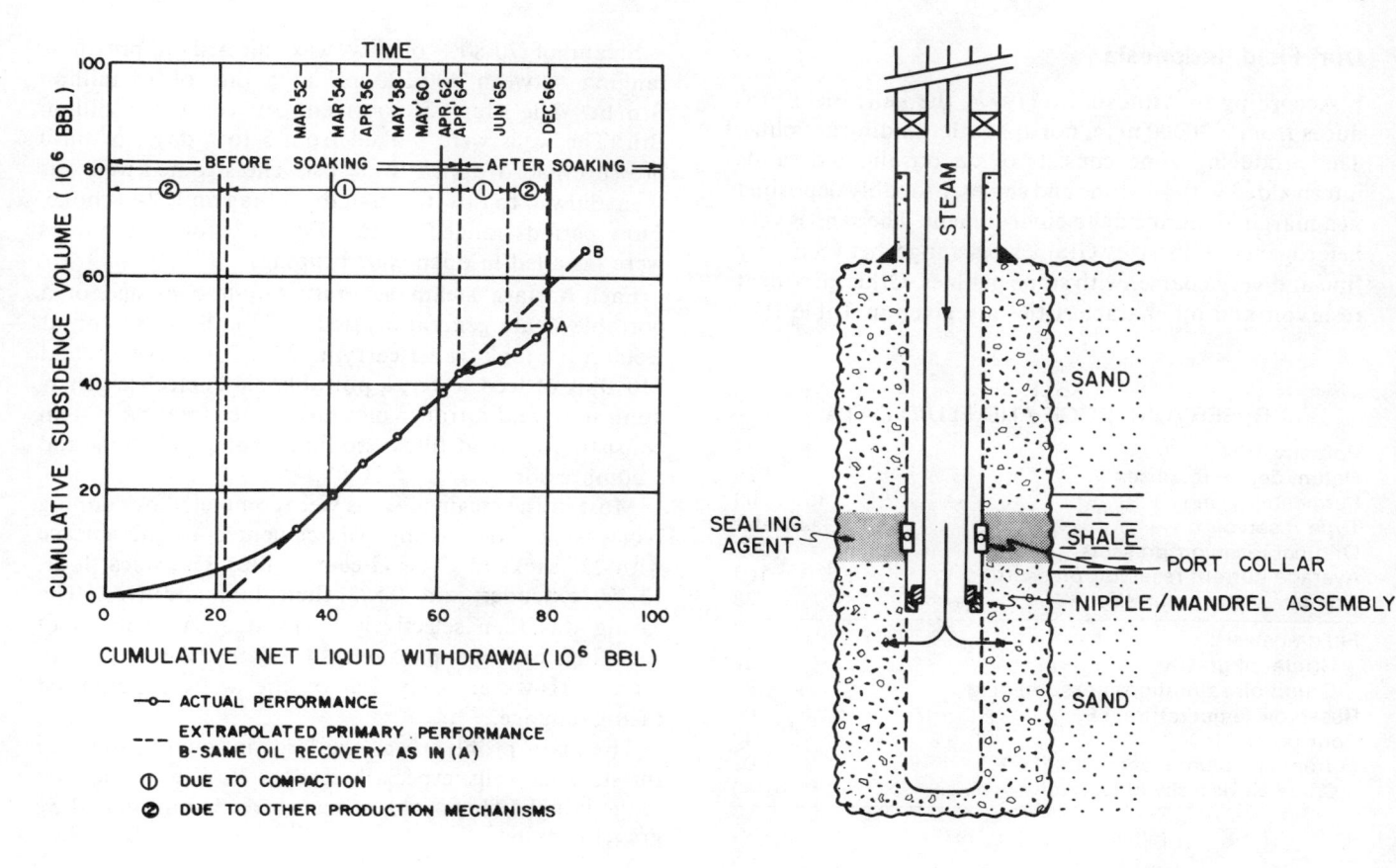

Fig. 6. Comparison of compaction with other production mechanisms.
(after deHaan and van Lookeren, 1969)

Fig. 7. Selective injection.
(after Giusti, 1974)

Steam of 90% surface quality was injected at wellhead pressures of 400-800 psi, and rates of 1200-3000 Bbls/day/well, through 4 1/2-in expandable tubing. The tubing was anchored to a packer to prevent the steam from entering the annulus. The size of the steam slug varied between 20000-70000 Bbls. As was expected, most of the steam penetrated the upper zone. However, by injecting a sealing agent as shown in Figure 7, it became possible to inject steam in the lower zone.

The wells were left to soak for a period of 2 to 4 days before putting them on production. In general, response to first-cycle stimulation was very good. Some wells produced 1000 BOPD for many months as compared with 100-150 BOPD during the primary production period. However, this favorable performance was not repeated in the second stimulation cycle (Figure 8). The water cut increased from 2-8% and remained constant throughout the project.

The overall oil-steam ratio was 0.3 Bbls/bbl. Between 1963 and 1966, 6% of the oil-in-place was recovered. This recovery would have taken some 20 years by primary means.

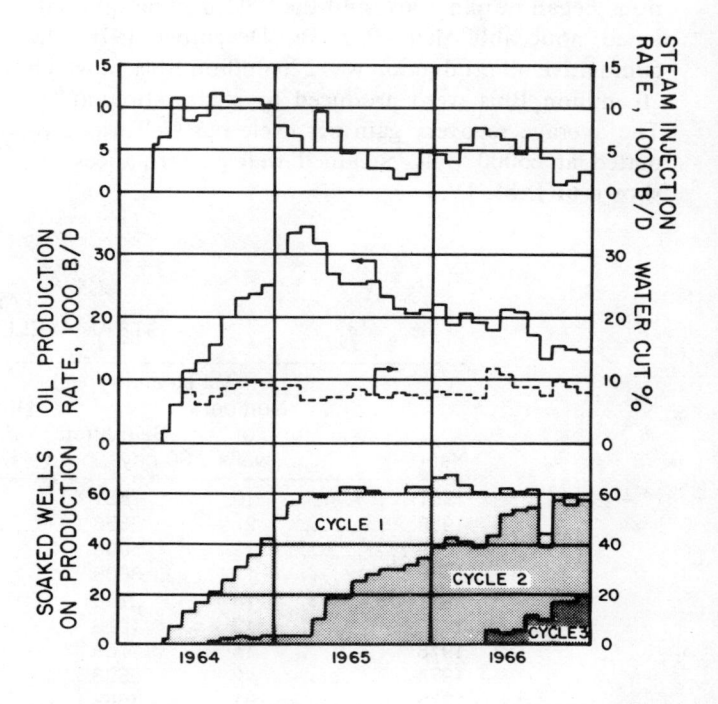

Fig. 8. Over-all project performance — Total steam injection and oil production rates for all soaked wells vs time.
(after deHaan and van Lookeren, 1969)

Duri Field, Indonesia

According to Atmosudiro (1977), the Duri Field produces from a 26000 acre, north-south trending anticline. The producing zone consists of unconsolidated sands interbedded with siltstone and shales, probably deposited in a marginal marine delta environment. The sand is very heterogeneous in size. Grain sizes range between very fine and very coarse, with some pebbles. Other pertinent reservoir and oil characteristics are given in Table III.

TABLE III
RESERVOIR, ROCK AND FLUID DATA

Porosity, %	37
Datum depth, ft subsea	525
Permeability, md	100-5000
Type reservoir drive	Gas/water
Original reservoir pressure, psig	267
Average current reservoir pressure, psig	100
Connate water saturation, %	38
Fluid analysis:	
Crude-oil gravity, °API	22
Crude-oil saturation pressure, psig	175
Reservoir temperature, °F	100
Pour point, °F	40
Formation volume factor rb/STB	1.02
Crude-oil viscosity at reservoir cond., cp	140-180

(after Atmosudiro, 1977)

The field was put on production in May 1958. By late 1963, production had peaked at 65000 BOPD. Thereafter, production declined steadily to 43000 BOPD (from 315 wells) by mid-1967. Steam stimulation operations began in mid-1967, and the field decline rate flattened noticeably thereafter. By December 1976, the cumulative oil production was 270 million Bbls, of which 20 million Bbls were produced by steam stimulation. The average recovery gain per cycle per well was estimated at 65000 Bbls. Steamed well performances are shown in Table IV.

Steam of 70-80% quality was injected at pressures ranging between 400-500 psi at a rate of 15 million Btu/hr. The slug size ranged between 2.5-5 billion Btu. The wells were soaked from 3 to 5 days, or until pressure had dropped to 50 psi. The steamed well was then allowed to flow through the tubing on 1/2-in choke. Flow periods ranged up to 10 days, before rod pumps were installed in open-ended tubings.

Each surface steam generating unit consisted of a portable steam generator rated at 22 million Btu/hr, at 1,500 psi; a tank trailer carrying 235 Bbls of fuel oil and 210 Bbls of feed water; a portable diesel electric generating unit; and a trailer mounted water treating system consisting of sand filters, sodium zeolite softeners and a compressor.

Most of the steamed wells were completed by running 7-in casing. Each string was cemented to the surface with 225 sacks of Class G cement plus 35% silica flour, 12½% extender and 0.8% fluid loss additives. The casing was then selectively perforated. As a result of steaming, casing expansions of 1½ to 2½-in were observed. However, only 7% of the wells experienced casing damage.

The main problem was the large inflow of sand into the steamed wells, especially if the wells were not soaked long enough. This problem was somewhat alleviated by gravel packing.

Huntington Beach Field, California (USA)

According to Adams and Khan (1969), this field lies both onshore and offshore. The reservoir is a broad, faulted anticline with an areal extent of 370 acres. The reservoir rock is 200-ft thick, fine-grained, consolidated sand, somewhat shaly, with interbedded shale stringers. The reservoir consists of an upper and lower zone. The lower zone contains a substantial amount of montmorillonite, and its oil is very viscous. The two zones

TABLE IV
STEAM WELL PERFORMANCE

	1st steaming		2nd steaming		3rd steaming	
Year	Number of wells	Gain after 90 days, bo/d	Number of wells	Gain after 90 days, bo/d	Number of wells	Gain after 90 days, bo/d
1967	10	821	....	......	.....	......
1968	27	6558	....	......	.....	......
1969	21	5008	3	395	.....	......
1970	17	3603	1	396	.....	......
1971	18	2748	6	728	.....	......
1972	12	1026	7	722	2	288
1973	14	1768	1	12	.....	......
1974	42	5593	7	478	.....	......
1975	31	1668	24	1410	3	183
1976	47	4591	12	380	5	529

(after Atmosudiro, 1977)

are separated by some 300 ft of water-bearing sands with interbedded shale. Oil characteristics and reservoir properties are given in Table V.

TABLE V
RESERVOIR AND CRUDE OIL CHARACTERISTICS, TAR POOL, HUNTINGTON BEACH, CALIF.

	Upper Tar	Lower Tar
Average depth, top of sand, ft	2000	2400
Average net sand, ft	100	50
Average permeability, md	2300	2300
Average porosity, %	38	38
Average water saturation, %	35	35
Original oil-in-place, Bbl/acre-ft	1800	1800
Oil gravity, °API	13	12
Present reservoir pressure, psig	600	1200
Reservoir temperature, °F	110	115
Oil viscosity at 80°F, cp	1800	11000
Oil viscosity at 100°F, cp	940	3700
Oil viscosity at 120°F, cp	530	1600

(after Adams and Khan, 1969)

For the purpose of cyclic steam stimulation, 86 new wells were drilled. The new wells were completed as follows (Figure 9): 1) a casing stub was cemented from the top of the lower zone to the base of the upper zone; 2) a casing string was run from the surface, set at the top of the upper zone, and cemented to a depth of 1500 ft with neat cement and 30% silica flour; and 3) a 60-mesh gravel flow packed liner equipped with expansion joints, high temperature hanger and hookwall packer was run and hung inside the casing.

Steam was injected through the tubing, which was hung 50 to 100 ft above the bottom of the well. Two techniques were followed in order to minimize heat stresses and reduce casing failures. In the old wells, nitrogen cushions were injected in the annulus between tubing and casing. In the 86 new wells, the annulus was packed with high temperature gel. Steam of 80% quality was injected at the rate of 1300 Bbls/day, at 800 to 1000 psi. The steam was supplied by seven gas fired generators. Each generator had a capacity of 22 million Btu/hr.

Before beginning steam stimulation, the daily production rate was 850 Bbls. After performing 125 first-cycle, 96 second-cycle, 34 third-cycle and 3 fourth-cycle operations, the daily rate increased to 3500 Bbls/day. It was estimated that 4 million additional Bbls were recovered as a result of steaming. The oil-steam ratio averaged about 0.3 Bbls/bbl. All the recovery was from the upper zone. The lower zone did not accept steam due to its high content of montmorillonite and the high viscosity of its oil. The optimum soak time was calculated at 9 days.

Figure 10 shows the production response per cycle, and Figure 11 shows the average gross production. It is evident from these figures that in each successive cycle, peak production declined and the cycle length decreased, which is in agreement with almost all published performances of steam stimulation operations.

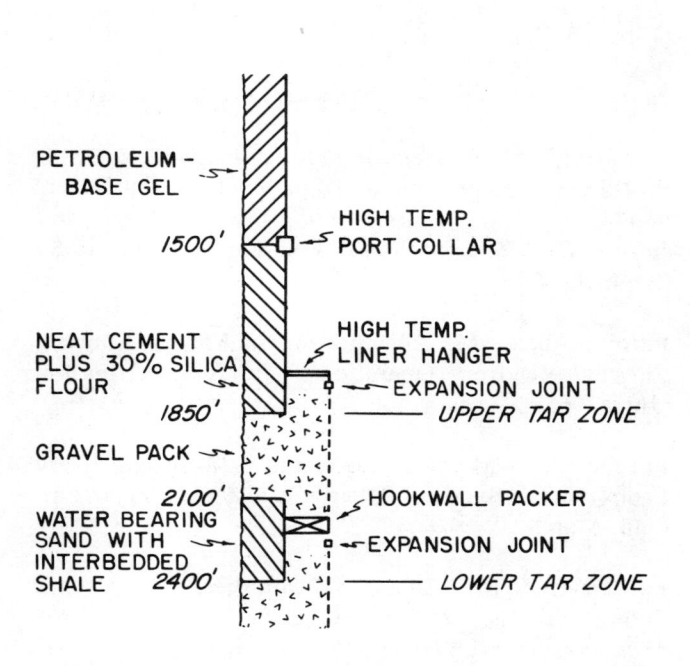

Fig. 9. Schematic of new well completion. (after Adams and Khan, 1969)

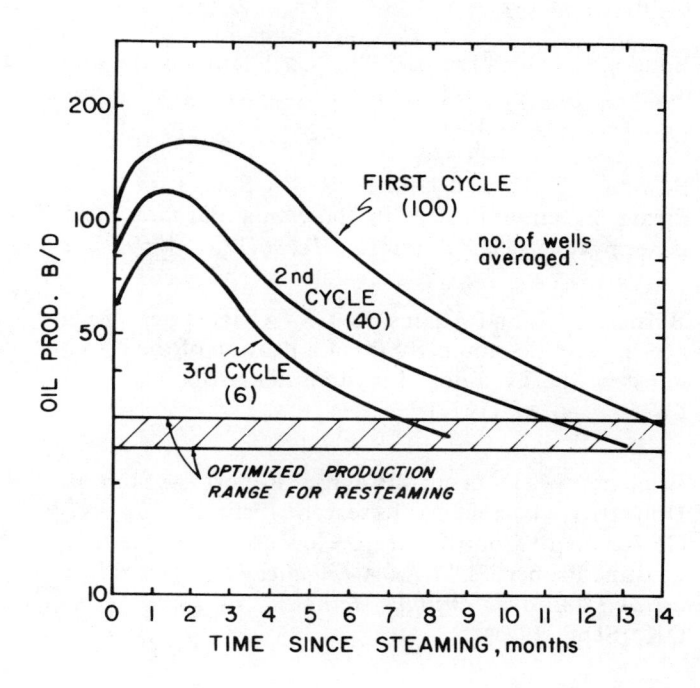

Fig. 10. Production response by cycles. (after Adams and Khan, 1969)

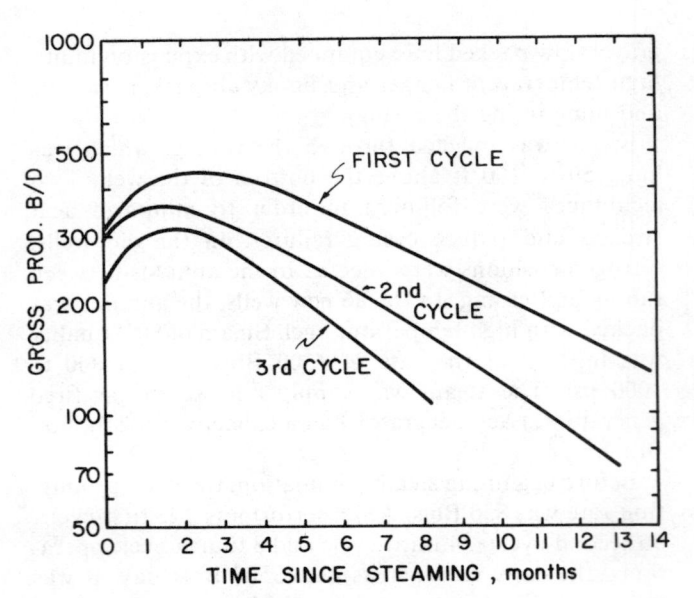

Fig. 11. Averaged cyclic gross production. (after Adams and Khan, 1969)

BIBLIOGRAPHY

Adams, B. H. and Khan, A. M.: "Cyclic Steam Injection Project Performance Analysis and Some Results of a Continuous Steam Displacement Pilot," *Jour. Pet. Tech.* (Jan., 1969).

"Appendix F, Thermal Process," *Enhanced Oil Recovery, National Petroleum Council.* (Dec., 1976).

Atmosudiro, H. W.: "Steam Soak Increases Recovery in Indonesia," *O. & G. Jour.* (Aug. 1, 1977).

Boberg, T. C. and Lantz, R. B.: "Calculation of the Production Rate of a Thermally Stimulated Well," *Jour. Pet. Tech.* (Dec., 1966).

Bowman, C. H. and Gilbert, S.: "A Successful Cyclic Steam Injection Project in the Santa Barbara Field, Eastern Venezuela," *Jour. Pet. Tech.* (Dec., 1969).

Burnett, A. I. and Adams, K. C.: "A Geological, Engineering and Economic Study of a Portion of the Lloydminster Sparky Pool, Lloydminster, Alberta," *Bull. Cdn. Pet. Geol.* (1977).

Brigham, W. E. et al: "Progress Report — Stanford University, Petroleum Research Institute (SUPRI); (1) Agents to Control Steam Flow and (2) Correlation of Brine Properties," *ERDA Enhanced Oil, Gas Recovery and Improved Drilling Methods,* Vol. 1 - Oil, Tulsa, OK. (Sept., 1977).

Closmann, P. J.: "Steam Zone Growth During Multiple-Layer Steam Injection," *Soc. Pet. Eng. Jour.* (Mar., 1967).

Closmann, P. J., Ratliff, N. W. and Truitt, N. E.: "A Steam-Soak Model for Depletion-Type Reservoirs," *Jour. Pet. Tech.* (June, 1970). *Trans.,* AIME (1970) 249.

Crawford, P. B.: "Thermal Recovery Guide Helps Select Projects," *World Oil.* (Aug. 1, 1971).

"Cyclic Steam Drive Set for Peace River," *O. & G. Jour.* (July 29, 1974).

Davidson, L. B., Miller, F. G. and Mueller, T. D.: "A Mathematical Model of Reservoir Response During the Cyclic Injection of Steam," *Soc. Pet. Eng. Jour.* (June, 1967).

de Haan, H. J. and van Lookeren, J.: "Early Results of the First Large-Scale Steam Soak Project in the Tia Juana Field, Western Venezuela," *Jour. Pet. Tech.* (Jan., 1969). *Trans.,* AIME (1969) 246.

de Swaan O., A.: "Improved Numerical Model of Steam-Soak Process," *O. & G. Jour.* (Jan. 3, 1972).

Dietz, D. N.: "Review of Thermal Recovery Methods," SPE 5558 presented at the SPE-AIME 50th Annual Fall Meeting, Dallas, TX. (Sept., 1975).

Dorfman, M. H., Oskay, M. M. and Gaddis, M. P.: "Self-Potential Profiling — A New Technique for Determination of Heat Movement in a Thermal Oil Recovery Flood," SPE 6790 presented at the SPE-AIME 52nd Annual Fall Meeting, Denver, CO. (Oct., 1977).

Dorsey, J. B. and Brinkley, T. W.: "Performance Review of Shoats Creek Unit Vaporizing Gas-Drive Project," *Jour. Pet. Tech.* (April, 1968). *Trans.,* AIME (1968) 243.

Doscher, T. M., Ershagi, I. and Herzberg, D. E.: "Solvent-Steam Stimulation of High Viscosity Oil Reservoirs," *ERDA Enhanced Oil, Gas Recovery and Improved Drilling Methods,* Vol. 1 - Oil, Tulsa, OK. (Sept., 1977).

Farouq Ali, S. M.: "Application of Solvent Slugs in Thermal Recovery Operations," *Producers Monthly.* (July, 1965).

Farouq Ali, S. M.: "Current Status of In-Situ Recovery From the Tar Sands of Alberta," *Jour. Cdn. Pet. Tech.* (Jan.-March, 1975).

Farouq Ali, S. M.: "Effects of Steam Pressure and Water Saturation On the Performance of a Steamflood," *Producers Monthly.* (Oct., 1966).

Farouq Ali, S. M.: "Steam Stimulation — An Evaluation," *Producers Monthly.* (Oct., 1967).

Farouq Ali, S. M.: "Stimulation by Thermal Methods," *Producers Monthly.* (April, 1968).

Franco, A.: "Shell de Venezuela Steams Multiple Pays," *O. & G. Jour.* (Sept. 16, 1974).

Froning, S. P. and Birdwell, B. F.: "Here's How Getty Controls Injectivity Profile in Ventura," *O. & G. Jour.* (Feb. 10, 1975).

Giusti, L. E.: "CSV Makes Steam Soak Work in Venezuela Field," *O. & G. Jour.* (Nov. 4, 1974).

Hagoort, J., Leijnse, A. and van Poelgeest, R.: "Steam-Strip Drive: A Potential Tertiary Recovery Process," *Jour. Pet. Tech.* (Dec., 1976).

Herbeck, E. F., Heintz, R. C. and Hastings, J. R.: "Fundamentals of Tertiary Oil Recovery — Part 8 — Thermal Recovery by Hot Fluid Injection," *Pet. Eng.* (Aug., 1976).

Herbeck, E. F., Heintz, R. C. and Hastings, J. R.: "Fundamentals of Tertiary Oil Recovery — Part 9 — Thermal Recovery by In-Situ Combustion," *Pet. Eng.* (Feb., 1977).

Johnson, F. S.: "ERDA In-House Research on Stimulating Heavy Oil Recovery by In Situ Combustion," *ERDA Enhanced Oil, Gas Recovery and Improved Drilling Methods,* Vol. 1 - Oil, Tulsa, OK. (Sept., 1977).

Jones, J.: "Cyclic Steam Reservoir Model for Viscous Oil, Pressure Depleted, Gravity Drainage Reservoirs," SPE 6544 presented at the SPE-AIME 47th Annual California Regional Meeting, Bakersfield, CA. (April, 1977).

Kuo, C. H., Shain, S. A. and Phocas, D. M.: "A Gravity Drainage Model for the Steam-Soak Process," *Soc. Pet. Eng. Jour.* (June, 1970). *Trans.,* AIME (1970) 249.

Kuuskraa, V. A., Muller, J. M. and Vipperman, O. T.: "Potential and Economics of Enhanced Oil Recovery," *The Future Supply of Nature-Made Petroleum and Gas,* Pergamon Press, New York (1976).

Martin, J. C.: "A Theoretical Analysis of Steam Stimulation," SPE 1579 presented at the SPE-AIME 41st Annual Fall Meeting, Dallas, TX. (Oct., 1966). Revised 1967. *SPE Reprint Series No. 10 Thermal Recovery Techniques.* 1972 Edition.

McIntyre, H.: "Imperial Plans Giant Syncrude Works," *Canadian Chemical Processing.* (Dec., 1977).

Miller, J. S. and Larman, J. H.: "ERDA In-House Research on Heavy Oil Recovery Using Unconventional Methods," *ERDA Symposium on Enhanced Oil and Gas Recovery,* Vol. 1 - Oil, Tulsa, OK. (Sept., 1976).

Myhill, N. A. and Stegemeier, G. L.: "Steam-Drive Correlation and Prediction," *Jour. Pet. Tech.* (Feb., 1978).

Niko, H. and Troost, P. J. P. M.: "Experimental Investigation of the Steam-Soak Process in a Depletion-Type Reservoir," SPE 2978 presented at the SPE-AIME 45th Annual Fall Meeting, Houston, TX. (Oct., 1970). *SPE Reprint Series No. 10 Thermal Recovery Techniques.* 1972 Edition.

Noran, D.: "Growth Marks Enhanced Oil Recovery," *O. & G. Jour.* (March 27, 1978).

Ramey, H. J., Jr. and Brigham, W. E.: "A Review of Thermal Recovery Applications in the United States," *The Future Supply of Nature-Made Petroleum and Gas,* Pergamon Press, New York (1976).

Rivero, R. T. and Heintz, R. C.: "Resteaming Time Determination — Case History of a Steam-Soak Well in Midway Sunset," *Jour. Pet. Tech.* (June, 1975).

Seba, R. D. and Perry, G. E.: "A Mathematical Model of Repeated Steam Soaks of Thick Gravity Drainage Reservoirs," *Jour. Pet. Tech.* (Jan. 1969). *Trans,* AIME (1969) 246.

Sperry, J. S.: "Development and Field Testing of the Vapor Therm Process, Iola, Kansas," *ERDA Enhanced Oil, Gas Recovery and Improved Drilling Methods,* Vol. 1 - Oil, Tulsa, OK. (Sept., 1977).

"Steam-Distillation Pilot Looking Good," *O. & G. Jour.* (Oct. 11, 1971).

Thurber, J. L. and Welbourn, M. E.: "How Shell Attempted to Unlock Utah Tar Sands," *Pet. Eng.* (Nov., 1977).

Tyvand, P. A.: "Heat Dispersion Effect on Thermal Convection in Anisotropic Porous Media," *Jour. of Hydrology.* (1977).

Williamson, A. S., Drake, L. P. and Chappelear, J. E.: "A Steam Soak Well Model for an Isothermal Reservoir Simulator," SPE 5739 presented at the 4th Annual Symposium on Numerical Simulation of Reservoir Performance, Los Angeles, CA. (Feb., 1976).

Wu, C. H.: "A Critical Review of Steamflood Mechanisms," SPE 6550 presented at the SPE-AIME 47th Annual California Regional Meeting, Bakersfield, CA. (April, 1977).

Yoelin, S. D.: "The TM Sand Steam Stimulation Project," *Jour. Pet. Tech.* (Aug., 1971). *Trans.,* AIME (1971) 251.

Chapter 2
STEAM FLOODING
(Including Hot Water Injection)

INTRODUCTION

The primary function of thermal recovery methods is to reduce the viscosity of the in-place oil. In the process, many reservoir volumes of hot water, steam or air are injected in the reservoir, which further enhances the driving forces. The capillary forces are not directly affected by the heat, but when the oil trapped by capillarity is heated, its light fractions are distilled and become mobile.

Steam injection is currently the principal enhanced oil recovery method. The number of steam injection wells in California alone has risen from 140 in 1966 to 1630 in 1975 (Ramey and Brigham, 1976). More impressive yet is the present daily oil recovery by steam injection (Prats, 1978) which amounts to 405 MBPD worldwide (U.S. 240, Venezuela 148, Canada 9 and others 8). This is compared to 55 MBPD attributed to steam stimulation and in situ combustion. All other enhanced recovery methods combined account for 221 MBPD.

The main advantage of steam injection over other enhanced oil recovery methods is that steam can be applied to a wide variety of reservoirs. Two limiting factors are: depth (less than 5000 ft) and reservoir thickness (greater than 10 ft). The depth limitation is imposed by the critical pressure of steam (3202 psia); the reservoir thickness is determined by the rate of heat loss to base and cap rock. Other reservoir parameters beneficial to steam injection are: 1) oil gravity above 12° API; 2) oil viscosity between 100-10000 cp at reservoir temperature; 3) permeability above 50 md; 4) porosity above 25%. However, these parameters should be considered as guidelines only (Prats, 1978).

Steam and water are both excellent heat carriers, but the heat content of a unit mass of steam is much higher than that of water at the same temperature and pressure. For example, the heat content of one pound of saturated steam at 401°F and 250 psi is 1133 Btu; the heat content of one pound of water at the same temperature and pressure is only 308 Btu. When the relative permeability to steam is considered, there are instances when water is a better carrier of heat than steam on a volumetric basis (Sarem and Hawthorne, 1966). But for a given amount of heat, steam introduces much less water into the formation. As a result, less water is produced with the oil, and the less water produced, the more heat that remains in the formation (Dietz, 1975).

Hot water injection may be preferred in shallow reservoirs containing oils in the viscosity range of 100-1000 cp, but because of the excessive heat losses in surface transmission, wellbore and reservoir rock, steam injection is generally preferred (Farouq Ali, 1974). Furthermore, field tests of hot water flooding have been hampered by viscous fingering and low volumetric sweep efficiency.

The temperature of the steam is determined by the injection pressure, but its quality is determined by the characteristics of the steam generating unit. Most units utilized in oil fields put out 80% quality steam and require very high quality feedwater. This is a disadvantage, especially in areas where potable water is in short supply. The amount of water needed depends on several design and performance factors, but the average oil-steam ratio is 0.2 Bbl/bbl of water converted into steam.

Steam flooding is a process similar to waterflooding. A suitable well pattern is chosen and steam is injected into a number of wells while the oil is produced from other wells. Ideally, injected steam forms a steam saturated zone around the injection well (Figure 1). The temperature of the steam zone is nearly equal to that of the injected steam. As the steam moves away from the well, its temperature drops as it continues to expand in response to pressure drop. At some distance from the well, which mainly depends on the initial temperature of the steam and rate of pressure change with distance from the well, the steam condenses and forms a hot water bank. In the steam zone, oil is displaced by steam

STEAM FLOODING

distillation and gas (steam) drive. In the hot water zone, physical changes in the characteristics of the oil and reservoir rock take place and result in oil recovery. These changes are: thermal expansion of the oil, reduction of viscosity and residual saturation and changes in relative permeabilities.

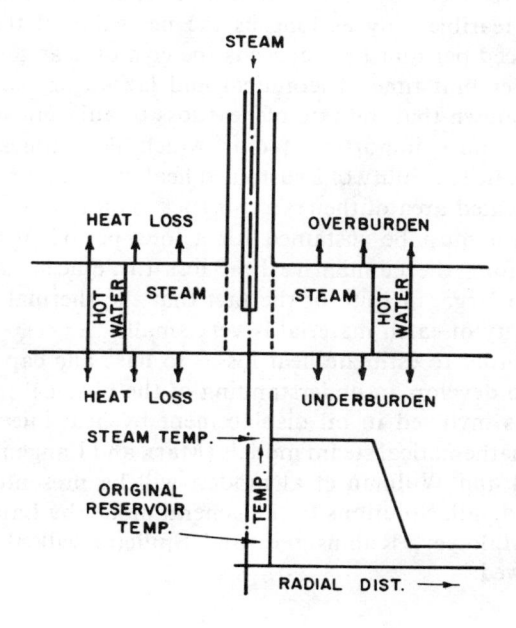

Fig. 1. Schematic diagram of steam injection and approximate distribution of formation temperature.
(after Farouq Ali, 1974)

Actual performance of the steam flood is considerably different from the ideal situation. When steam is injected, it usually forms a finger-like channel through the easiest conduit and quickly reaches the producing well. With time and continued injection, the steam finger, being less dense than the surrounding oil, travels upward in the reservoir and blankets the oil. This gravity override by the steam results in the upper one-third of the reservoir being swept by steam and the remaining two-thirds being swept by hot water, thus resulting in uneven vertical sweep efficiencies.

Gravity overrides are aggravated by the presence of a gas zone (Farouq Ali and Meldau, 1978). Injection of steam at the bottom of the reservoir may be effective in reducing override severity, but only when the reservoir properties and oil viscosity throughout the reservoir are homogeneous and no bottom water zone is present (since bottom water forms an easy conduit for the steam). In multilayered reservoirs, steam injection must take place at different intervals in order to ensure even distribution of the steam throughout the oil zone. For heterogeneous reservoirs, chemicals and high temperature gel have been developed to plug steam thief zones (Fitch and Minter, 1976 and Knapp and Welbourn, 1978).

It is often necessary to steamsoak the producing wells before steam injection is begun. This is done in order to reduce the backpressure that would develop at the injection well when the cold, viscous oil near the producing well moves in response to the steam injection.

The design of steam flood projects requires clear understanding of steam properties and the physical mechanisms involved in oil displacement by both steam and water. It is also necessary to be able to estimate heat losses in order to properly calculate the capacities of needed steam generating equipment.

TECHNICAL DISCUSSION

Properties of Steam

When one pound of water at an initial temperature t_i(°F) is heated at a constant pressure p_s (psia), it will attain a maximum temperature t_s, called the saturation temperature, before it is converted into steam. The amount of heat absorbed by the water h_w is given by:

$$h_w = C_w (t_s - t_i), \; t_i \geq 32°F$$

where:

C_w is the specific heat of the water (Btu/lb − °F) in the temperature range t_i to t_s.

With continued supply of heat, the water temperature does not change until all the water is converted into steam. The amount of heat ℓ (Btu/lb) required to change the water from a liquid at temperature and pressure of t_s and p_s, respectively, to steam at the same temperature and pressure is called the enthalpy of vaporization or the latent heat of steam. The steam at t_s and p_s is called saturated steam. Its heat content h_s is the enthalpy of the steam and is given by:

$$h_s = h_w + \ell$$

Further heating of the steam to a temperature t_{sup} above t_s, while maintaining the pressure at p_s, converts the steam from saturated to superheated steam. The heat content (enthalpy) h_{sup} of the superheated steam is given by:

$$h_{sup} = h_s + C_s (t_{sup} - t_s)$$

where:

C_s is the specific heat of steam in the temperature range of t_s to t_{sup}.

If the amount of heat applied to the water at the saturation temperature t_s is $X\ell$ (where X is a fraction) only the fraction X (lb) of the water will be converted into steam. The steam in this case would be a mixture of saturated water and saturated steam. This steam is termed wet of quality X. Its heat content h_s, or enthalpy of the mixture, is given by:

$$h_s = h_w + X \ell$$

The volume of one pound of wet steam of quality X is given by:

$$V = (1 - X) V_w + X V_s$$

where:

V_w and V_s are the volumes of saturated liquid (water) and saturated steam, respectively

Figure 2 shows that the enthalpy of 20% quality steam at 400 psia and 450°F is about 580 Btu/lb. The figure also shows that at this same pressure and temperature, the enthalpy of saturated water (X = 0) is about 425 Btu/lb, and the enthalpy of saturated steam (X = 1) is 1200 Btu/lb. This means that the energy content of steam is about 2.8 times that of water at the same temperature and pressure.

When steam is superheated, its heat energy content increases. However, it is interesting to note that at a pressure of 300 psia, the energy content of superheated steam increases only about 0.05% per 1°F rise in temperature above the saturation temperature, and the heat content increases by 0.1% per 1°F at 1500 psia.

Model Studies

Oil recovery by sustained heat injection is economically feasible only as long as the net value of the oil displaced per unit time exceeds the cost of heat generation per unit time. Theoretical and laboratory studies have shown that the rate of heat loss to adjacent strata is the most important factor which determines the economic feasibility of a sustained heat injection project. The heated area of the reservoir rock is quite large, and the heat must be sustained for a long period of time. Therefore, the cumulative heat loss to adjacent strata is also large, in spite of the fact that the thermal conductivity of earth material is very small.

In order to estimate heat losses to base and cap rock and to develop an understanding of the physical mechanisms involved in oil displacement by heat injection, two mathematical steam models (Marx and Langenheim model and Willman et al model) will be presented in some detail. Solutions to the general model by Lauwerier, Malofeev, Rubinshtein and Spillette will also be reviewed.

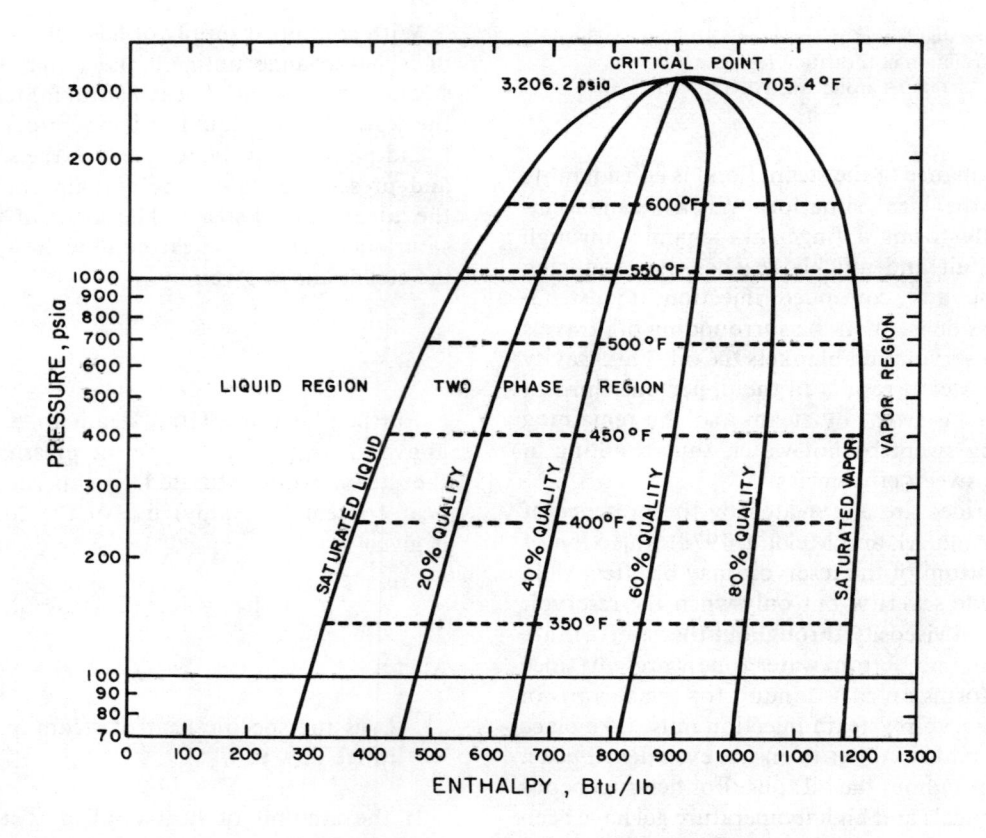

Fig. 2. Pressure-enthalpy chart for saturated steam showing steam quality lines.
(from the Oil and Gas Journal)

Marx and Langenheim Model

This model was first introduced by Marx and Langenheim in 1959 and was further clarified by Farouq Ali in 1966. Although the authors did not discuss the basic assumptions upon which the model was formulated, they have implicitly assumed that the reservoir base and cap rock are geometrically, hydrologically and thermally homogeneous and isotropic, and that radial heat conduction can be ignored. In addition, they have assumed that only steam is displacing the oil without a hot water bank ahead of it, and that the fluids are incompressible. Figure 3 shows the reservoir and base and cap rock at an initial temperature T_r. The thicknesses of the base and cap rock are assumed infinite. At time $t = 0$, heat is applied to the face of the reservoir rock and the temperature T_s is sustained. Consider the origin of the y coordinate to be at the contact between reservoir and cap rock. The differential equation which describes the heat flow in the y direction is given by:

$$\frac{\partial T}{\partial t} = D \frac{\partial^2 T}{\partial y^2} \qquad (1)$$

The initial and boundary conditions are: $T(y,0) = T_r$ for $0 \leq y \leq \infty$; and $T(0,t) = T_s$. D is the thermal diffusivity of the cap rock, defined as $K/\rho_c C_c$, where K is the thermal conductivity (Btu/ft-hr-°F), and ρ_c and C_c are the density (lb/ft³) and specific heat (Btu/lb-°F), respectively, of the cap rock.

The solution to Equation 1 is given by:

$$T(y,t) = T_s - \frac{\Delta T}{2\sqrt{Dt}} \, \text{erf}(x) \qquad (2)$$

where:

$$x = y^2/4Dt$$

$$\Delta T = (T_s - T_r)$$

$$\text{erf}(x) = \frac{1}{\sqrt{\pi}} \int_0^x e^{-t^2} dt$$

Equation 2 gives the temperature at any point y in the cap rock at any time t following application of sustained heat at the face of the reservoir rock. The heat H_t, conducted in the vertical direction is given by:

$$H_t = -K \left[\frac{\partial T}{\partial y} \right]_{y=0}$$

or

$$H_t = \frac{K \, \Delta T}{\sqrt{\pi Dt}} \qquad (3)$$

| CAP ROCK, Tr |
| RESERVOIR, Tr |
| BASE ROCK, Tr |

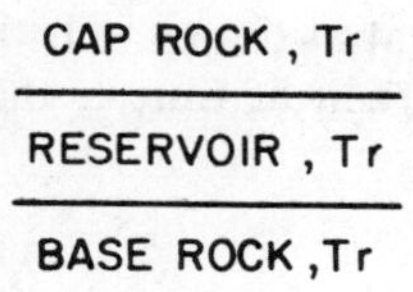

Fig. 3. Initial temperature of reservoir, base and cap rock.

If a continuous supply of steam or hot water is injected in the reservoir, then the heat will propagate into the reservoir and hence, the area of the cap rock through which heat is lost will expand continuously with time. Fig. 4 shows three stages of heat propagation, assuming no temperature gradients in the sand in the vertical direction, and that the temperature distribution is a step function. Figure 4A shows the heat wave occupying an area ΔA, at time $\tau = 0$.

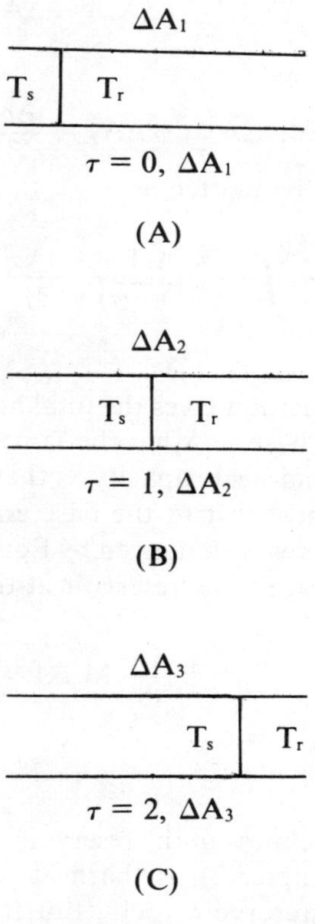

ΔA_1

| T_s | T_r |

$\tau = 0, \Delta A_1$

(A)

ΔA_2

| T_s | T_r |

$\tau = 1, \Delta A_2$

(B)

ΔA_3

| T_s | T_r |

$\tau = 2, \Delta A_3$

(C)

Fig. 4. Propagation of unit step heat wave.

where:

τ = time step

t = total time since beginning of injection

$\tau < t$.

From Equation 3, it is evident that at time t, the heat loss to the cap rock is given by:

$$H_t = \frac{K \, \Delta T}{\sqrt{\pi D(t-0)}} \, \Delta A_1$$

Figure 4B shows that at time $\tau = 1$, the heat wave occupies an area ΔA_2, then at time t, the total heat loss is given by:

$$H_t = \frac{K \Delta T}{\sqrt{\pi D(t-0)}} \Delta A_1 + \frac{K \Delta T}{\sqrt{\pi D(t-1)}} (\Delta A_2 - \Delta A_1)$$

Likewise, the total heat loss when the heat wave covers an area ΔA_3 (Figure 4C) is given by:

$$H_t = \frac{K \Delta T}{\sqrt{\pi D(t-0)}} \Delta A_1 + \frac{K \Delta T}{\sqrt{\pi D(t-1)}} (\Delta A_2 - \Delta A_1) + \frac{K \Delta T}{\sqrt{\pi D(t-2)}} (\Delta A_3 - \Delta A_2)$$

Therefore, as the heat wave travels in the reservoir, the total heat loss to the cap rock is given by:

$$H_t = \sum_{\tau=0}^{t} \frac{K \Delta T}{\sqrt{\pi D(t-\tau)}} (\Delta A_{\tau+1} - \Delta A_\tau) \text{ and } \Delta A_0 = 0$$

Since in the limit $\Delta A_{\tau+1} - \Delta A_\tau = \frac{\partial A_\tau}{\partial \tau} d\tau$, the above expression may be written as:

$$H_t = \int_0^t \frac{K \Delta T}{\sqrt{\pi D(t-\tau)}} \frac{\partial A_\tau}{\partial \tau} d\tau \quad (4)$$

The above equation gives the total heat loss to the cap rock, and if the base rock has the same thermal conductivity, density and heat capacity as that of the cap rock, then the total heat lost to the base and cap rock combined would be twice that given by Equation 4. The heat U utilized in heating the reservoir at time t is given by:

$$U = h \frac{dA}{dt} M \Delta T$$

where:

h is the thickness of the reservoir, and M (see Equation 2, Chapter 3), is the heat capacity per ft^3 of saturated reservoir rock (Btu/ft^3-°F). Therefore, the constant heat injection rate H_o (Btu/hr) is given by:

$$H_o = 2 \int_0^t \frac{K \Delta T}{\sqrt{\pi D(t-\tau)}} \frac{\partial A}{\partial \tau} d\tau + h \frac{dA}{dt} M \Delta T \quad (5)$$

The solution to Equation 5 yields:

$$A(t) = \left[\frac{H_o M h D}{4 K^2 \Delta T}\right]\left[e^{x^2} \text{erfc}(x) + \frac{2x}{\sqrt{\pi}} - 1\right] \quad (6)$$

where:

A(t) is the cumulative area in ft^2

$$\text{erfc}(x) = 1 - \text{erf}(x)$$

$$x = \left[\frac{2K}{Mh\sqrt{D}}\right] t^{1/2} \text{ (dimensionless)}$$

The oil displacement rate V_o (Bbl/day) for this idealized reservoir is given by:

$$V_o = 4.273 \left[\frac{H_o \phi (S_o - S_{or})}{M \Delta T}\right] (e^{x^2} \text{erfc } x) \quad (7)$$

where:

ϕ = porosity

S_o = oil saturation

S_{or} = residual oil saturation

The heat loss (W_o^*) to cap and base rock as a fraction of total heat injected was given by Ramey (1965) for the Marx and Langenheim model, as follows:

$$W_o^* = 1 - \frac{1}{t_D}\left[e^{t_D} \text{erfc}(\sqrt{t_D}) + 2\sqrt{t_D/\pi} - 1\right] \quad (8)$$

where:

$t_D = 4Dt/h^2$ (dimensionless)

Equations 6, 7 and 8 constitute the set of equations needed to evaluate a heat injection project according to the Marx and Langenheim model. The authors gave the following example. Suppose that 5000 lb/hr of saturated steam are to be continuously injected in a single well at 500 psig into a reservoir under the following conditions: h = 20 ft; ϕ = 0.25; S_w = 0.20; S_o = 0.60; S_{or} = 0.10; T_r = 80 °F; T_s = 470° F at 514.7 psia; steam energy cost $\$_h$ = \$2.00/million Btu; net value of displaced oil $\$_o$ = \$8.00; specific heat of rock C_r = 0.21 Btu/lb-°F; specific heat of water C_w = 1.0 Btu/lb-°F; specific heat of oil C_o = 0.5 Btu/lb-°F; rock grain density ρ_r = 167 lb/ft^3; water density ρ_w = 62.4 lb/ft^3; oil density ρ_o = 50.0 lb/ft^3; thermal conductivity of base and cap rock K = 1.50 Btu/ft-hr-°F; thermal diffusivity of base and cap rock D = 0.0482 ft^2/hr; available heat of steam at 470°F, 500 psia (Figure 2) = 1150 Btu/lb. It is required to estimate: area swept out in the first 1000 hrs; oil displacement rate at 1000 hrs; economic areal limit for sustained heat application; and time required to reach economic areal limit.

The solution is as follows:
1. Thermal energy input per hour =
   $H_o \times 1150 = 5.74$ MMBtu/hr
2. $\Delta T = T_s - T_r = 390°F$
3. $M = (1 - \phi)\rho_r C_r + S_w \phi \rho_w C_w + S_o \phi \rho_o C_o$
   $= 33.2$ Btu/ft^3-°F
4. $x = \left[\dfrac{2K}{Mh\sqrt{D}}\right] t^{1/2} = \left[\dfrac{2 \times 1.50}{33.2 \times 20 \sqrt{0.0482}}\right] 1000^{1/2}$
   $= 0.651$
5. From Table I, e^{x^2} erfc$(x) = 0.545$, and
   $\left[e^{x^2} \text{erfc}(x) + \dfrac{2x}{\sqrt{\pi}} - 1\right] = 0.281$
6. Substituting into Equation 6, A (t = 1000 hr) = 14700 ft^2, or 0.338 acres.
7. Substituting into Equation 7, $V_o = 130$ BOPD at t = 1000 hrs.
8. The economic limit is reached when the net value of the displaced oil per unit time is equal to the cost of heat per unit time. From Equation 7, it follows that the economic limit is reached when:

$$\left[e^{x^2} \text{erfc}(x)\right]_l = (5.618 \times 10^{-6}) \left[\dfrac{\$_h M \Delta T}{\$_o \phi (S_o - S_{or})}\right]$$

where:

$\left[e^{x^2} \text{erfc}(x)\right]_l$ = practical limit of $e^{x^2}\text{erfc}(x)$

$\$_h$ = cost of heat energy (dollars/million Btu)

$\$_o$ = net value of oil (dollars/bbl)

Accordingly, for this example $\left[e^{x^2}\text{erfc}(x)\right]_l = 0.182$, and with the help of Table I, $x_l = 2.948$. Inserting this value of x in Equation 6, we get $A_l = 112000$ ft^2 or 2.57 acres.

9. From the definition of x as given in (4) above, we have: $2.948 = 0.0206$ $t_l^{1/2}$; or $t_l = 20500$ hrs, or 854 days.

If it is desired to know the rate of heat loss at the economic limit, Equation 8 can be used. Instead, Ramey (1965) prepared the graph shown in Figure 5. For this example, $\lambda = M/M_{ob} = 33.2/31.12 = 1.07 \cong 1.0$ and,

$$\log_{10}\left(\dfrac{4Dt}{h^2}\right) = \log_{10}\left(\dfrac{4 \times 0.0482 \times 20500}{20^2}\right) = 0.99$$

which, from Figure 5, gives a value of 72%; i.e., when the economic limit is reached, 72% of the heat injected into the reservoir is being lost to the base and cap rock. It should be noted that Figure 5 can be used to estimate heat losses at any time during heat injection.

According to Ramey (1959), a variable injection rate can be expressed as a series of step functions such that $H = H_1 + H_2 + \ldots H_n$, H_1. The terms H_1 H_2 ... denote injection rates (Btu/hr) at successive time periods. Thus, the heated area is given by:

$$A(t) = \dfrac{MDh}{4K^2 \Delta T} \left\{ H_n \left[e^{x_n^2} \text{erfc } x_n + \dfrac{2x_n}{\sqrt{\pi}} - 1\right] + \right. \quad (9)$$

$$\left. \sum_{m=1}^{M=n-1} (H_m - H_{m+1}) \left[e^{x_m^2} \text{erfc } X_m + \dfrac{2x_m}{\sqrt{\pi}} - 1\right] \right\}$$

Note that for constant injection rates ($H_m = H_{m+1}$) the rates within the summation are zero, and the expression simplifies to Equation 6. From Equation 9, the change in the steam area is given by:

$$A(t_{n+1}) = A(t_n) + \dfrac{MDh\, H_{n+1}}{4K^2(\Delta T)_{n+1}} \left[f(t_{n+1}) - f(t_n)\right] \quad (9')$$

where:

$\Delta T_{n+1} = T_s(t_{n+1}) - T_r$

$f(t_n) = \left[e^{x_n^2} \text{erfc } x_n + \dfrac{2x_n}{\sqrt{\pi}} - 1\right]$

Equation 9 can be used when steam injection is followed by cold water, and Equation 9' is used in oil recovery calculations (Farouq Ali, 1970).

Willman et al Model

The authors presented an equation to calculate the size of the swept area at any time since the beginning of steam injection. Figure 6 is the graphical representation of their equation as given by the authors. Table II shows the correspondence between the symbols used by Marx and Langenheim and by Willman et al. As an application of Figure 6, we consider the previous data given by Marx and Langenheim, except that h in this example is 22 ft.

Therefore, $\dfrac{h}{8} \sqrt{\pi/\alpha}\, \dfrac{(\rho C_p)_r}{(\rho C_p)_{OB}} = 25.02$

and, $\dfrac{K(T_s - T_r)}{14.6\, i_{st} h_{fg}} \sqrt{\pi/\alpha} = 821.534 \times 10^{-6}$

From Figure 6, at 2.34 years (854 days), the ordinate is approximately equal to 38, from which R = 215 ft and the swept area (πR^2) is 145314 ft^2. It is evident that Figure 6 yielded an area which is somewhat larger than that given by Marx and Langenheim's model (112,000 ft^2 for h = 20 ft). Figure 5 shows that vertical heat loss as predicted by the Willman et al equation is slightly less than that given by Marx and Langenheim.

TABLE I

x	$e^{x^2}\text{erfc}(x)$	$e^{x^2}\text{erfc}(x)\dfrac{2x}{\sqrt{\pi}} - 1$	x	$e^{x^2}\text{erfc}(x)$	$e^{x^2}\text{erfc}(x)\dfrac{2x}{\sqrt{\pi}} - 1$	x	$e^{x^2}\text{erfc}(x)$	$e^{x^2}\text{erfc}(x)\dfrac{2x}{\sqrt{\pi}} - 1$
0.00	1.00000	0.00000	1.00	0.42758	0.55596	4.50	0.12248	4.20019
.02	.97783	.00039	.05	.41430	.59910	.60	.11994	.31048
.04	.95642	.00155	.10	.40173	.64295	.70	.11749	.42087
.06	.93574	.00344	.15	.38983	.68746	.80	.11514	.53136
.08	.91576	.00603	.20	.37854	.73259	.90	.11288	.64194
0.10	0.89646	0.00929	1.25	0.36782	0.77830	5.00	0.11070	4.75260
.12	.87779	.01320	.30	.35764	.82454	.20	.10659	4.97417
.14	.85974	.01771	.35	.34796	.87127	.40	.10277	5.19602
.16	.84228	.02282	.40	.33874	.91847	.60	.09921	.41814
.18	.82538	.02849	.45	.32996	0.96611	.80	.09589	.64049
0.20	0.80902	0.03470	1.50	0.32159	1.01415	6.00	0.09278	5.86305
.22	.79318	.04142	.55	.31359	.06258	.20	.08986	6.08581
.24	.77784	.04865	.60	.30595	.11136	.40	.08712	.30874
.26	.76297	.05635	.65	.29865	.16048	.60	.08453	.53184
.28	.74857	.06451	.70	.29166	.20991	.80	.08210	.75508
0.30	0.73460	0.07311	1.75	0.28497	1.25964	7.00	0.07980	6.97845
.32	.72106	.08214	.80	.27856	.30964	.20	.07762	7.20195
.34	.70792	.09157	.85	.27241	.35991	.40	.07556	.42557
.36	.69517	.10139	.90	.26651	.41043	.60	.07361	.64929
.38	.68280	.11158	.95	.26084	.46118	.80	.07175	7.87311
0.40	0.67079	0.12214	2.00	0.25540	1.51215	8.00	0.06999	8.09702
.42	.65912	.13304	.05	.25016	.56334	.20	.06830	.32101
.44	.64779	.14428	.10	.24512	.61472	.40	.06670	.54508
.46	.63679	.15584	.15	.24027	.66628	.60	.06517	.76923
.48	.62609	.16771	.20	.23559	.71803	.80	.06371	8.99344
0.50	0.61569	0.17988	2.25	0.23109	1.76994	9.00	0.06231	9.21772
.52	.60588	.19234	.30	.22674	.82201	.20	.06097	.44206
.54	.59574	.20507	.35	.22255	.87424	.40	.05969	.66645
.56	.58618	.21807	.40	.21850	.92661	.60	.05846	9.89090
.58	.57687	.23133	.45	.21459	1.97912	.80	.05727	10.11539
0.60	0.56780	0.24483	2.50	0.21081	2.03175	10.00	0.05614	10.33993
.62	.55898	.25858	.60	.20361	.13740			
.64	.55039	.27256	.70	.19687	.24350			
.66	.54203	.28676	.80	.19055	.35001			
.68	.53387	.30117	.90	.18460	.45690			
0.70	0.52593	0.31580	3.00	0.17900	2.56414			
.72	.51819	.33062	.10	.17372	.67169			
.74	.51064	.34564	.20	.16873	.77954			
.76	.50328	.36085	.30	.16401	.88766			
.78	.49610	.37624	.40	.15954	2.99602			
0.80	0.48910	0.39180	3.50	0.15529	3.10462			
.82	.48227	.40754	.60	.15127	.21343			
.84	.47560	.42344	.70	.14743	.32244			
.86	.46909	.43950	.80	.14379	.43163			
.88	.46274	.45571	.90	.14031	.54099			
0.90	0.45653	0.47207	4.00	0.13700	3.65052			
.92	.45047	.48858	.10	.13383	.76019			
.94	.44455	.50523	.20	.13081	.87000			
.96	.43876	.52201	.30	.12791	3.97994			
.98	.43311	.53892	.40	.12514	4.09001			

(after Marx and Langenheim, 1959)

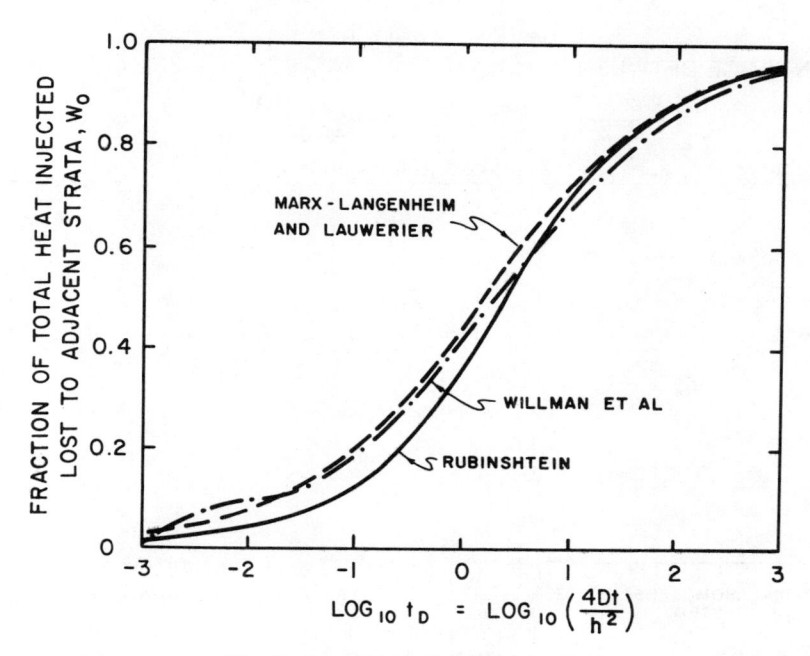

Fig. 5. Vertical heat loss versus time.
(after Ramey, 1965)

This difference could account for the larger swept area calculated from Figure 6 as compared with that calculated by Equation 6.

Finally, Willman et al used the Buckley-Leverett saturation model to predict oil recovery. Figures 7 and 8 show the saturation distribution after one year of injection and oil recovery prediction for the following data:

$S_{oi} = 85\%$; $S_w = 15\%$; $T_r = 60°F$; $H_o = 681000$ Btu/hr; $T_s = 340°F$; μ_o at $60°F = 900$ cp; μ_o/μ_w at $340°F = 16.8$; μ_o/μ_w at $60°F = 900$; $\phi = 23\%$; steam saturation in steam zone $(S_s)_{sz} = 50\%$; and residual oil saturation $= 18\%$.

The calculations were made assuming a radial system having the same area as a 0.625 acre five-spot pattern.

The General Model

This model (Spillette, 1965) is based on the assumption that the heat is transferred by two mechanisms: 1) the physical movement of the injected fluids and 2) thermal conduction. Convective heat transfer between the injected fluid and the original reservoir fluids and the permeable reservoir rock is assumed to be accounted for by the additional assumption of instantaneous thermal equilibrium in the reservoir. With these assumptions, the heat transfer model is given by:

$$\frac{\partial}{\partial t}(MT) + \nabla \cdot (\rho_f C_f \bar{V} T) = \nabla \cdot K \nabla T \qquad (10)$$

where:

$$M = (1-\Phi)\rho_r C_r + \phi \rho_o C_o S_o + \phi \rho_w C_w S_w$$

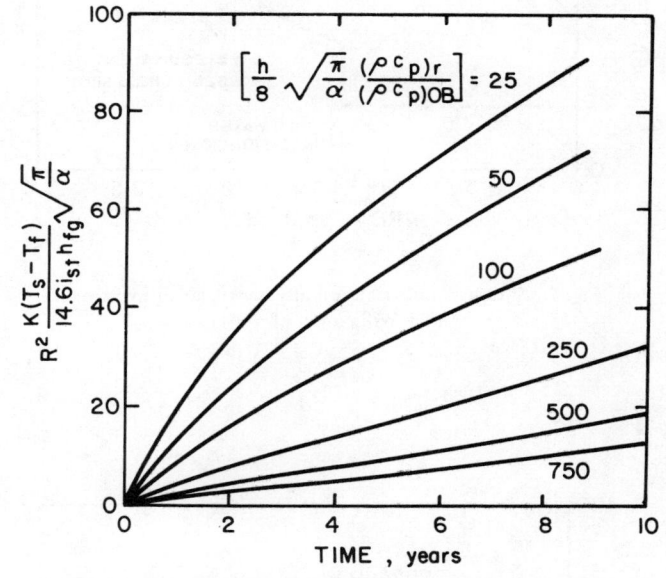

Fig. 6. Variation of R^2 with time and reservoir parameters.
(after Willman et al, 1961)

$\rho_f, C_f, \bar{V}$ respectively are: density, specific heat of fluids flowing in the reservoir and the velocity vector of the fluids.

Then for an oil-water system:

$$\rho_f C_f \bar{V} = \rho_o C_o \bar{V}_o + \rho_w C_w \bar{V}_w$$

When Equation 10 is applied to the base or cap rock, $\bar{V} = 0$, and the equation reduces to:

$$\frac{\partial}{\partial t}(MT) = \nabla \cdot K \nabla T$$

TABLE II
CORRESPONDENCE BETWEEN SYMBOLS: (WILLMAN ET AL) and (MARX AND LANGENHEIM)

Symbol		
Willman et al	Marx and Langenheim	Example
h	h	22
γ	D	.0482
K	K	1.5
$(\rho C_p)_r$	$\rho_r C_r$	35.07
$(\rho C_p)_{OB}$	K/D	31.12
T_s	T_s	470
T_r	T_r	80
$i_{st} h_{fg}$	$(H_o \times 24)/(350)$	393757.50

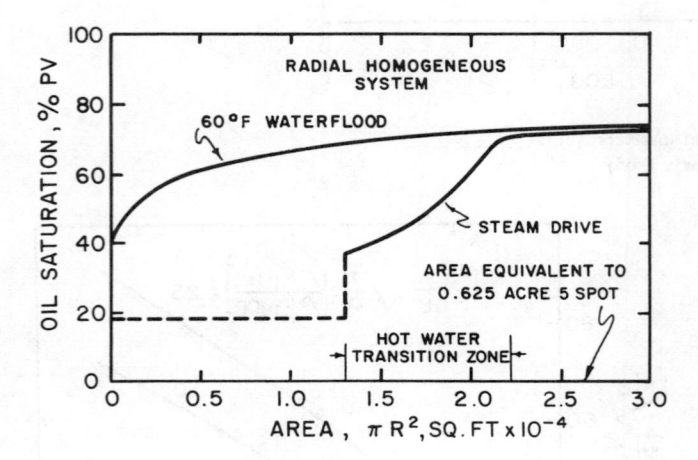

Fig. 7. Saturation distribution after one year of injection. (after Willman et al, 1961)

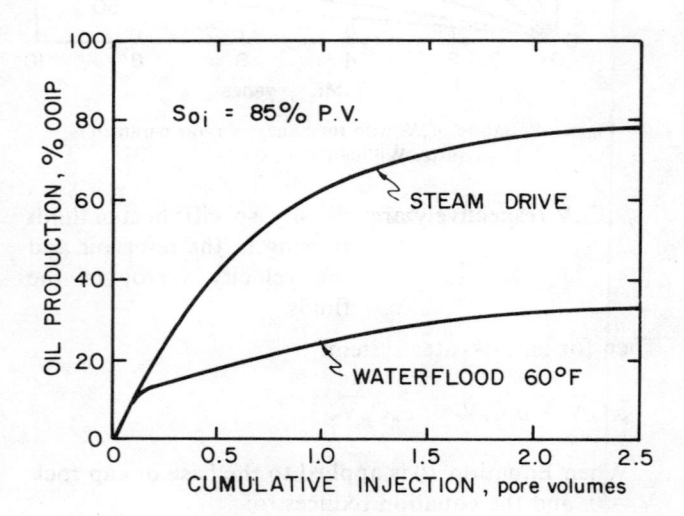

Fig. 8. Oil recovery prediction, radial homogeneous system. (after Willman et al, 1961)

For a constant M and K and one-dimensional flow, the preceding equation reduces to:

$$\frac{\partial T}{\partial t} = D \frac{\partial^2 T}{\partial y^2}$$

which is the same as Equation 1.

Equation 10 is the general energy balance equation. It is applicable to any heat injection system regardless of geometry or physical properties. Lauwerier (1955) solved this equation with the following assumptions: 1) linear, incompressible flow in a homogeneous sand; 2) constant physical properties and fluid saturations; 3) infinite thermal conductivity within the sand in the vertical direction; 4) zero horizontal thermal conductivity in reservoir, base and cap rock; and 5) constant injection temperature. The solution given by Lauwerier is as follows:

$$\bar{T} = \text{erfc} \frac{\xi/\Theta}{2\sqrt{t_D - \xi/\Theta}} \tag{11}$$

where:

$$T = \frac{\bar{T} - T_r}{\Delta T}, \text{ and T is the temperature at time t}$$

$$\xi/\Theta = t_D (\rho_r C_r / \rho_f C_f)$$

Figure 5 shows that the vertical heat loss as derived from the Marx and Langenheim model is identical to that derived from Equation 11. Figure 9 shows the sand temperature distribution as derived by Equation 11.

Malofeev (1960) showed that Equation 11 is valid for radial flow if:

$$\xi/\Theta = t_D (\pi r^2 h / Qt) (\rho_r C_r / \rho_f C_f)$$

where:

Q is the volume rate of flow (ft^3/hr).

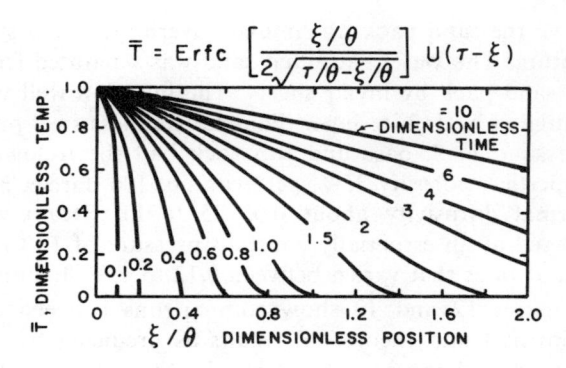

Fig. 9. Sand temperature distribution, Lauwerier's solution. (after Spillette, 1965)

Figure 10 compares Lauwerier's solution as modified by Malofeev with other solutions that will be briefly discussed later on in this section. Figure 10 is based on the following data: $\rho_f C_f = 62.4$ Btu/ft^3-°F; $\rho_r C_r = 42.45$ Btu/ft^3-°F; $D = 0.03857$ ft^2/hr; $K = 1.4$ Btu/hr-ft-°F.

Rubinshtein (1959) solved Equation 10 with the same set of assumptions given by Lauwerier except that he assumed constant, isotropic thermal conductivities in the reservoir and in the impermeable strata surrounding the reservoir. He obtained the following solution only for the vertical heat losses:

$$W_o^*(t_D) = 2/3 \sqrt{t_D/\pi} \left[1 - (1 + 1/t_D) e^{-1/t_D}\right]$$
$$+ (1 + 2/3 t_D) \text{erfc } 1/\sqrt{t_D}$$

The above equation was given by Ramey (1965) and is depicted in Figure 5.

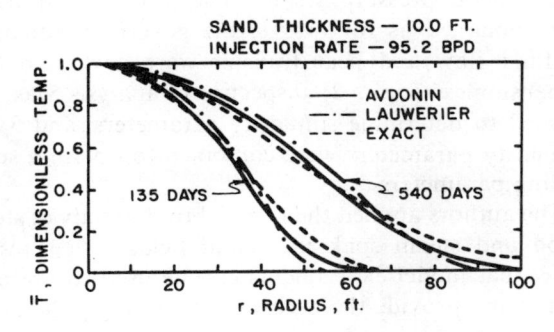

Fig. 10. Sand temperature distribution radial flow. (after Spillette, 1965)

Avdonin (1964) and Rubinshtein (1960, 1962) examined the effects of the various assumptions made by Lauwerier in regard to the thermal conductivities in the reservoir. Avdonin then solved Equation 10 for the case of a non-zero value for the thermal conductivity in the horizontal direction within the reservoir, and Rubinshtein obtained the solution for the case of radial flow in which both vertical and horizontal heat conduction were considered in the base and cap rock. Rubinshtein also included horizontal conduction in the reservoir, but the vertical conductivity remained infinite.

Spillette (1965) obtained numerical solutions of Equation 10 without the restrictive assumptions of Lauwerier. He used the method of characteristics and an alternating direction implicit (ADI) finite difference technique. His solution, which he called "exact solution" is depicted in Figure 10.

In conclusion, Figure 10 shows that the solutions given by Avdonin, Lauwerier (or Malofeev) and Spillette are very close, and thus an elaborate numerical solution such as the one given by Spillette seems unnecessary. Figure 5 shows the coincidence of Lauwerier's solution and that of Marx and Langenheim. The deviation of Rubinshtein's solution from that of Marx and Langenheim is significant only when t is small or when the reservoir is very thick. According to Flock et al (1967), the effect of neglecting the horizontal thermal conductivity of the base and cap rock is approximately 2% for injection rates equal to or higher than 7.5×10^6 Btu/hr and for thicknesses greater than 10 ft.

Whether the mathematical models can predict the performance of actual heat-injection projects is open to question. The main deficiency of these models is that they ignore the effect of heat on the flow characteristics of the porous media. In fact, little is known about the thermal behavior of these media under simultaneous flow of condensing steam and reservoir fluids. Even the effect of heat on absolute and relative permeabilities is still being debated in the literature. Weinbrandt et al (1975) conducted experiments on cores at elevated temperatures. They concluded that when the rock is heated, tight pore openings are sealed by thermal expansion, which could be the cause of the observed changes in permeabilities. Their data show that, in response to an increase in temperature, irreducible water saturation and relative permeability to oil increase; but residual oil saturation and relative permeability ratio (k_w/k_o) as well as absolute permeability decrease.

Experimental Studies

Experimental studies are aimed at estimating field behavior of the heat injection process. These studies are superior to theoretical model studies in that no restrictive, a priori assumptions need to be made in regard to the thermal parameters of the reservoir. On the other hand, experimental studies are subject to severe scaling problems, and they cannot consider all the heterogeneities of the reservoir. Shutler (1970) stated that improper scaling of capillarity in laboratory models of thick sand can lead to optimistic oil recovery predictions, and failure to account for vertical permeability can lead to pessimistic recovery predictions.

While numerical models are often restricted to 1000 grid blocks, a physical model can attain fine detail, as the number of beads that are used in constructing a modern physical model may attain 10 million (Stegemeier and Laumbach, 1977). In addition, unless the different parameters have been determined experimentally, numerical models cannot adequately predict the performance of heat injection processes. Accordingly, numerical and physical models serve a complementary role.

Willman et al Experimental Study

The authors conducted a series of steam and hot water flooding tests on cores saturated with oils of different viscosities and distillation properties. They concluded that:

1. Steam or hot water injection recovers more oil than ordinary waterflooding.
2. Increased recovery by hot water injection is caused by viscosity reduction and thermal expansion of the oil.
3. Hot water flooding is particularly applicable to viscous oils since it significantly reduces the oil-water viscosity ratio.
4. Recovery by steam injection is significantly higher than recovery by hot water flooding because, in addition to its effect on the viscosity and thermal expansion of the oil, steam distills and gas-drives the oil.
5. Because of steam distillation, oil produced shortly before steam breakthrough has a lower API gravity than the original oil-in-place.
6. Residual oil saturation after steam flooding is independent of the initial oil saturation.
7. The percentage increase in oil recovery with high pressure (and temperature) steam is less than the percentage increase in heat required to raise the temperature of the high pressure saturated steam.
8. Because of the greater amount of heat content per unit mass of steam as compared to the heat content of a unit mass of water at the same temperature, the mass of water needed as steam to heat the reservoir to a given temperature is much less than if the water is injected as a liquid.
9. To minimize heat requirements, injection rates should be high, flood patterns should be small and formations should be thick.
10. When the initial oil saturation is high, oil recovery per barrel of injected steam would also be high.

Baker's Study

This model (Figure 11) consisted of a 4 in thick circular sand pack, 72 in. in diameter. The sand pack was placed in the center of a 58-in high cylindrical pressure vessel. Water-saturated sand was placed both below and above the sand pack, to simulate overburden and substratum. The water saturated sand was separated from the sand pack by mylar sheets. The injection well was simulated by 1/2-in. inner diameter stainless steel pipe. The sand pack base and cap rock had the following properties: porosity 35%, permeability 100 darcies and thermal diffusivity about 0.03125 ft^2/hr. Steam was injected at an essentially constant pressure of 17 to 22 psia at rates that varied between 0.12 to 1.61 lb/min.

Figures 12 and 13 show comparisons between experimental and theoretical results as predicted by the Marx and Langenheim model. In these figures, the steamed volume is the volume swept by steam, and the heated volume is defined by a heated radius, which is the farthest point at which a temperature rise of 5°F was observed. The difference in volume between the heated and steamed zones is equal to the volume of the hot water zone.

Based on this study, the following conclusions were made:

1. For a given reservoir thickness and thermal diffusivity, vertical heat loss is a function of time alone. This is in agreement with theoretical conclusions.
2. The extent of the hot water zone is significant.
3. Gravity override by the steam is important at all tested rates.

Stegemeier and Laumbach's Vacuum Model

This is a modern, physical model that operates under sub-atmospheric pressures and low temperatures. The principle involved implies that, if the ratio of latent to sensible heat is properly scaled, comparable model results could be obtained at low temperatures with sub-atmospheric pressure steam. The scaling parameters were obtained as follows: 1) the governing equations of fluid flow and heat transfer were converted to a dimensionless form; 2) inspectional analysis was performed to determine similarity parameters; and 3) the similarity parameters were combined to obtain a set of scaling parameters.

The authors applied their model to the study of steam flood and steam soak in several fields. Their results show that model experiments can be used to optimize design and provide the necessary operating policies.

Wellbore Heat Loss

Unless a heat injection well is properly designed, considerable amounts of heat can be lost from the injected fluid as it travels down the wellbore. In order to appreciate the magnitude of this loss, Figure 14 shows the temperature profiles associated with water injection at a rate of 500 Bbl/day through a 7-in casing at a wellhead temperature of 400°F. After one week, the tem-

STEAM FLOODING

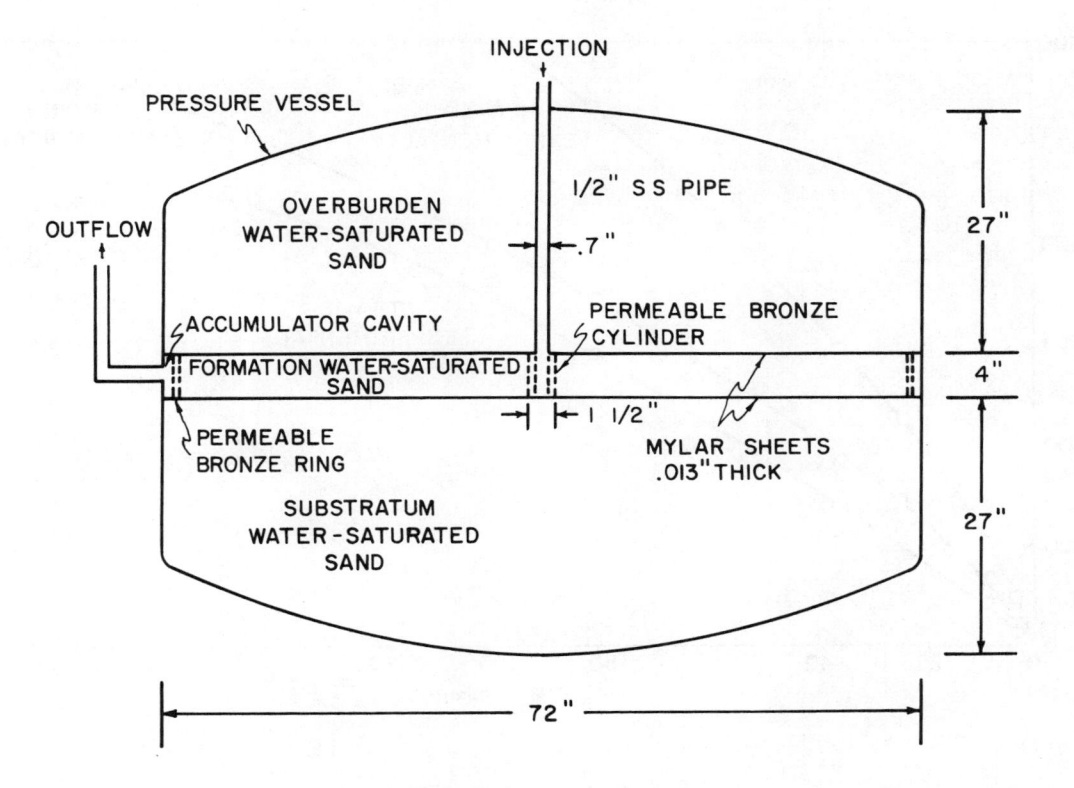

Fig. 11. Schematic drawing of flow model.
(after Baker, 1969)

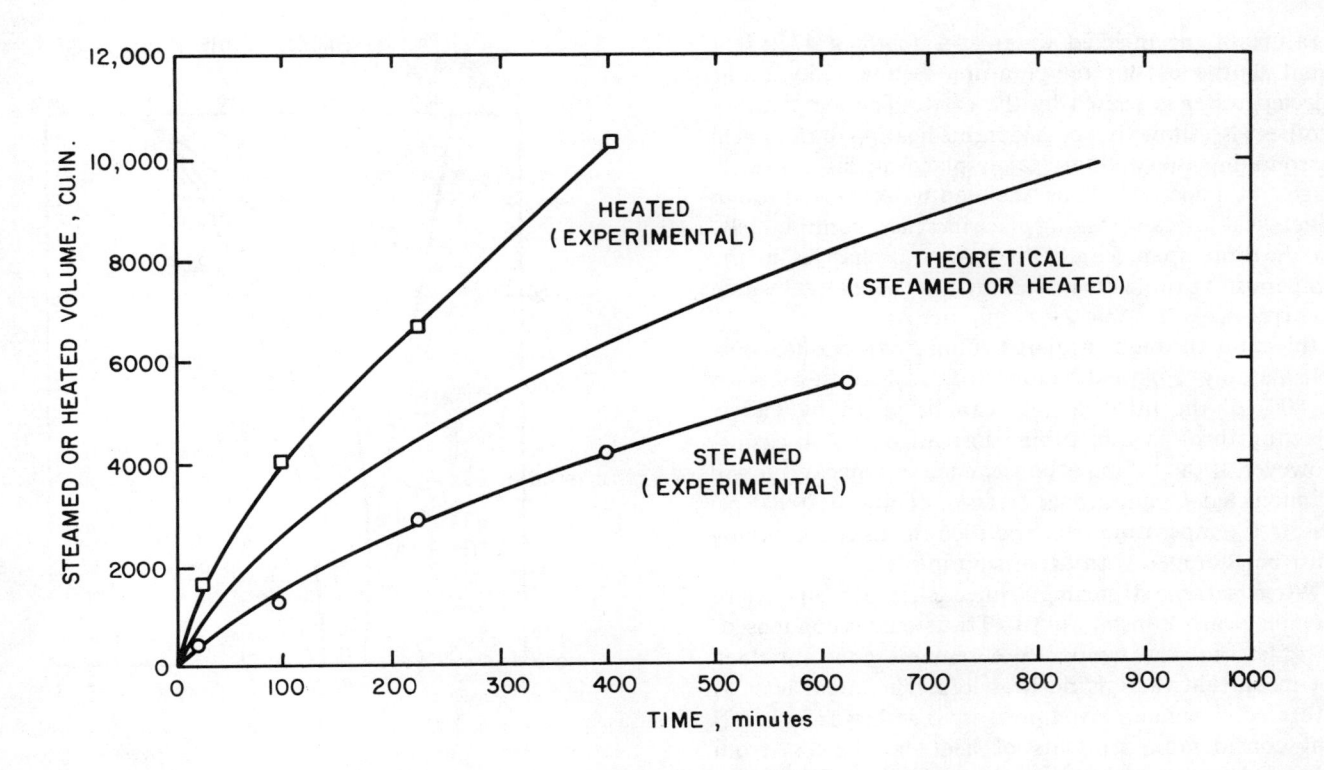

Fig. 12. Comparison of swept volumes (for steam and heat) determined from experimental and theoretical models by Marx and Langenheim theory. Injection rate 0.15 lb/min.
(after Baker, 1969)

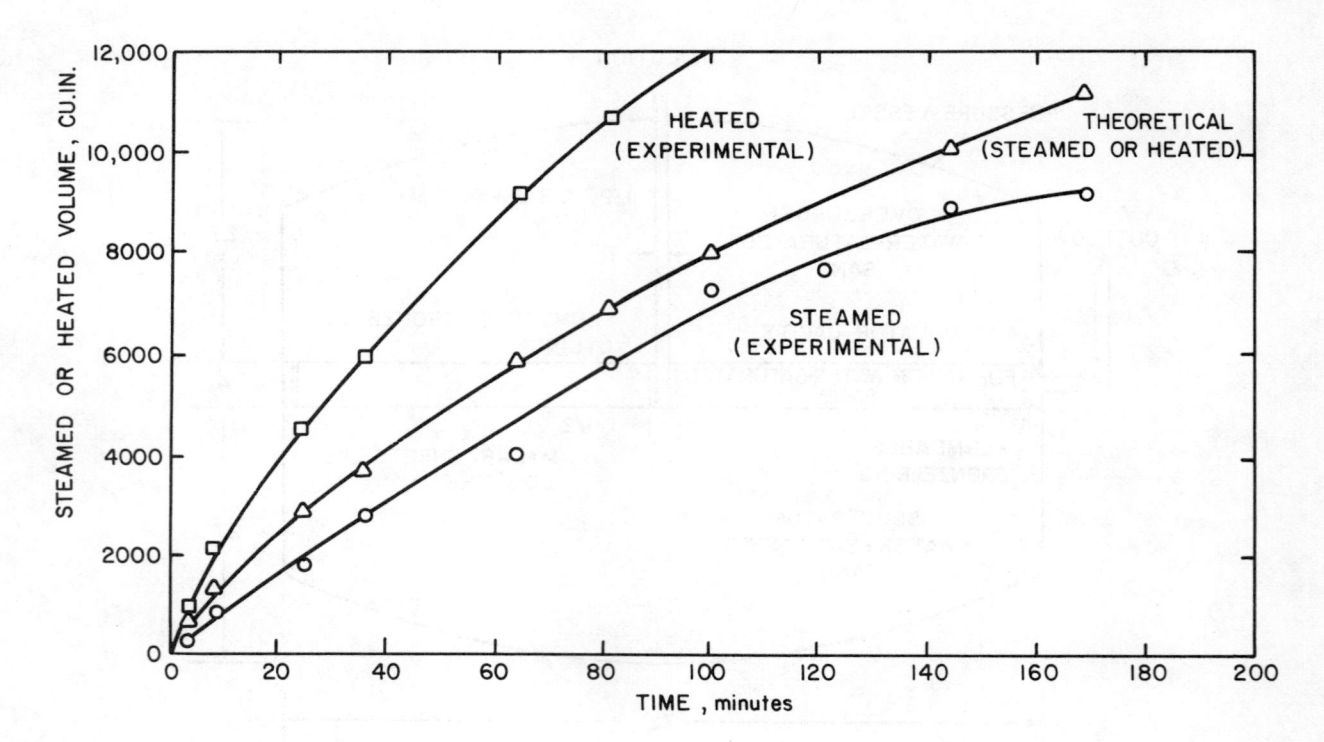

Fig. 13. Comparison of swept volumes (for steam and heat) determined from experimental and theoretical models by Marx and Langenheim theory. Injection rate 0.49 lb/min.
(after Baker, 1969)

perature of the injected water at a depth of 4750 ft is equal to the earth's temperature. Below 4750 ft, the injected water is heated by the earth. The temperature profiles also show that considerable heating of the strata surrounding the wellbore takes place during the early stages of injection. Thus the depth interval through which heat losses are taking place increases continuously. As the time span lengthens, however, changes in the temperature profile are not as large as the changes which occurred when injection was initiated.

Injection through insulated tubing reduces heat loss considerably. Figure 15 shows that at 4000 ft, as much as 80% of the injected heat can be saved by merely injecting through 2-in tubing instead of a 7-in casing. However, if the tubing is constrained in a packer, it will be subjected to additional stresses of about 10000 psi per 50°F temperature rise, and then the injection tubing must be equipped with expansion joints.

When saturated steam is injected, the temperature remains nearly constant until all the steam is condensed. The fact that the temperature remains constant does not mean that there is no heat loss. The latent heat of saturated steam at a given pressure is so large (Figure 2) that considerable amounts of heat can be lost from saturated steam without any drop in its temperature. Once all the steam has been condensed into hot water, the temperature will drop as heat is lost from the water.

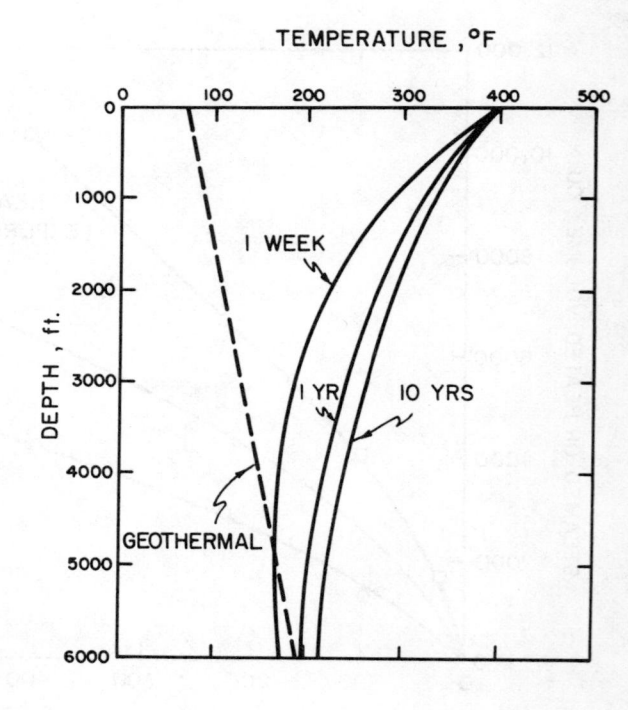

Fig. 14. Computed temperatures resulting from injection of 500 b/d of hot water down 7-in. casing.

(after Ramey, 1965)

STEAM FLOODING

CONDITIONS: SURFACE INJECTION TEMPERATURE – 397°F
STEAM INJECTED AT SATURATION PRESSURE – 223°F
HOT WATER INJECTED AT 1000 psi
HEAT LOSS BASED ON TEMPERATURE OF 130°F
TOTAL HEAT INJECTION RATES:
STEAM = 191.0 MM BTU/D
HOT WATER = 44.9 MM BTU/D

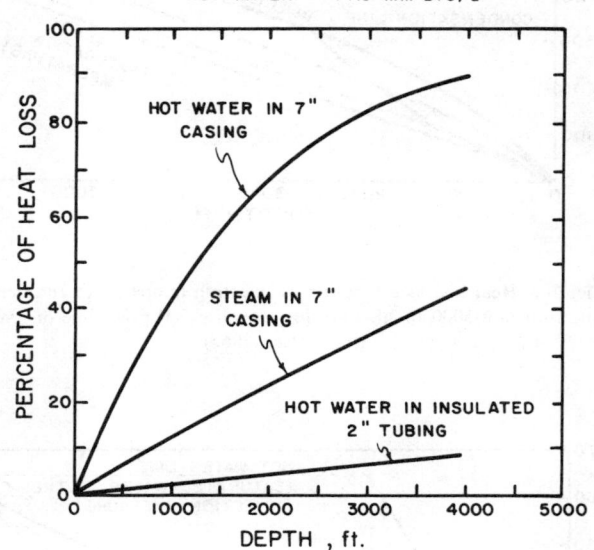

Fig. 15. Computed heat loss vs depth for injection of 500 b/d of steam or hot water for one week.
(after Ramey, 1965)

With a given set of geothermal properties, casing and tubing sizes, the main factors which affect heat losses are: 1) injection time; 2) injection rate; 3) depth of oil-bearing horizon; and 4) injection pressure (for saturated steam) and injection pressure and temperature (for superheated steam) (Satter, 1965). Figures 16-23 demonstrate the effects of these factors for the given set of parameters shown in Table III.

According to Ramey (1965), temperature in the tubing which results from injection of hot water is given by:

$$T_{tub}(z,t) = az + b - aA + (T_i + aA - b)e^{-z/A}$$

where:

$T_{tub}(z,t)$ = temperature of tubing at a depth z and time t, days

$A = Qc[K + Uf(t_D)] / (2\pi r_1 UK)$

a = geothermal gradient, °F/ft

b = surface temperature, °F

c = specific heat of water (for fresh water $c \cong 1$ Btu/lb-°F)

r_1 = inside radius of tubing, ft

r_2 = outside radius of casing, ft

$f(t_D)$ = a time function controlling heat loss to earth, evaluated from Figure 24.

U = overall heat transfer coefficient between inside tubing and outside of casing, Btu/day ft²-°F

T_i = surface injection temperature, °F

Q = water injection rate, Bbl/day

K = earth's thermal conductivity, Btu/day ft °F

D = thermal diffusivity of earth ft²/day

Example: Calculate the water temperature at 4000 ft after 7 days given the following data: Q = 500 Bbl/day; injection through 2-in. I.D. tubing; annulus filled with insulating material and U = 2.4 Btu/day ft²-°F; a = 0.02°F/ft; b = 70°F; T_i = 400°F; D = 0.96 ft²/day; K = 33.60 Btu/day ft-°F; c = 1 Btu/lb-°F.
Solution:

$$Q = 500 \times 350 = 175000 \text{ lb/day}$$
$$r_1 = 1/12 \text{ ft}$$

$$\log_{10}(967.88) = 2.99$$

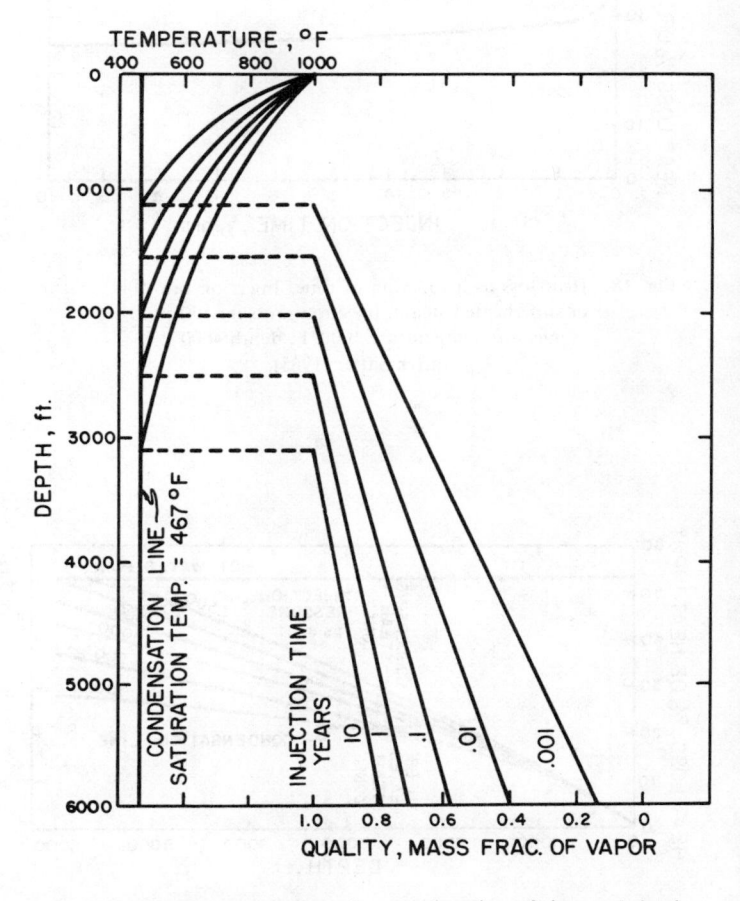

Fig. 16. Temperature and quality as a function of time and depth. Injection rate 5000 lb/hr of superheated steam, injection temperature 1000°F, injection pressure 500 psia.
(after Satter, 1965)

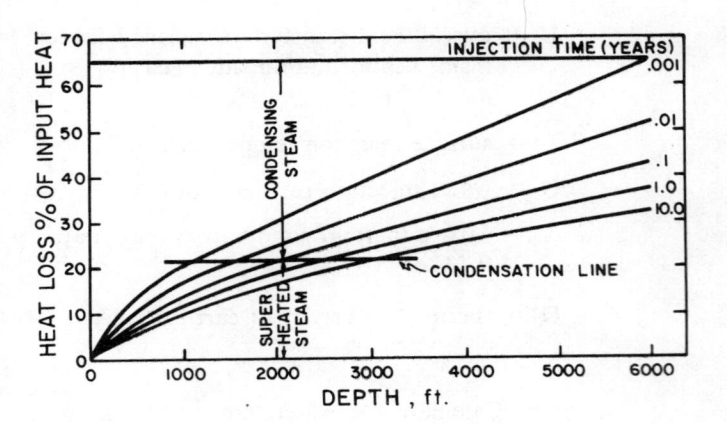

Fig. 17. Heat loss as a function of time and depth. Injection rate 5000 lb/hr of superheated steam, injection pressure 500 psia, injection temperature 1000°F.
(after Satter, 1965)

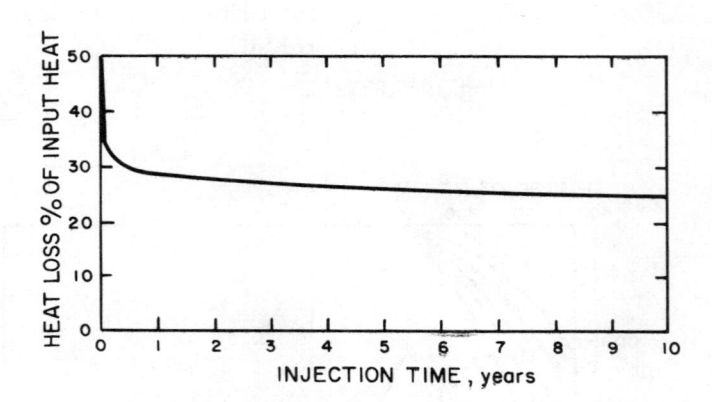

Fig. 18. Heat loss as a function of time. Injection rate 5000 lbs/hr of superheated steam, injection pressure 500 psia, injection temperature 1000°F, depth 4000 ft.
(after Satter, 1965)

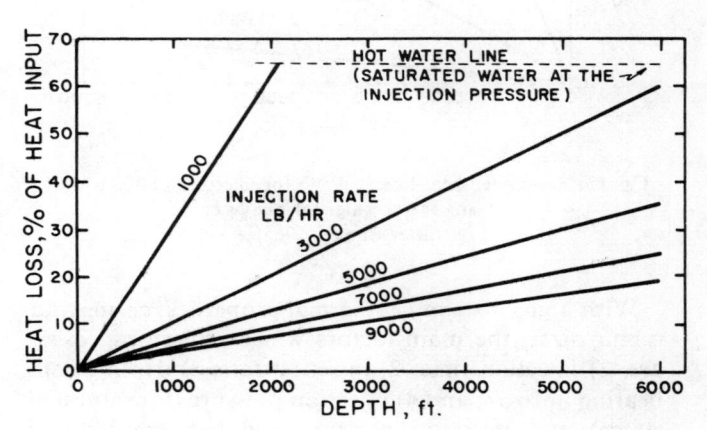

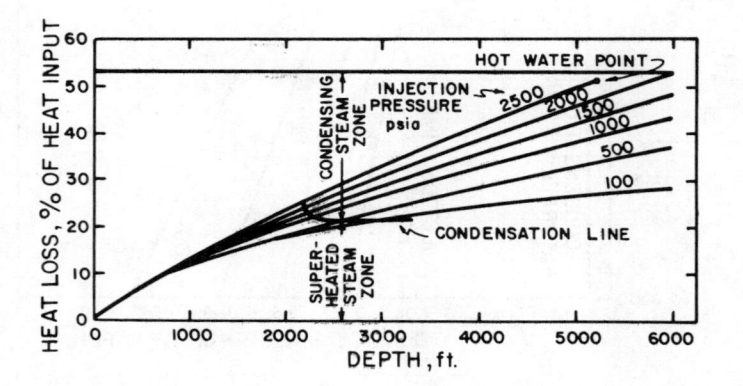

Fig. 19. Heat loss as a function of injection pressure and depth. Injection rate 5000 lb/hr, time one year, temperature 1000°F.
(after Satter, 1965)

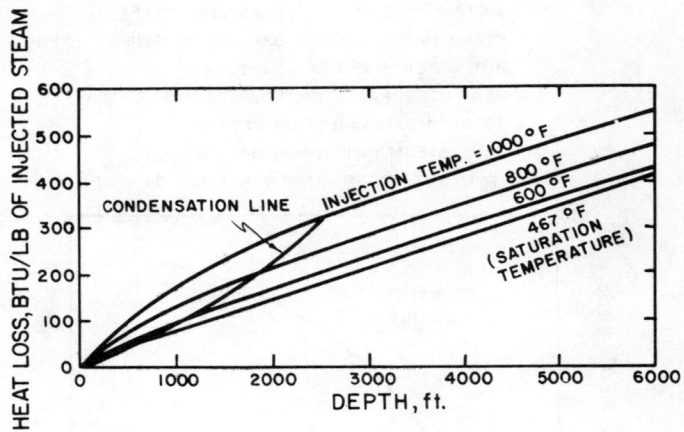

Fig. 20. Heat loss as a function of injection temperature and depth. Injection rate 5000 lb/hr, injection pressure 500 psia, time one year.
(after Satter, 1965)

Fig. 21. Heat loss as a function of injection rate and depth. Wellhead condition of steam, saturated steam at 500 psia and 467°F; injection time one year.
(after Satter, 1965)

From Figure 24, $\log_{10} f(t_D) = 0.6$

$$f(t_D) = 3.98$$

$$A = \frac{175000 \times 1 [33.6 + 1/12 \times 2.4 \times 3.98]}{2\pi \times 1/12 \times 2.4 \times 33.60} \cong 142560$$

$$T(7 \text{ Days})_{4000'} = 0.02 \times 4000 + 70 - 0.02 \times A + (400 + 0.02A - 70)\exp(-4000/A) = 392°F$$

Suppose that in the above example, the injection took place through a 7-in casing, then:

$$U = \infty, \quad A = Qcf(t_D)/2\pi K,$$

$$Dt/r_e^2 = 78.8$$

$$f(t_D) = 2.75$$

STEAM FLOODING

$$A = 2{,}280 \text{ ft}^{-1}$$

$$T(7 \text{ Days})_{4000'} = 0.02 \times 4000 + 70 - 0.02 \times A + (400 + 0.02A - 70)\exp(-4000/A) = 169°F$$

The above procedure can also be used to calculate the temperature of cold water injections. For example, injection of 4790 Bbl/day of water of temperature 58.5°F for 75 days will reduce the reservoir temperature (at 6605 ft) from 125° to 65°F.

When steam is injected through an insulated tubing, the total heat loss rate q (Btu/day) is given by:

$$q = \frac{2\pi r_1 UK}{K + r_1 U f(t_D)} \left[(T_i - b)z - \frac{az^2}{2} \right]$$

If steam is injected through the casing, $U = \infty$, then the above equation reduces to:

$$q = \frac{2\pi K}{f(t_D)} \left[(T_i - b)z - \frac{az^2}{2} \right]$$

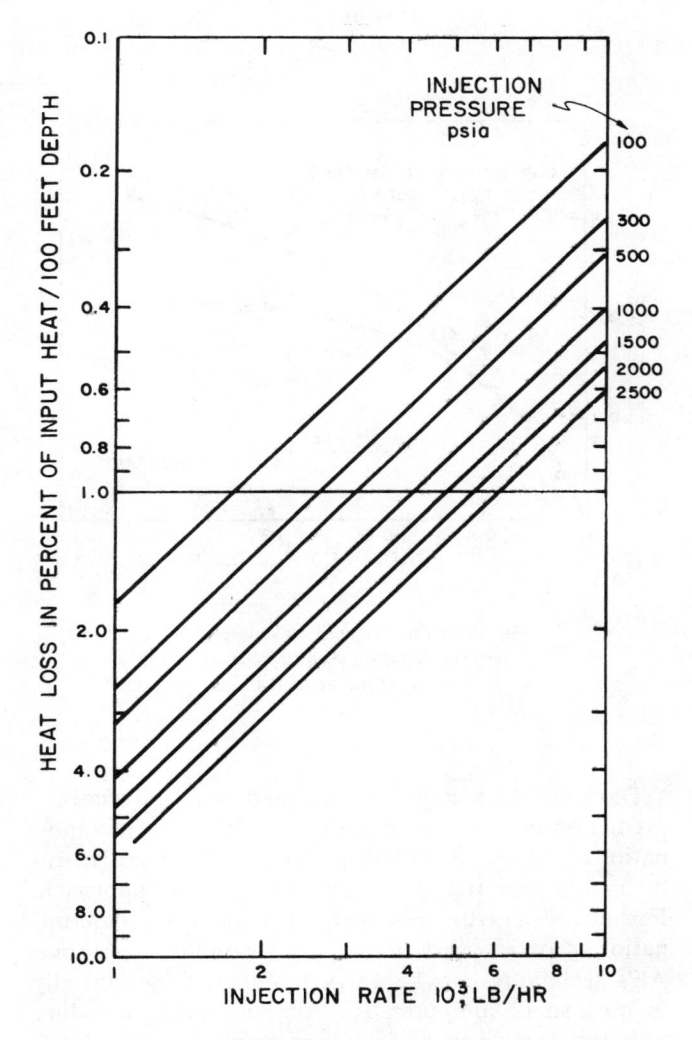

Fig. 22. Heat loss as a function of injection rate and pressure in case of injecting saturated steam for one year. (after Satter, 1965)

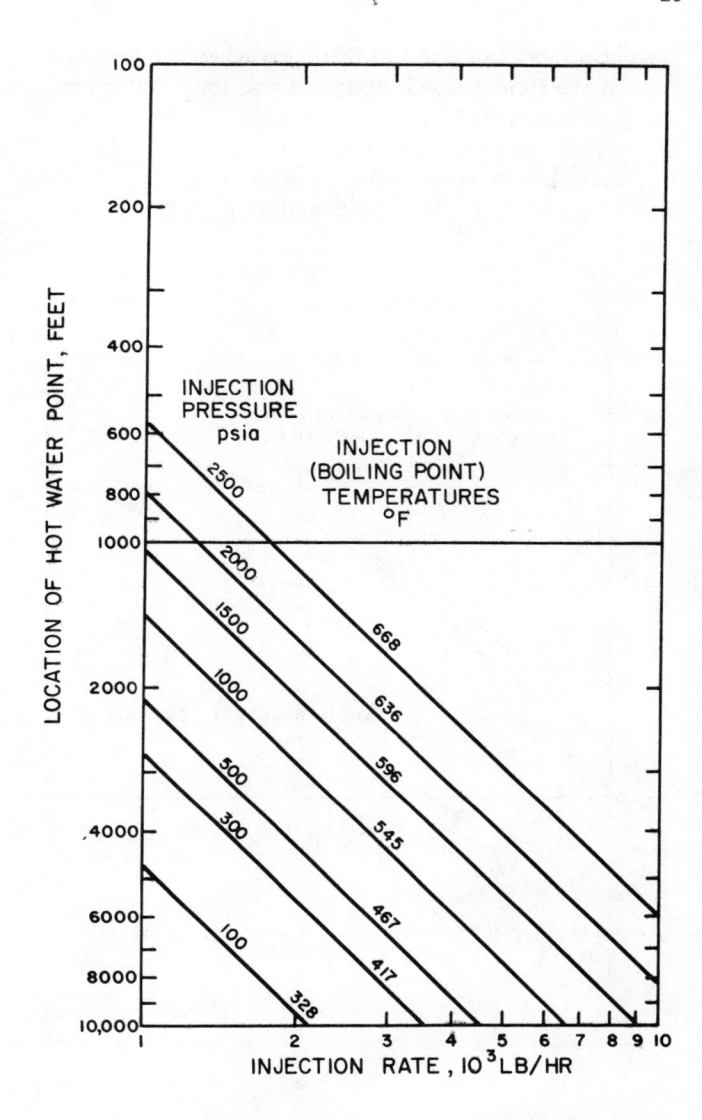

Fig. 23. Location of hot water point as a function of injection rate and pressure in case of injecting saturated steam for one year. (after Satter, 1965)

TABLE III
PARAMETERS FOR FIGURES 16-23

Surface temperatures:	75°F
Geothermal gradient:	0.011°F/ft
Earth's thermal conductivity:	1.0 Btu/hr. ft°F
Earth's thermal diffusivity:	0.046 ft²/hr
Casing's inside diameter	5.989 in
Casing's outside diameter	6.625 in
Tubing's inside diameter	2.441 in
Tubing's outside diameter	2.875 in

Surface Heat Loss

Heat losses from transmission lines connecting the wellhead and the thermal unit can be adequately estimated from Table IV (Ramey, 1965). For example, Table IV shows that a 3-in pipe with 1 1/2-in of magnesia insulation would loose 115 Btu/hr ft when the temperature of the fluid inside is 400°F. Thus, for a 100 ft pipe, the heat loss would be:

Heat loss = 115 x 100 x 24 x 10^{-6}
= 0.276 MM Btu/day (per 100 ft.)

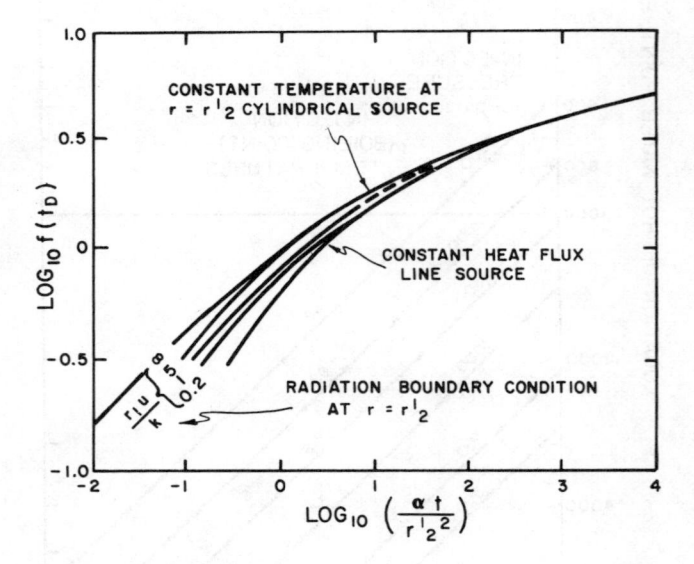

Fig. 24. Transient heat conduction in an infinite radial system. (after Ramey, 1962)

Oil Recovery Estimates

Oil recovery estimates can be obtained either from laboratory studies or from numerical models. It has been pointed out earlier that results obtained from laboratory models are uncertain because of dimensional scaling problems. In numerical models, the physical phenomena are constrained by specified relationships, and there are uncertainties associated with the different parameters and the effect of heat on these parameters.

Willman et al (1961) conducted laboratory experiments and gave oil recovery estimates based on a Buckley-Leverett approach. Their results have been presented in the preceding sections.

Shutler (1970) formulated a two-dimensional, three-phase numerical model. His results (Figure 25) agreed quite well with experimentation and emphasized the effects of gravity override by steam in thick reservoirs. This was also reported by Baker (1969) in his experimental studies. Shutler concluded that for thin, homogeneous reservoirs, one-dimensional numerical models are adequate for process variable studies.

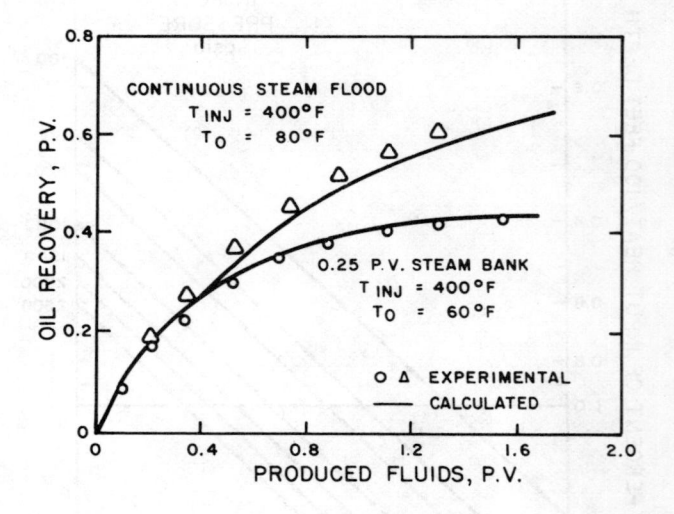

Fig. 25. Comparison of calculated and experimental oil recovery curves. (after Shutler, 1970)

Davies et al (1968) presented a method for oil recovery prediction in five-spot steam floods. Based on a combination of the Davies et al method, the Willman et al method and a modified Marx and Langenheim approach, Farouq Ali (1970) presented a technique for determination of oil recovery in five-spot steam floods. Farouq Ali's technique can be carried out either graphically or on a small computer. His computational procedure followed the scheme of Higgins and Leighton (1962) which is known in waterflooding as the streamline-channel flow method.

TABLE IV
APPROXIMATE HEAT LOSS FROM PIPE

Insulation	Conditions	Heat Loss, Btu/hr-sq ft surface area for inside Temperatures of		
		200 F	400 F	600 F
Bare metal pipe	Still air, 0 F	540	1560	3120
	Still air, 100 F	210	990	2250
	10-mph wind, 0 F	1010	2540	4680
	10-mph wind, 100 F	440	1710	3500
	40-mph wind, 0 F	1620	4120	7440
	40-mph wind, 100 F	700	2760	5650

		Heat Loss, Btu/hr-ft of linear length of pipe at inside temperatures of			
Magnesia pipe insulation, air temperature 80 F		200 F	400 F	600 F	800 F
	Standard on 3-in pipe	50	150	270	440
	Standard on 6-in pipe	77	232	417	620
	1½-in on 3-in pipe	40	115	207	330
	1½-in on 6-in pipe	64	186	335	497
	3-in on 3-in pipe	24	75	135	200
	3-in on 6-in pipe	40	116	207	322

(after Ramey, 1965)

Coats (1974) described a three-dimensional, three-phase simulator of the steam injection process which accounted for steam distillation, solution gas and temperature dependent relative permeability. His results showed that oil recovery increased with increasing steam quality and decreasing initial oil saturation. The distillation properties of the oil had only moderate effects on the recovery. For a 63 ft thick reservoir, oil recovery was insensitive to the location of the steam injection point, but for a 189 ft thick reservoir, both recovery and recovery rate were high when the steam was injected at the bottom of the reservoir. Increased formation compressibility had a strong effect on injection volume regardless of initial water saturation.

Based on the same principles of Coats' model, Ferrer and Farouq Ali (1977) formulated a three-phase, two-dimensional model. They examined the sensitivity of predictions to the input parameters. Their studies showed that recoveries were higher when relative permeabilities were considered temperature dependent. They concluded that steam distillation and gravity segregation are important mechanisms in steam injection, and that reservoir compaction enhances steam soak response but can be detrimental to steam drive.

CONCLUSIONS

Steam flooding has emerged as the major thermal recovery method. Its main advantages are:

1. Steam floods are easier to control than in situ combustion, and for the same pattern size, the response time is 25-50% of the response time with in situ combustion.
2. The steam flood process does not cause cracking of the oil; therefore, no environmentally objectionable flue gases are produced and no exceptionally corrosive waters are formed, as in the case of in situ combustion.
3. Steam injection wells and oil producing wells are subject to much lower temperatures than the temperatures encountered in in situ combustion. Farouq Ali and Meldau (1978) present a survey of thermal well completion practices for various fields. (See Table V.)
4. Steam flooding can be applied to reservoirs containing high API gravity oils. Usually these oils cannot produce sufficient coke to sustain combustion.

The main disadvantages are:

1. Steam flooding requires large supplies of high quality fresh water.
2. Steam floods cannot be performed at depths below 5000 ft, nor to reservoirs containing freshwater-sensitive clays.
3. The capital investment is lower than that needed for in situ combustion, but the fuel consumption per barrel of oil produced is higher than in the case of in situ combustion.

CASE HISTORIES

In this section, four case histories are presented: 1) South Belridge Field, California; 2) Tia Juana Field, Western Venezuela; 3) Schoonebeek Field, the Netherlands; and 4) Cat Canyon Field, California. Each of these cases was selected because it included some novel technique. In the South Belridge Field, a suitable tech-

TABLE V
SURVEY OF THERMAL WELL COMPLETION PRACTICES

No.	Field	Operator	Depth Ft	Max. Steam Inj. Pres. PSIG	Casing Size In	Casing Wt. Lb/Ft	Casing Grade	Casing Type Coupling	Pre-Stress?	Annulus During Cylic	Liner Length Ft	Liner Size In	Liner Wt. Lbs	Liner Grade	Liner Type Coupling
1	Kern River	Getty	900	350	7	23	K-55	STC	No	Steam	None	—	—	—	—
2	Midway-Sunset	Chevron	1200	450	8⅝	36	K-55	LTC	No	Steam	400	6⅝	28	K-55	BUTT
3	Cold Lake	Imperial	1500	1600	5½	20	N-80	BUTT	No	Steam	80	4	15	H-40	FJ (5)
4	Tia Juana	Maraven	1600	1500	9⅝	43.5	N-80	BUTT	No	Vent	100	7	29	N-80	BUTT (6)
5	Mt. Poso	Shell	1800	600	8⅝	32	?	BUTT	No	Steam	—	—	—	—	—
6	San Ardo	Texaco	2350	800	8⅝	32	K-55	STC	No	Vent	150	6⅝	28	K-55	SF
7	Guadalupe	Union	3000	1800	7	26	N-80	BUTT	No	Gas Pack	100	5½	17	N-80	BUTT
8	E. Cat Canyon	Conoco	3500	2500	7	23	S00-95	BUTT	Yes	Steam	400	5½	17	N-80	X-Line
9	Morichal	Phillips	3500	1500	7	23	N-80	BUTT	Yes	Crude	250	3½	13	N-80	FJ

Note: (1) All operators cement to the surface.
(2) All use 50% silica flour except Chevron uses 30%.
(3) All use a lead or brass sealed liner hanger, all except Chevron gravel pack the liner, and
(4) None of the operators used an expansion joint in the liner.
(5) Plan to use S00-95 and prestress in new wells, modified n-80. Do not gravel pack.
(6) Openhole gravel pack; expansion joint in liner.

(after Farouq Ali and Meldau, 1978)

nique was found and resulted in improvement of injection profiles. In the Tia Juana Field, air was pumped in order to map the steam flow pattern. In the Schoonebeek Field, gas was continuously injected in the tubing-casing annulus, new well-head equipment design was presented, and oil displacement by steam drive was calculated by material balance. The Cat Canyon Field was selected because of the particular well completion methods used and the nature of the problems that evolved.

South Belridge Field, California (USA)

The South Belridge Field structure consists of a broad, southeasterly plunging anticline, about nine miles long and two miles wide (Gates and Brewer, 1975). The producing sands thicken in the flanks but become shaley in the direction of the anticlinal axis (Figures 26 and 27). The sands are locally known as the D and E Zones of the Tulare formation of Pleistocene age. The reservoir characteristics are given in Table VI.

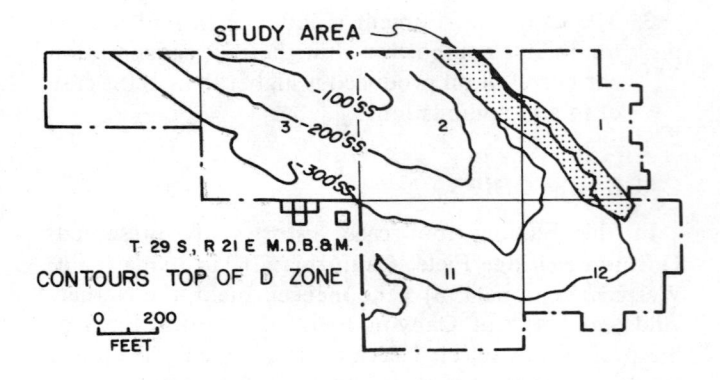

Fig. 26. Location of study area on Mobil properties, South Belridge field.
(after Gates and Brewer, 1975)

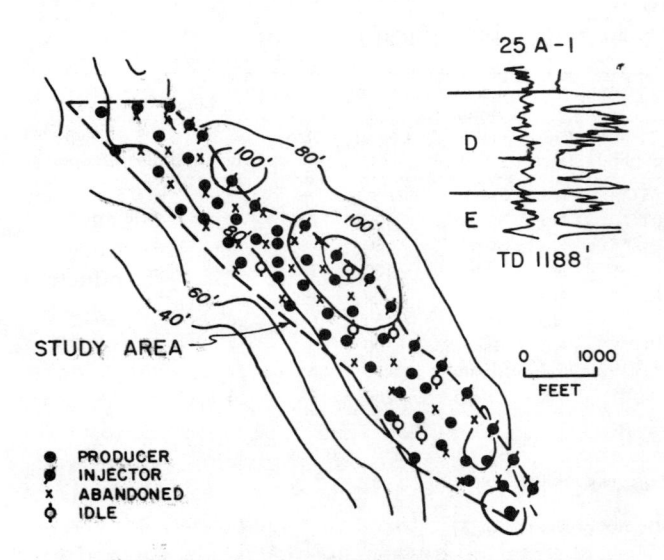

Fig. 27. Isopach of net sand in the D and E zones.
(after Gates and Brewer, 1975)

TABLE VI
STUDY-AREA RESERVOIR CHARACTERISTICS

Area (D and E zone), acres	204
Average depth, ft	1009-1197
Gross formation thickness, ft	210
Average net sand (D and E), ft	91
Effective porosity, %	35
Original oil saturation, %	76
Original water saturation, %	24
Formation volume factor rb/STB	1.03
Initial oil in place, STB/acre-ft	2000
Initial oil in place, total, MMSTB	37.1
Permeability, md	3000
Initial formation pressure, psig	420
Reservoir temperature, °F	95
Oil gravity, °API	13.0
Dead oil viscosity at 95°F, cp	1600
Dead oil viscosity at 250°F, cp	16
Sand character	Unconsolidated
Geologic age	Pleistocene
Formation	Tulare

(after Gates and Brewer, 1975)

The field was discovered in 1911, but development drilling did not begin until 1943. By the end of 1944, 40 wells were drilled on a 5 acre spacing. By 1952, oil production had peaked at 1000 BOPD. By early 1965, oil production had declined to 235 BOPD.

Cyclic steam injection was initiated in 1965. As a result, the daily production rate increased and peaked at 1600 BOPD by mid-1967. Thereafter, the rate declined rapidly, mainly because of diminished effectiveness of successive steam injection cycles and because of well failures. Eventually all but two of the original wells failed.

Well failures during cyclic steam injection had a peculiar effect on water production. The water cut reached 85% in 1964. However, it was found that this high water cut was not caused by reservoir water influx; rather, it was caused by casing failures at shallow intervals opposite water-bearing horizons.

A continuous steam injection pilot study was begun in 1966 and continued for three years. Based on this study, 15 new injectors and 39 new producers were drilled along parallel lines as shown in Figure 27. Continuous steam injection began in 1969, and within 4 1/2 years the oil recovery exceeded the total oil recovered during the preceding 25 years by both primary means and cyclic steam injection. By 1973, the cumulative oil-steam ratio was about 0.28.

From the pilot study it was learned that the injected steam tended to flow through thin sections of the reservoir sands, thus resulting in poor injection profiles. To improve the injection profiles, some injectors were completed by running and cementing two 3 1/2-in tubing strings. Then by using oriented jets, one string was perforated opposite the D Zone and the other string was perforated opposite the E Zone. Steam was injected at a rate of 300 Bbls/day into each string of 1 1/4-in tubing that was set in a packer above the zone. This type of completion proved to be unsatisfactory. Additional efforts were made and were also found unsuccessful. These efforts included well cleaning and washing of perforations, increasing injection rates and selective plugging by chemical agents.

A suitable technique was found and resulted in improvement of injectivity profiles. This technique consisted of running and cementing 5 1/2-in O.D. casing in an 8 3/4-in hole to a depth of 50 ft below the lowest sand. The casing was then jet perforated opposite the sand. The number and diameter of the perforations were determined such that the sum of the maximum injection rates of the perforations was equal to the total desired injection rate in the well. (For example, if it was desired to inject 600 Bbls/day of water converted into steam, and the maximum injection capacity of each 0.25-in perforation at the injection pressure was 25 Bbls/day, then 24 perforations were made throughout the sand section.) Steam was injected through a 2 7/8-in O.D. tubing set in a packer above the injection interval.

Tia Juana Field, Venezuela

This field is located on the east coast of Lake Maracaibo (Figure 28), which is known as the Bolivar Coast. It is one of several fields on the Bolivar Coast that produce oils of gravities in the range of 10-15° API (de Haan and Schenk, 1969). The initial oil-in-place in all the Bolivar Coast fields was estimated at 20 billion Bbls. The combined daily production in 1967 was about 400000 BOPD. Reservoir depths range between 1000 and 5000 ft, but most reservoir depths are less than 3000 ft. Net oil sand thicknesses are 50-300 ft, porosities range between 30-40%, permeabilities are greater than one darcy, initial oil saturations are about 80% and oil viscosities are in the range of 100-10000 cp.

A steam flood test was made in the northern part of the field (Figure 29). Pertinent reservoir data at the beginning of the test are given in Tables VII and VIII. The field produces from two reservoirs (Upper and Lower zone) and steam flood was undertaken in both zones simultaneously. Primary recovery was greater in the upper zone due to the presence of lower viscosity oil. Because of differences in cumulative primary prouction, a significant pressure differential existed between zones. Separate steam injection wells were drilled for each zone. Each production and injection well was completed with 7-in WSO casing and cemented with a special cement-silica powder mixture which could sustain elevated temperatures. Injection was made through tubing and a 7-in packer set at the top of the producing horizon. The annulus was kept dry and open to the atmosphere in order to reduce wellbore heat losses. A 4 1/2-in saw slotted liner packed with 10-14 Tyler mesh gravel was set through the producing horizon. (Although most of these liners failed, they were considered to be the most suitable for this operation.) Wellhead connections were made flexible to allow for thermal expansion.

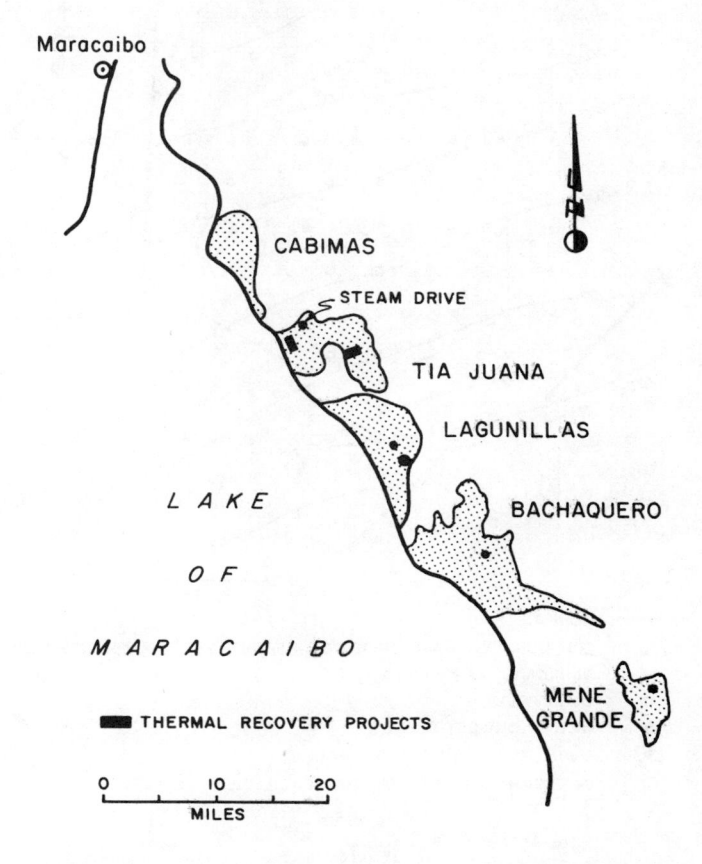

Fig. 28. Shell's main heavy oil fields in the Maracaibo Basin. (after de Haan and Schenk, 1969)

Steam injection began in September 1961, and continued for five years at a rate of 1000 Bbls/day. With injectivities of up to 60 Bbls/day-psi, there were no difficulties in injecting steam at the desired rates.

In order to map the flow pattern, air was injected with the steam. The volume of steam that was flowing towards a well was measured in terms of the volume of nitrogen produced from that well. The results showed that each injection well did not influence more than two wells, mainly because of gravity override and heterogeneity of the sands. In addition, as soon as a production well became influenced by an injection well, the production well became pressurized and steam approaching the well from other injection wells was diverted.

Unsuccessful attempts were made to improve steam flow patterns by restricting production of individual wells. Steam soaking of some of the production wells resulted in further increased production. However, by 1966, production had declined primarily because of steam and water bypassing. Nonetheless, the cumulative oil-steam ratio was 0.6 Bbl/bbl. The test performance data in 1966 are given in Table IX.

Fig. 29. Test area.
(after de Haan and Schenk, 1969)

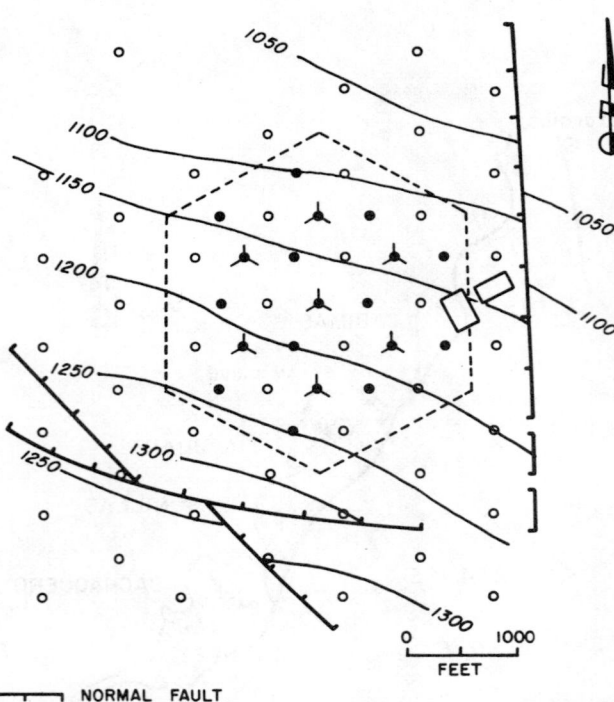

NORMAL FAULT
—1300— CONTOURS ON BASE 'HARD BED' (ABOUT 250' ABOVE TOP SAND)
------ BOUNDARY TEST AREA
○ OLD WELL
● NEWLY DRILLED WELLS
▲ TWIN STEAM INJECTION WELLS
□ INJECTION AND PRODUCTION FACILITIES

TABLE VII
RESERVOIR PROPERTIES AND INITIAL CONDITIONS

Project area*, acres	137
Well spacing, ft	435
Depth to top of sand, ft	Upper zone — 1450
	Lower zone — 1700
Gross open interval, ft	Upper zone — 150
	Lower zone — 100
Thickness separation shaly zone, ft	90
Total net oil sand, ft	200
Dip, degrees	3
Temperature, °F	113
Permeability, darcies	1-3
Porosity, %	33
Oil saturation, %	75
Water saturation, %	25
Oil formation volume factor, rb/STB	1.04
Stock-tank oil-in-place, STB	50 × 10^6

*As defined in Fig. 29, the area indicated is equal to the theoretical drainage area of the 12 old wells.

(after deHaan and Schenk, 1969)

TABLE VIII
STATUS AT START OF TEST

Cumulative oil production, STB	
12 old wells	5.7 × 10^6
12 new wells	0.3 × 10^6
24 wells	6.0 × 10^6
Recovery, % STOIIP	12.0
Oil production rate 24 wells, STBOPD	1840
Water cut, %	1-2
Oil saturation, %	71
Gas saturation, %	4
Pressure at top of sand, psig	Upper zone — 175
	Lower zone — 350
Approximate tank oil viscosity at 100°F, cp	Upper zone — 1300
	Lower zone — 5000
Oil formation volume factor rb/STB	1.01

(after deHaan and Schenk, 1969)

TABLE IX
TEST PERFORMANCE DATA AS OF END OF 1966

Cumulative steam injection (water equivalent), Bbl	
Injection wells	16.73 × 10^6
Production wells (for steam soaking)	0.30 × 10^6
Cumulative oil production during test, STB	10.3 × 10^6
STOIIP, %	20.6
Cumulative primary plus test production, Bbl	16.3 × 10^6
STOIIP, %	32.6
Cumulative water production, Bbl	11.4 × 10^6
Percent of steam injection	67.0

(after deHaan and Schenk, 1969)

Schoonebeek Field, the Netherlands

The Schoonebeek Field (van Dijk, 1968) is located in eastern Holland near the German border (Figure 30). It was discovered in 1943 and developed shortly after the war. The field consists of two pressure isolated reservoirs that produce from the same sand. A sealing fault divides the reservoir into a southwestern area where solution-gas drive is the primary production mechanism, and a northwestern area where active water drive is the prevailing mechanism. Details of the steam-drive project area and reservoir data are shown in Figure 31 and Table X, respectively.

The injection wells were completed by cementing 9 5/8-in casing above the oil sand. The tubing was hung from a support so that thermal expansion could occur only in the downward direction. Two braden heads were installed (Figure 32) to allow for relative movements of tubing and casing. Gas, at a small rate, was continuously injected in the casing-tubing annulus in order to reduce wellbore heat losses. Production wells were completed with 7-in casings instead of the 9 5/8-in, and both production and injection wells were completed with liners slotted over the entire sand interval. The surface completions did not fail in spite of producing well temperatures of more than 300°F and casing expansions of 4.5 to 7 ft.

STEAM FLOODING

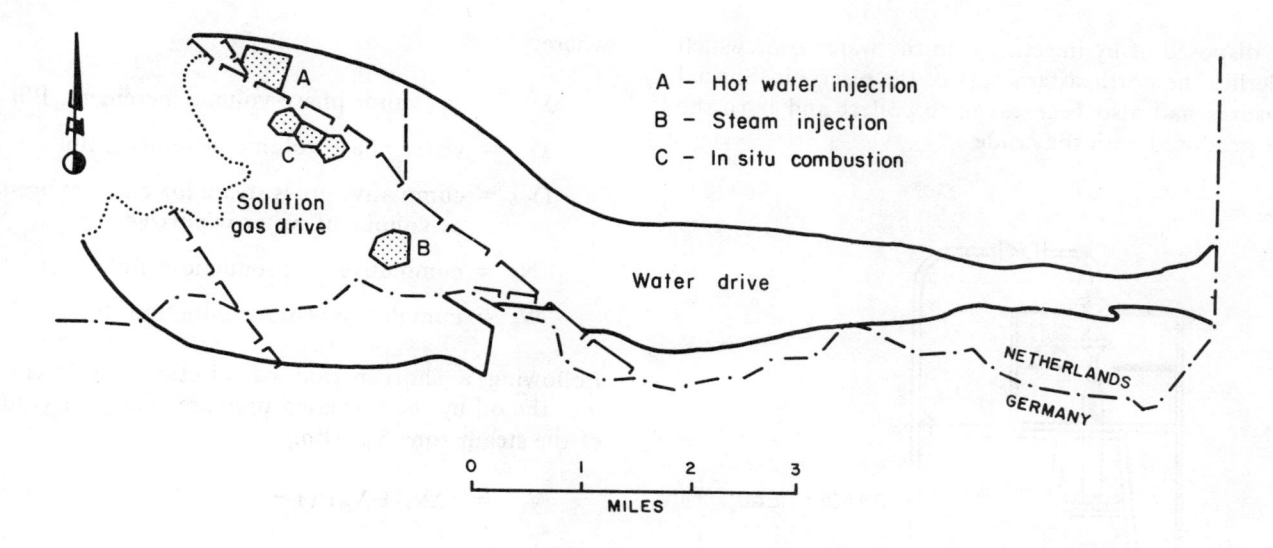

Fig. 30. Schoonebeek oil field, the Netherlands.
(after van Dijk, 1968)

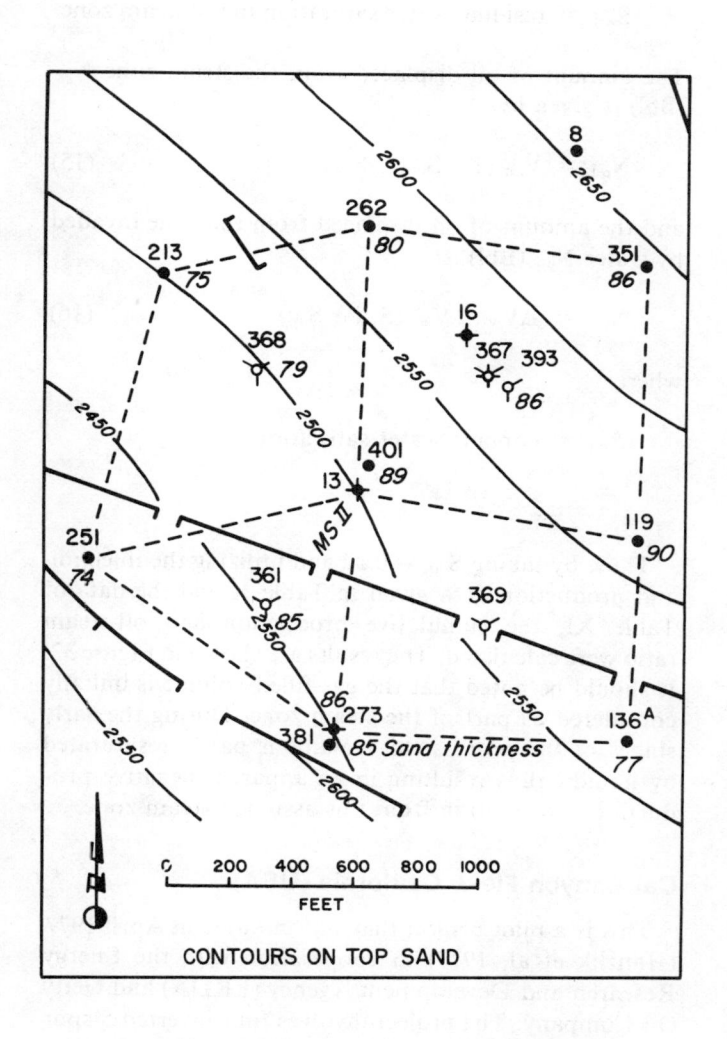

Fig. 31. Steam-drive project area, Schoonebeek Field.
(after van Dijk, 1968)

TABLE X
RESERVOIR DATA STEAM INJECTION AREA, SCHOONEBEEK FIELD

Pattern area, acres	65.5
Sand thickness, ft	74-90
Dip (average), degrees	6.5
Porosity (average), %	30
Permeability, md	1000-10000
Reservoir pressure at project start, approximate, psi	120
Reservoir temperature, °F	100
Oil saturation, %	85
Water saturation, % (approximate value at project start)	11
Gas saturation, %	4
Tank oil viscosity, cp approximate	200
STOIIP within pattern, STB	12.6×10^6
Cumulative production at project start, %	4

(after van Dijk, 1968)

Injection began in September 1960. The maximum injection rate of 1500 Bbls/day was attained by mid-1962. Frequent well cleaning and acidizing were necessary in order to maintain injectivity. One reason was that the gas injected in the annulus contained CO_2 which formed scale. The scale then dropped and plugged the liner. Casing leaks at shallow depths also occurred. The leaks were caused by heavy external casing corrosion, which was induced by the effects of high temperature and sodium bichromite that was used as an inhibitor in the mud left in the annulus between casing and conductor string. This problem was alleviated by injecting an oil gel in the annulus. The injected steam reacted with the sulfur contained in the oil and resulted in the formation of H_2S. Thus, the produced water was isolated and disposed of by injecting it in the water zone which underlies the northeastern part of the reservoir. Special measures had also been taken to collect and burn the H₂S produced with the crude.

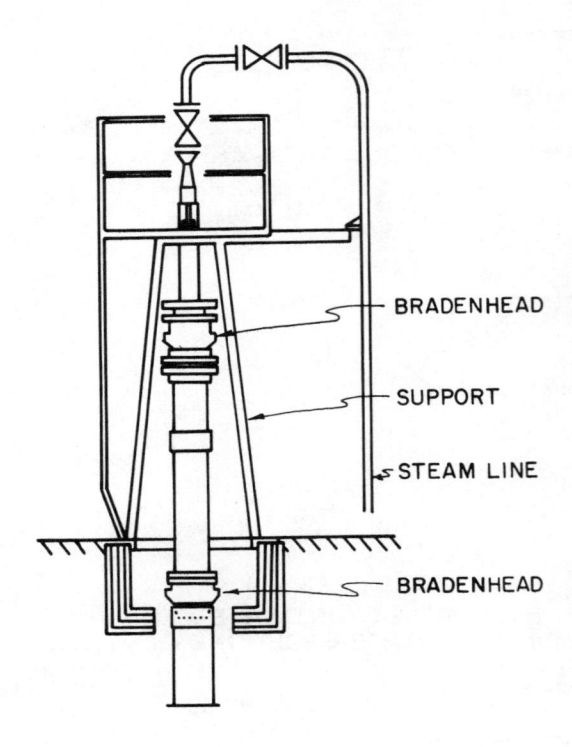

Fig. 32. Surface completion injection well. (after van Dijk, 1968)

The steam drive project was both technologically and economically successful. The overall oil-steam ratio was 0.37 Bbl/bbl. However, individual well performances showed that distribution of steam did not conform to a regular pattern. For example, after 6 years of steam injection, Well Nos. 213 and 136-A (Figure 31) were still cold, whereas Well No. 8, which lies outside the pattern, strongly responded to steam injection in Well No. 367. In addition, analysis of cores obtained during drilling of replacement wells gave conclusive evidence of gravity override by the steam.

Oil displacement by the steam drive was obtained by using simple material balance equations, and by neglecting compression or expansion of fluids that might have taken place under varying pressures and temperatures. This approach is justified since residual saturations are not usually known with accuracy.

By neglecting the mass of vapor, the volume of the steam zone was estimated as follows:

$$\Delta V_g = N_p + W_p - Q_{tot} \qquad (12)$$

$$\Delta V_w = Q_{TOT} - W_P \qquad (13)$$

where:

ΔV_g = gas vapor phase volume increment, Bbl

ΔV_w = water phase volume increment, Bbl

Q_{tot} = cumulative gross steam injection, expressed as volume of water vaporized, Bbl

N_p = cumulative oil production, Bbl

W_p = cumulative water production, Bbl

Following a short period of injection, gas dissolves into the oil by the increased pressure. The pore volume of the steam zone V_{pst} (Bbl) is:

$$V_{pst} = (\Delta V_g + V_{gi})/(1 - S_{orst} - S_{wst}) \qquad (14)$$

where:

V_{gi} = volume of gas at start of project, Bbl

S_{orst} = residual oil saturation in the steam zone

S_{wst} = residual water saturation in the steam zone.

The amount of oil displaced from the steam zone N_{pst} (Bbl) is given by:

$$N_{pst} = V_{pst} (1 - S_{orst} - S_{ow}) - V_{gi} \qquad (15)$$

and the amount of oil displaced from the zone invaded by water N_{pw} (Bbl) is:

$$N_{pw} = \Delta V_w - V_{pst} (S_{wst} - S_{ow}) \qquad (16)$$

where:

S_{ow} = connate water saturation.

Thus, by taking $S_{orst} = 0.22$ and utilizing the injection and production data given in Table X and the data of Table XI, the cumulative production and oil-steam ratio were calculated. The results are shown in Figure 33. It should be noted that the gas-filled volume is initially considered as part of the steam zone. During the early stages of the process this volume is partly resaturated by liquids, thus resulting in an apparent negative production contribution from this assumed steam zone.

Cat Canyon Field, California (USA)

This is a pilot project that was initiated in April 1977 (Hanzlik et al, 1977) in cooperation with the Energy Research and Development Agency (ERDA) and Getty Oil Company. The project involves four inverted 5-spot pilot patterns covering an area of about 20 acres. The reservoir is the S1-B sand of the Sisquoe formation.

STEAM FLOODING

The reservoir contains 9° API oil and viscosity ranges between 8000-600000 cp at the reservoir temperature of 110°F. The average stock-tank oil viscosity is 60000 cp. The average reservoir depth is 2500 ft, porosity 31%, permeability 5000 md, thickness 80 ft and water saturation 35%.

TABLE XI
INJECTION AND PRODUCTION DATA USED IN CALCULATIONS

Date	Cumulative Gross Steam Injection (Bbl)	Cumulative Net Steam Equivalent Injection (Bbl)	Cumulative Oil Production (Bbl)	Cumulative Water Production (Bbl)
1-1-61	106521	91350	17618	6594
7-1-61	638114	543211	132883	87114
1-1-62	1457663	1225537	523654	526875
7-1-62	2350309	2032909	993825	1251146
1-1-63	3312603	2889997	1370961	2189059
7-1-63	4212054	3612781	1634021	2978064
1-1-64	5080370	4459056	1947355	3829919
7-1-64	6023291	5288739	2247117	4763919
1-1-65	7046001	6188397	2619886	5824616
7-1-65	7966379	6974811	2964101	6788131
1-1-66	8923918	7796423	3338878	7657045
7-1-66	9767772	8521723	3676468	8436964
1-1-67	10558891	9214472	3962570	9236325

(after van Dijk, 1968)

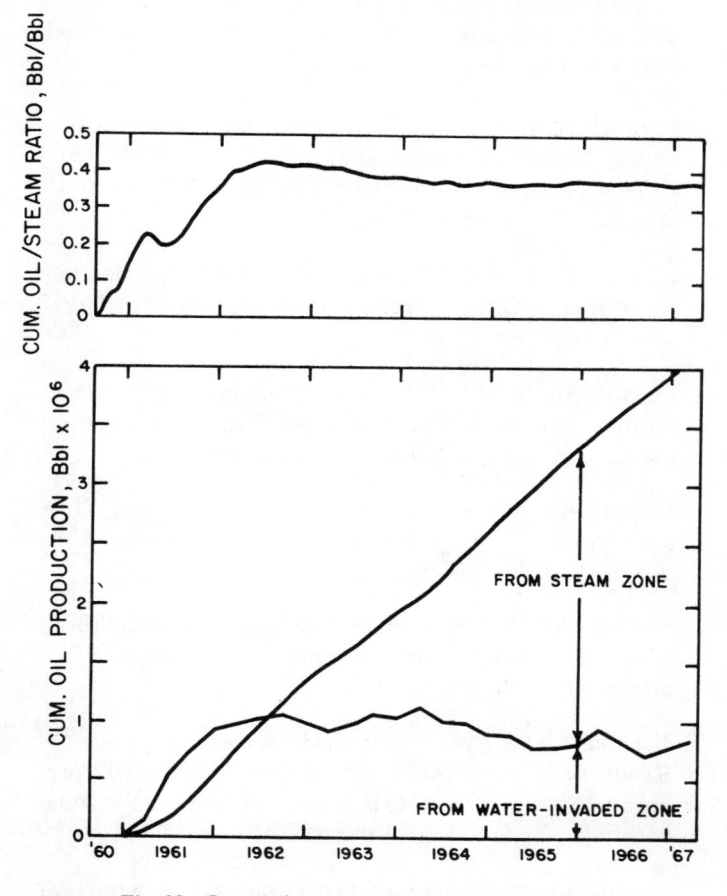

Fig. 33. Cumulative production and oil-steam ratio. (after van Dijk, 1968)

Although no meaningful performance data have been made available, this project was selected for inclusion here because of the particular well completion methods used and the nature of the problems encountered since the inception of the project.

Thirteen new wells were drilled specifically for this project. The details of completing production and injection wells are given in Figure 34. The luminite cement was treated to achieve 12 hours setting time. After 12 hours, tension of approximately 275000 pounds was applied to the casing. This tension was held for an extended period of time before the casing was landed and the shoe was drilled. (The authors did not indicate the length of time during which tension was applied.) It is believed that this method of prestressing will prevent casing failures that might be induced by compressive thermal stresses. The productive interval was reamed to 12-in diameter. A 5 1/2-in liner was then run and gravel packed using pressure packing methods. Getty Oil used 40 mesh slots in the liner and 10 x 18 mesh gravel. Steam injection was conducted through 2 7/8-in tubing and thermal packer. It was planned to insulate the tubing with foamed-in-place sodium silicate foam.

Between April and June 1977, the project was afflicted with packer failure. Initially, all injectors were equipped with Baker C-2 LOK-SET Thermoseal packers. These packers had to be replaced. The replacements were manufactured by Brown and Guiberson, which also failed. It appears that the injection pressure and temperature of 1375 psi and 590°F, respectively, are beyond the operational limits of all currently available thermal packers.

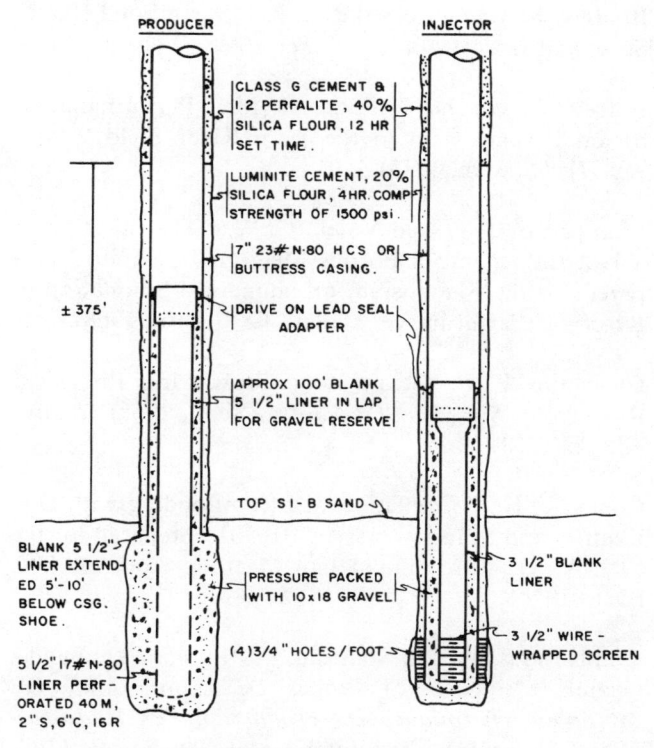

Fig. 34. SI-B sand completion techniques. (after Hanzlik and Birdwell, 1977)

BIBLIOGRAPHY

Alford, W. O.: "The '200' Sand Steamflood Demonstration Project," *ERDA Enhanced Oil, Gas Recovery and Improved Drilling Methods,* Vol. 1 - Oil, Tulsa, OK. (Sept., 1977).

Avdonin, N. A.: "On the Different Methods of Calculating the Temperature Fields of a Stratum During Thermal Injection," *Neft'i Gaz,* Vol. 7, No. 8. (1964).

Avdonin, N. A.: "Some Formulas for Calculating the Temperature Field of a Stratum Subject to Thermal Injection," *Neft'i Gaz,* Vol. 7, No. 3, (1964).

Baker, P. E.: "An Experimental Study of Heat Flow in Steam Flooding," *Soc. Pet. Eng. Jour.* (March, 1969). *Trans.,* AIME (1969) 246.

Bleakley, W. B.: "Penn Grade Crude Oil Yields to Steam Drive," *O. & G. Jour.* (March 25, 1974).

Blevins, T. R., Aseltine, R. J. and Kirk, R. S.: "Analysis of a Steam Drive Project, Inglewood Field, California," *Jour. Pet. Tech.* (Sept., 1969).

Blevins, T. R. and Billingsley, R. H.: "The Ten-Pattern Steamflood, Kern River Field, California," *Jour. Pet. Tech.* (Dec., 1975).

Bradley, B. W. and Gatzke, L. K.: "Steamflood Heater Scale and Corrosion," *Jour. Pet. Tech.* (Feb., 1975).

Bursell, C. G. and Pittman, G. M.: "Performance of Steam Displacement in the Kern River Field," *Jour. Pet. Tech.* (Aug., 1975).

Chappelear, J. E. and Volek, C. W.: "The Injection of a Hot Liquid into a Porous Medium," SPE 2013 prepared for the Symposium on Numerical Simulation of Reservoir Performance, Dallas, TX. (April, 1968).

Closmann, P. J.: "Steam Zone Growth in a Preheated Reservoir," *Soc. Pet. Eng. Jour.* (Sept., 1968). *Trans.,* AIME (1968) 243.

Coats, K. H.: "Simulation of Steamflooding with Distillation and Solution Gas," SPE 5015 prepared for the SPE-AIME 49th Annual Fall Meeting, Houston, TX. (Oct., 1974).

Combarnous, M. and Sourieau, P.: "Injection De Fluides Chauds. Principes Et Etudes De Laboratoire," *Les Methodes Thermiques De Production Des Hydrocarbures.* Revue De L'Institut Francais Du Petrole (Juil-Aout, 1976).

Cook, D. L.: "Influence of Silt Zones on Steam Drive Performance Upper Conglomerate Zone, Yorba Linda Field, California," *Jour. Pet. Tech.* (Nov., 1977).

Crookston, R. B., Culham, W. E. and Chen, W. H.: "Numerical Simulation Model For Thermal Recovery Processes," SPE 6724 presented at the SPE-AIME 52nd Annual Fall Meeting, Denver, CO. (Oct., 1977).

Davies, L. G., Silberberg, I. H. and Caudle, B. H.: "A Method of Predicting Oil Recovery in a Five-Spot Steamflood," *Jour. Pet. Tech.* (Sept., 1968). *Trans.,* AIME (1968) 243.

deHaan, M. J. and Schenk, L.: "Performance Analysis of a Major Steam Drive Project in the Tia Juana Field, Western Venezuela," *Jour. Pet. Tech.* (Jan., 1969).

Dietz, D. N.: "Review of Thermal Recovery Methods," SPE 5558 prepared for the SPE-AIME 50th Annual Fall Meeting, Dallas, TX. (Sept., 1975).

Doscher, T. M.: "Tertiary Recovery of Crude Oil" *The Future Supply of Nature-Made Petroleum and Gas,* Pergamon Press, New York (1976).

Duerksen, J. H. and Gomaa, E. E.: "Status of the Section 26C Steamflood, Midway-Sunset Field, California," SPE 6748 presented at the SPE-AIME 52nd Annual Fall Meeting, Denver, CO. (Oct., 1977).

Earlougher, R. C., Jr.: "Some Practical Considerations in the Design of Steam Injection Wells," SPE 2202 prepared for the SPE-AIME 43rd Annual Fall Meeting, Houston, TX. (Sept., 1968).

Farouq Ali, S. M.: "Current Status of Steam Injection As a Heavy Oil Recovery Method," *Jour. Cdn. Pet. Tech.* (Jan.-March, 1974).

Farouq Ali, S. M.: "Effects of Differences in the Overburden and the Underburden on Steamflood Performance," *Producers Monthly.* (Dec., 1966).

Farouq Ali, S. M.: "Effect of Temperature and Pressure on the Viscosity of Steam and Other Gases," *Producers Monthly.* (March, 1967).

Farouq Ali, S. M.: "Fluid Viscosity, Its Estimation, Measurement, and Role in Thermal Recovery," *Producers Monthly.* (Jan., 1967).

Farouq Ali, S. M.: "Graphical Determination of Oil Recovery in a Five-Spot Steamflood," SPE 2900 prepared for the SPE-AIME Rocky Mountain Regional Meeting, Casper, WY. (June, 1970).

Farouq Ali, S. M.: "Marx and Langenheim's Model of Steam Injection," *Producers Monthly.* (Nov., 1966).

Farouq Ali, S. M.: "Oil Recovery by Hot Waterflooding," *Producers Monthly*. (Feb., 1968).

Farouq Ali, S. M.: "Oil Recovery by a Steamflood in a Waterflooded Reservoir," *Producers Monthly*. (Sept., 1966).

Farouq Ali, S. M.: "Practical Considerations in Steamflooding," *Producers Monthly*. (Jan., 1968).

Farouq Ali, S. M.: "Recovery of Bradford Crude By Continuous Steam Injection," *Producers Monthly*. (Aug., 1966).

Farouq Ali, S. M.: "Thermal Recovery of Oil by Steam Injection - General Considerations," *Producers Monthly*. (Dec., 1965).

Farouq Ali, S. M.: "Viscosity of Steam and Steam-Gas Mixtures," *Producers Monthly*. (April, 1967).

Farouq Ali, S. M.: "Wet Steam For Thermal Recovery." *Producers Monthly*. (Feb., 1966).

Farouq Ali, S. M. and Meldau, R. F.: "A Current Appraisal of Steam Injection Field Projects," SPE 7183 presented at the SPE-AIME Rocky Mountain Regional Meeting, Cody, WY. (May, 1978).

Ferrer, J. and Farouq Ali, S. M.: "A Three-Phase, Two-Dimensional Compositional Thermal Simulator for Steam Injection Processes," *Jour. Cdn. Pet. Tech*. (Jan.-March, 1977).

Fitch, J. P. and Minter, R. B.: "Chemical Diversion of Heat Will Improve Thermal Oil Recovery," SPE 6172 presented at the SPE-AIME 51st Annual Fall Meeting, New Orleans, LA. (Oct., 1976).

Flock, D. L., Quon, D., Leal, M. A. and Thachuk, A. R.: "Modelling of a Thermally Stimulated Oil Reservoir — An Evaluation of Theoretical and Numerical Methods," *Jour. Cdn. Pet. Tech*. (Oct.-Dec., 1967).

Gates, C. F. and Brewer, S. W.: "Steam Injection Into the D and E Zone, Tulare Formation, South Belridge Field, Kern County, California," *Jour. Pet. Tech*. (March, 1975).

Hanzlik, E. J. and Bridwell, B. F.: "Steamflood of Heavy Oil, Cat Canyon Field," *ERDA Enhanced Oil, Gas Recovery and Improved Drilling Methods*, Vol. 1 - Oil, Tulsa, OK. (Sept., 1977).

Hartline, B. K. and Lister, C. R. B.: "Thermal Convection in a Hele-Shaw Cell," *Jour. Fluid Mech*., Vol. 79, Part 2. (1977).

Higgins, R. V. and Leighton, A. J.: "Computer Prediction of Water Drive of Oil and Gas Mixtures Through Irregularly Bounded Porous Media — Three-Phase Flow," *Jour. Pet. Tech*. (Sept., 1962).

Houpeurt, A. H.: "Tertiary Oil Recovery Processes: Physical Principles and Potentialities," *The Future Supply of Nature-Made Petroleum and Gas*, Pergamon Press, New York (1976).

Jones, J. R. and Shelton, J. L.: "Winkleman Dome Nugget Field," *Amoco Production Company Update Report*. (June, 1975).

Knapp, R. H. and Welbourn, M. E.: "An Acrylic/Epoxy Emulsion Gel System for Formation Plugging: Laboratory Development and Field Testing for Steam Thief Zone Plugging," SPE 7083 presented at the SPE-AIME 5th Symposium on Improved Methods for Oil Recovery, Tulsa, OK. (April, 1978).

Lauwerier, H. A.: "The Transport of Heat in an Oil Layer Caused by the Injection of Hot Fluid," *Appl. Sci. Res*., Sec. A. (1955).

Leighton, A. J.: "Enhanced Recovery of Heavy Oil in California," *The Future Supply of Nature-Made Petroleum and Gas*, Pergamon Press, New York (1976).

Malofeev, G. E.: "Calculation of the Temperature Distribution in a Formation When Pumping Hot Fluid into a Well," *Neft'i Gaz*, Vol. 3, No. 7. (1960).

Martin, W. L., Dew, J. N., Powers, M. L. and Steves, H. B.: "Results of a Tertiary Hot Waterflood in a Thin Sand Reservoir," *Jour. Pet. Tech*. (July, 1968) *Trans.,* AIME (1968) 243.

Marx, J. W. and Langenheim, R. H.: "Reservoir Heating by Hot Fluid Injection," *Trans.,* AIME (1959) 216.

Moss, J. T.: "Minitests Evaluate Thermal - Drive Variables," *O. & G. Jour*. (April 1, 1974).

O'Dell, P. M. and Rogers, W. L.: "Use of Numerical Simulation to Improve Thermal Recovery Performance in the Mount Poso Field, California," SPE 7078 prepared for the SPE-AIME 5th Symposium on Improved Methods for Oil Recovery, Tulsa, OK. (April, 1978).

Prats, M.: "A Current Appraisal of Thermal Recovery," SPE 7044 presented at the SPE-AIME 5th Symposium on Improved Methods for Oil Recovery, Tulsa, OK. (April, 1978).

Ramey, H. J., Jr.: "Further Discussion of 'Reservoir Heating by Hot Fluid Injection'," *Trans.,* AIME (1959) 216.

Ramey, H. J., Jr.: "How to Calculate Heat Transmission In Hot Fluid Injection," *Fundamentals of Thermal Oil Recovery,* Petroleum Engineer Publishing Co., Dallas, TX. (1965).

Ramey, H. J., Jr.: "Wellbore Heat Transmission," *Jour. Pet. Tech.* (April, 1962).

Ramey, H. J., Jr. and Brigham, W. E.: "A Review of Thermal Recovery Applications in the United States," *The Future Supply of Nature-Made Petroleum and Gas,* Pergamon Press, New York (1976).

Rubinshtein, L. I.: "An Asymptotic Solution of an Axially Symmetric Contact Problem in Thermal Convection for High Values of the Convection Parameter," *Dan SSSR,* Vol. 146, No. 5. (1962).

Rubinshtein, L. I.: "A Contact Thermal Conduction Problem," *Dan SSSR,* Vol. 135, No. 4. (1960).

Rubinshtein, L. I.: "The Total Heat Losses in Injection of a Hot Liquid into a Stratum," *Neft'i Gaz,* Vol. 2, No. 9. (1959).

Sarem, A. M. and Hawthorne, R. G.: "A Theoretical Comparison of Heat Transfer into Porous Media by the Injection of Steam or Hot Water," *Jour. Cdn. Pet. Tech.* (April-June, 1966).

Satter, A.: "Heat Losses During Flow of Steam Down a Wellbore," *Jour. Pet. Tech.* (July, 1965).

Shutler, N. D.: "Numerical Three-Phase Model of the Two-Dimensional Steam Flood Process," SPE 2798 prepared for the 2nd Symposium on Numerical Simulation of Reservoir Performance, Dallas, TX. (Feb., 1970).

Spillette, A. G.: "Heat Transfer During Hot Fluid Injection Into An Oil Reservoir," *Jour. Cdn. Pet. Tech.* (Oct.-Dec., 1965).

Squier, D. P., Smith, D. D. and Dougherty, E. L.: "Calculated Temperature Behavior of Hot-Water Inejction Wells," *Jour. Pet. Tech.* (April, 1962).

Stegemeier, G. L., Laumbach, D. D. and Volek, C. W.: "Representing Steam Processes with Vacuum Models," SPE 6787 presented at the SPE-AIME 52nd Annual Fall Meeting, Denver, CO. (Oct., 1977).

Stokes, D. D., Brew, J. R., Whitten, D. G. and Wooden, L. G.: "Steam Drive As a Supplemental Recovery Process in an Intermediate Viscosity Reservoir, Mount Poso Field, California," SPE 6522 prepared for the SPE-AIME 47th Annual California Regional Meeting, Bakersfield, CA. (April, 1977).

Stokes, D. D. and Doscher, T. M.: "Shell Makes a Success of Steam Flood at Yorba Linda," *O. & G. Jour.* (Sept. 2, 1974).

Valleroy, V. V., Willman, B. T., Campbell, J. B. and Powers, L. W.: "Deerfield Pilot Test of Recovery by Steam Drive," *Jour. Pet. Tech.* (July, 1967).

van Dijk, C.: "Steam-Drive Project in the Schoonebeek Field, The Netherlands," *Jour. Pet. Tech.* (March, 1968).

Weinbrandt, R. M., Ramey H. J., Jr. and Cassi, F. J.: "The Effect of Temperature on Relative and Absolute Permeability of Sandstones," *Soc. Pet. Eng. Jour.* (Oct., 1975).

Willhite, G. P. and Dietrich, W. K.: "Design Criteria for Completion of Steam Injection Wells," *Jour. Pet. Tech.* (Jan., 1967).

Willman, B. T. et al: "Laboratory Studies of Oil Recovery by Steam Injection," *Trans.,* AIME (1961) 222.

Wilson, L. A. and Root, P. J.: "Cost Comparison of Reservoir Heating Using Steam or Air," *Jour. Pet. Tech.* (Feb., 1966).

Yoelin, S. D.: "The TM Sand Steam Stimulation Project," *Jour. Pet. Tech.* (Aug., 1971).

Chapter 3
IN SITU COMBUSTION

INTRODUCTION

There are two fundamentally different processes of in situ combustion: forward combustion (also known as dry combustion) and reverse combustion. In forward combustion, the reservoir is ignited in the vicinity of an air injection well, and the combustion front propagates away from the well. Continued injection of air maintains the combustion and drives the combustion front through the reservoir in the general direction of air flow. Figure 1 is a schematic diagram of the in situ combustion process (Nelson and McNeil, 1961). The lower portion of the diagram (Figure 1B) shows the temperature distribution from the injection well to the

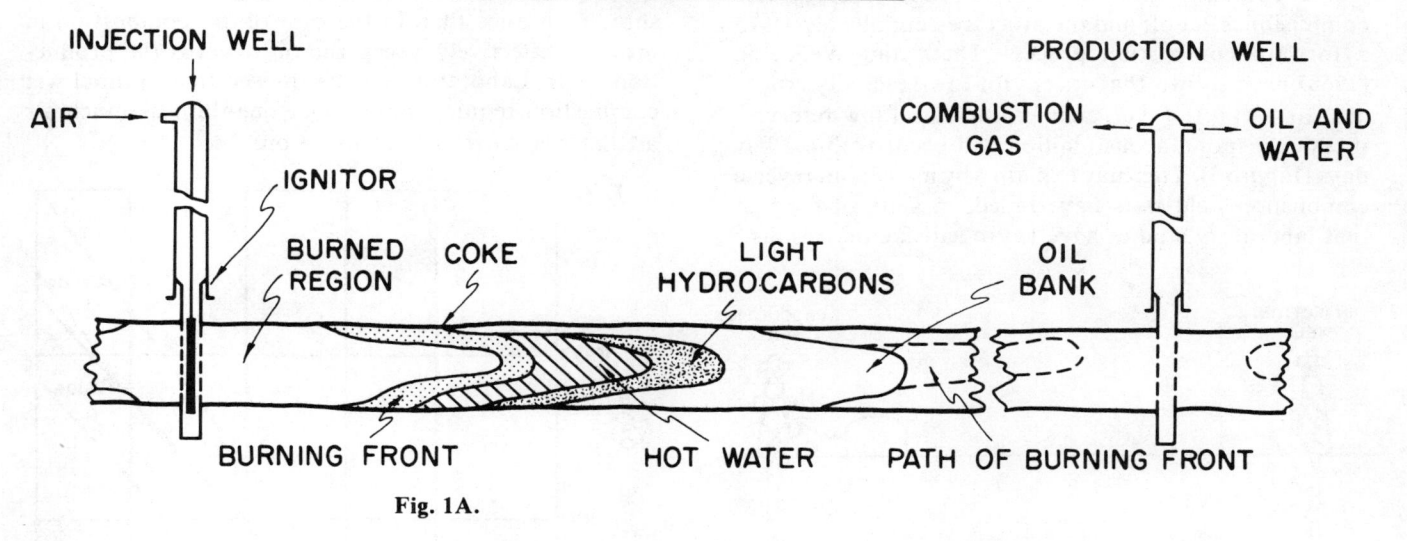

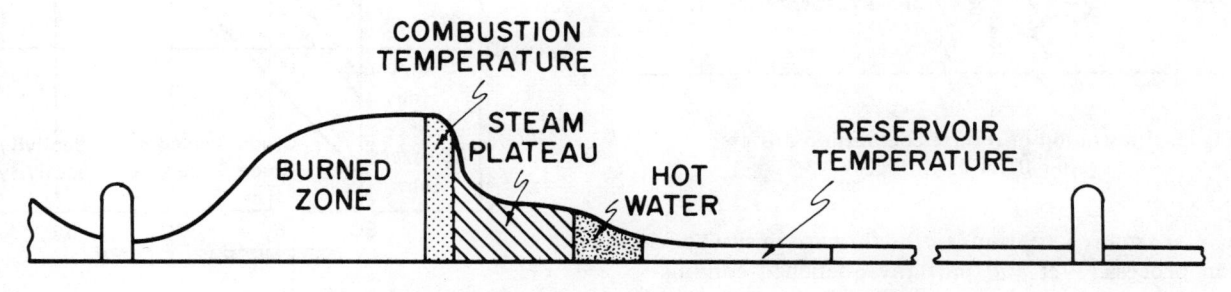

Fig. 1B. Schematic diagram of in situ combustion process. (modified from Nelson and McNeil, 1961)

production well. It is evident that if the injection is reversed, i.e., if air injection at the injection well is stopped and then switched to the production well, the oil bank will be forced to move in the reverse direction and through the burned zone (Figure 1A), while the combustion front continues to move without any change in direction. Since the temperature and heat content of the burned zone are quite large, the oil will be heated to temperatures ranging between 500-700°F. This high temperature reduces the viscosity of the oil by several orders of magnitude, thereby allowing high viscosity crudes to flow freely towards the producing well.

The reverse combustion process is first started as a forward combustion process by injecting air through wells that will eventually become oil producing wells. After burning out a short distance from the ignited wells, air injection is switched to adjacent wells. Continued injection of air in the adjacent wells drives the oil towards the previously ignited wells while the combustion front travels in the opposite direction towards the adjacent wells (Figure 2). This oxygen supply can only be provided by the air which is injected in the adjacent wells. Thus if the oil about the adjacent wells ignites spontaneously, the oxygen supply for reverse combusion is cut off and the process essentially reverts to a forward combustion process. Dietz and Weijdema (1968) have shown that except for low-reactivity crudes at near arctic reservoir temperatures and low reservoir pressures, spontaneous ignition will occur within a few days (Figure 3). This may explain why many pilot reverse combustion field tests have failed, in spite of the fact that laboratory studies have led to satisfactory results.

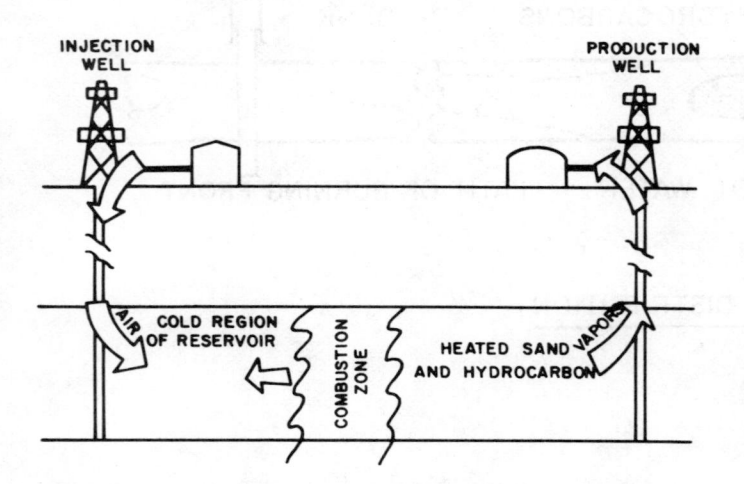

Fig. 2. Illustration of reverse-combustion process. (after Berry et al, 1960)

There are many variations of the forward in situ combustion process. Wet and partially quenched combustion are two of these variations. Although these methods have not progressed beyond laboratory and pilot field studies, the consensus is that they offer significantly higher potential than dry combustion. Referring to Figure 1, one can readily see that much heat is left in the burned zone behind the combustion front. In dry combustion, this heat is left to dissipate through the reservoir cap and base rock. However, when water in moderate amounts is injected simultaneously with the air, it flashes into superheated steam a short distance from the injection well; the evaporation front remains close to the injection well (Figure 4A). As the superheated steam mixed with air reaches the combustion front, only the oxygen is utilized in the burning process. Upon crossing the combustion front, the superheated steam mixes with nitrogen from the air and flue gas consisting mainly of CO and CO_2. This mixture of gases displaces the oil in front of the combustion zone and condenses as soon as its temperature drops to about 400°F. The length of the steam zone is determined by the amount of heat recovered from the burned zone upstream. Ideally, water can be injected in sufficient amounts in order to bring the evaporation front as close as possible to the combustion front (Figure 4B). This would result in transporting nearly all the excess heat of the burned zone to the steam zone. Thus, the steam zone would increase in size, and for a given well spacing the combustion front would travel a much shorter distance than in the case of dry combustion in order to effectively sweep the oil towards the production wells. Laboratory results showed that optimal wet combustion requires only about one-third as much air as that needed for dry combustion.

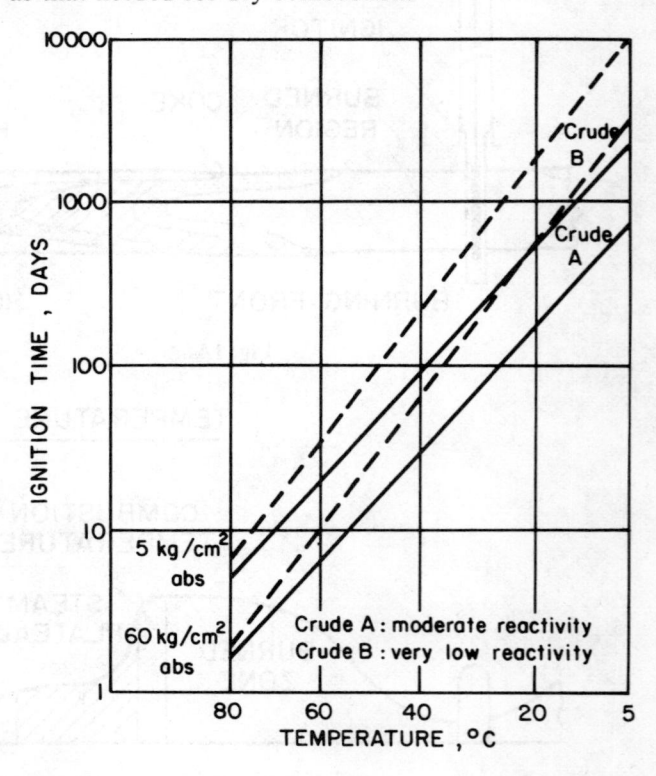

Fig. 3. Ignition time as a function of temperature at various air-injection pressures. (after Dietz, 1968)

Partially quenched combustion occurs when water is introduced in sufficient amounts into the combustion zone. Laboratory experiments showed that at a temperature of 400°F, oxygen is consumed within 0.4 to 3.3 ft. Therefore, introduction of water in the combustion zone would partially quench the combustion, and thus the oxygen must travel until it comes in contact with oil at 400°F. In this way, the combustion zone moves at the speed of the cooling water and all the processes (heating and partial evaporation of water, heat recovery from the formation and heat generation by combustion) occur in one fast-traveling front (Figure 4C).

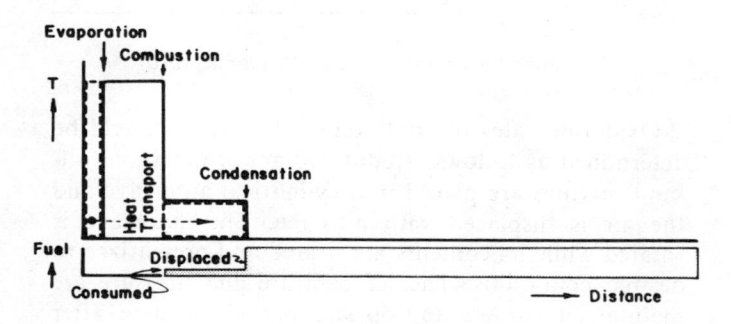

Fig. 4A. Normal wet combustion.

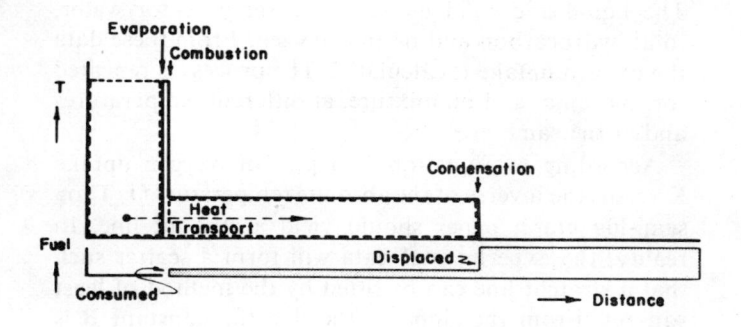

Fig. 4B. Optimal normal wet combustion.

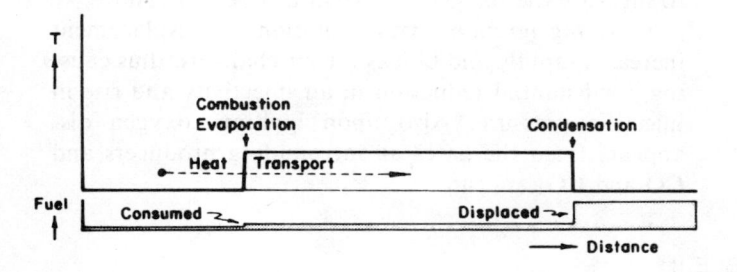

Fig. 4C. Partially quenched combustion.
(after Dietz and Weijdema, 1968)

In situ combustion is one of the most difficult enhanced oil recovery methods to control in the field or simulate on a computer. However, this technique may offer the highest percentage recovery of the thermal techniques.

TECHNICAL DISCUSSION

Ignition

In a forward combustion process, oil is ignited spontaneously. The process is as follows. Upon injection of air, some oxidation of the crude oil takes place at reservoir temperature and results in a slow temperature rise. The rise in temperature has the effect of increasing the rate of oxidation of the crude which, in turn, leads to further rise in temperature. The process continues until the temperature of the crude has risen to the point at which spontaneous ignition of the crude occurs. In some cases, heating of the air may reduce ignition time.

During the early stages of development of in situ combustion, numerous and sometimes unnecessary measures were taken in order to ignite the oil. In one method, high-powered heaters were lowered in the air injection wells. This resulted in the formation of coke around the wells, a substance that is much more difficult to ignite than crude oil. An attempt was also made to use concentrated nitric acid, a strong reactive and oxidizing agent, but violent explosions occurred and resulted in destroying the tubular equipment in the well.

Moderate heating of the injected air may be desired in order to reduce the ignition time. Normally, raising the reservoir temperature to 200°F will result in an ignition period of 1 or 2 days.

Preheating may also be required to control the initial location of ignition. Without preheating, ignition will occur at some distance beyond the face of the injector screen and a reverse combustion effect may follow. Extremely high temperatures occur in the interval where combustion direction is reversed and damage to injectors could result. Moderate preheating will cause ignition near the wellbore and the high combustion reversal temperatures can be avoided.

According to Tadema and Weijdema (1970), the ignition time can be estimated by the following formula:

$$t_i = \frac{\rho_1 c_1 T_o \, (1 + 2T_o/B)}{86400 \; \phi S_o \rho_o H A_o P_x^n B/T_o} \, e^{B/T_o} \qquad (1)$$

where:

t_i = ignition time (days)

ρ_1 = density of oil-bearing formation (kg/m³)

ρ_o = density of oil (kg/m³)

c_1 = specific heat of oil-bearing formation (kcal/kg°C)

T_o = initial temperature (°K)

A_o = constant (sec^{-1} atm^{-n})

B = constant (°K)

n = pressure exponent

S_o = oil saturation

ϕ = porosity

H = heat of reaction (kcal/kgO$_2$)

p_x = partial pressure of oxygen (atm abs)

p_x = 0.209 p, where p is the air injection pressure (atm abs)

The specific heat of the formation $\rho_1 c_1$ is determined from:

$$\rho_1 c_1 = (1 - \phi)\rho_s c_s + \phi S_o \rho_o c_o + \phi S_w \rho_w c_w \qquad (2)$$

where:

ρ_s = formation grain density (kg/m^3)

c_s = specific heat of the formation grain (k cal/kg°C)

c_o = specific heat of oil (K cal/kg°C)

S_w = water saturation

ρ_w = water density (kg/m^3)

c_w = specific heat of water (k cal/kg°C)

The constants A_o and B_o and n are determined by measuring the oxidation rates of different crude oil-sand mixtures at different temperatures and pressures. The oxidation rate K (mg O$_2$/kg oil, sec) is related to the partial pressure of the oxygen p_x and the oil temperature T_o by the following Arrhenius-type equation:

$$K = A_o p_x^n e^{-B/T_o} \qquad (3)$$

Tadema and Weijdema (1970) gave the following representative values for A_o, B and n (Table I).

Upon substitution of the field data shown in Table II in Equation 1, ignition times of 99 and 49 days are obtained for the South Belridge Field, and a Venezuelan field, respectively. These figures compare very favorably with the observed ignition times of 106 and 35 days, respectively.

TABLE I
REPRESENTATIVE VALUES OF A_o, B and n

Crude	A_o	B	n
A	3080	8860	0.46
B	925	8640	0.57
C	498	8880	0.79
D	84800	10270	0.48
E	1210	8680	0.45
F	7380	9480	0.31

(Modified from Tadema and Weijdema, 1970)

Oxidation rates of crude oil-sand mixtures could be determined as follows. About 200 gm of measured oil-sand mixture are placed in a cylindrical autoclave and the air is displaced with nitrogen. The autoclave is rotated while its contents are heated and pressurized to desired conditions. The temperature and pressure are maintained for an appropriate period of time after which the autoclave is allowed to cool. The gas content of the autoclave is then analyzed for O$_2$, CO$_2$ and CO. The liquid and solid material are analyzed for water, total hydrocarbon and bound oxygen. From these data the oxygen uptake is calculated. The process is repeated for the same sand-oil mixture, at different temperatures and at the same pressure.

According to Equation 3, a plot of oxygen uptake K versus the inverse of the absolute temperature (1/T) on semi-log graph paper should yield a straight line. In reality, the experimental data will form a scatter such that a straight line can be fitted by the method of least squares. From the slope of the line the constant B is determined, and the product $A_o p_x^n$ can then be evaluated. To determine each of the values A_o and n it is necessary to measure the oxygen uptake at different pressures.

Detecting ignition: Upon ignition, oil displacement increases rapidly and blocks the air channels, thus causing a substantial reduction in air injectivity and rise in injection pressure. Also upon ignition, oxygen disappears from the gases at surrounding producers and CO and CO$_2$ appear.

TABLE II
FIELD DATA FOR CALCULATION OF SPONTANEOUS IGNITION TIME

Field	T_o	P	P_x	A_o	B	n	H	Φ	S_o	S_w	ρ_o	$\rho_1 C_1$	Ignition Time Calculated	Observed
South Belridge	303.8	15.3	3.20	3080	8860	0.46	2940	0.37	0.60	0.37	970	553	99	106
Venezuela	312.2	29.6	6.18	1210	8680	0.45	2940	0.40	0.56	0.34	980	527	49	35

(Modified from Tadema and Weijdema, 1970)

Air Requirements

The main factors which govern the volume of air required for in situ combustion are: amount of fuel supply (coke) of the oil being burned and the efficiency of oxygen utilization. As the combustion zone moves through the reservoir, it continuously emits heat. The heat moves in the forward direction by conduction, as sensible heat in liquid and gas and as heat of vaporization in vapor. The hot fluids flush out volatile and mobile substances from the path of the combustion zone leaving behind residual hydrocarbon (coke), which is the fuel supply for combustion. The combustion zone can only move as fast as it depletes the coke. If the amount of coke deposited by the crude oil is excessive, the rate of advance of the combustion zone would be slow and the air requirement would be large. On the other hand, if the oil is paraffin-base and of high API gravity it could be completely flushed out and combustion would not be sustained in the absence of coke. Therefore, asphaltic and naphthenic-base crude oils are normally the best candidates for the in situ combustion process.

Efficiency of oxygen utilization depends on the following parameters: 1) carbon/hydrogen ratio (C/H) of the coke; 2) amount of CO produced; and 3) amount of O_2 which appears in the exhaust. When the C/H ratio of the crude oil is low, more air will be required since it takes more O_2 to oxidize hydrogen than to oxidize carbon. The amount of CO produced is a measure of the relative amount of carbon which is not completely oxidized. Since the amount of O_2 required to produce CO_2 is twice that required to produce CO, it follows that the larger the amount of CO produced the less the air required. Finally, the air requirement must be increased in order to supply an amount of O_2 which is equal to the amount that appears in the exhaust.

Considerable uncertainty exists in the chemical properties of coke deposition and combustion. Alexander, Martin and Dew (1962) found that fuel availability depends on:

1. Initial oil saturation and rock quality (Figure 5)
2. Residual oil saturation (Figure 6)
3. Produced gas flux (Figure 7)
4. C/H ratio (Figure 8)
5. API gravity (Figures 9A and 9B)
6. Conradson carbon residue (Figure 10)
7. Viscosity (Figure 11)
8. Low temperature oxidation (LTO) (Figure 12).

These authors concluded that: a) fuel availability decreased as the C/H ratio, viscosity, Conradson carbon residue and oil saturation decreased, and the API gravity increased; b) fuel availability increased when natural reservoir rock was used and decreased when clean porous media were used; c) LTO prior to high temperature combustion had a greater effect on fuel availability and composition than any other process variable such that it is possible to maximize fuel availability for a given system by proper LTO treatment; d) the C/H ratio of the fuel is lower for low temperature and higher for high temperature than the C/H ratio of the crude oil, and as the combustion temperature increased, the C/H ratio of the fuel burned increased.

Showalter (1963) showed that varying the pressure had little effect on air requirement and fuel consumption (Figure 13). Likewise, burning rate seems to have very little effect on fuel consumption, but the air requirement was higher at the higher burning rates. The volume of air required to burn 1 lb of hydrocarbon can be estimated by the following equation:

$$V = \frac{\left(\frac{R}{R+1}\right)\left(\frac{2.667(A + 0.5 B + C)}{A + B}\right) + \left(\frac{8}{R+1}\right)}{0.01873} \quad (4)$$

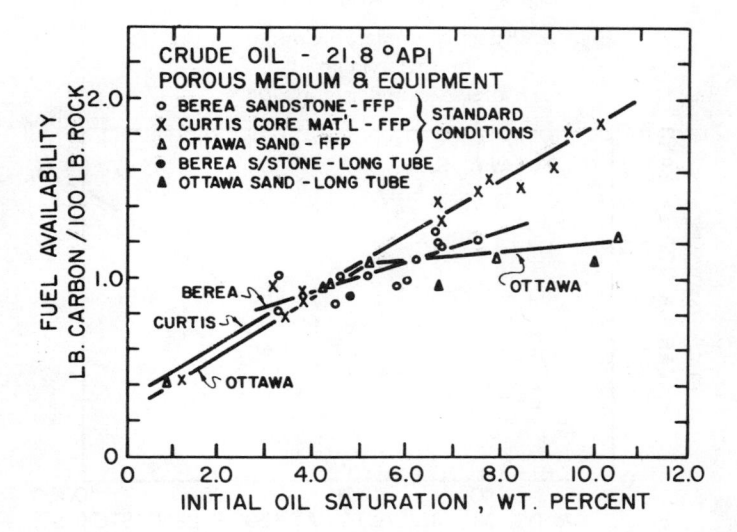

Fig. 5. Effect of initial oil saturation on fuel availability. (after Alexander et al, 1962)

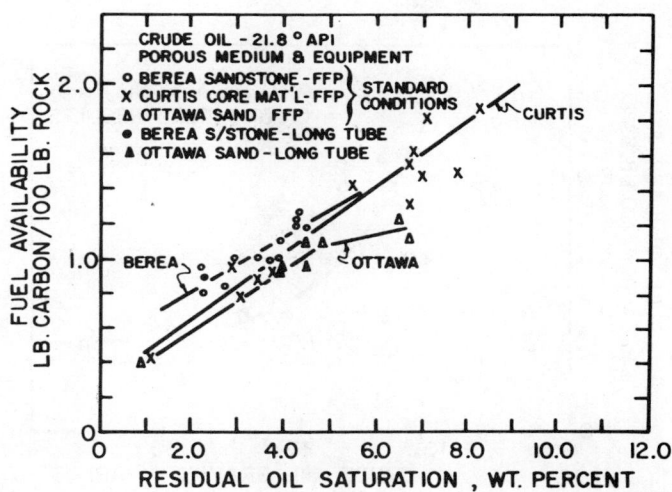

Fig. 6. Effect of residual oil saturation on fuel availability. (after Alexander et al, 1962)

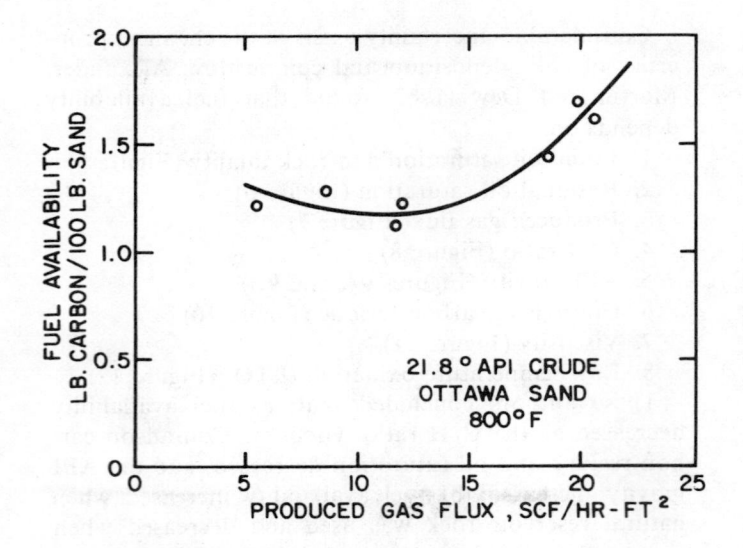

Fig. 7. Effect of produced-gas flux on fuel availability.
(after Alexander et al, 1962)

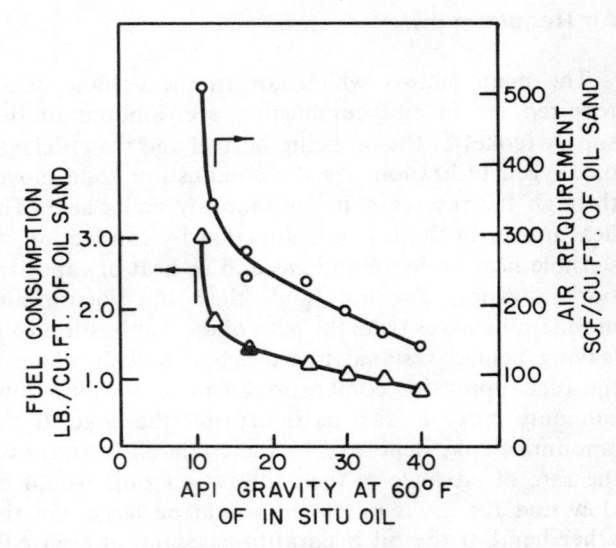

Fig. 9B. Combustion-drive fuel consumption and air requirement vs oil gravity.
(after Showalter, 1963)

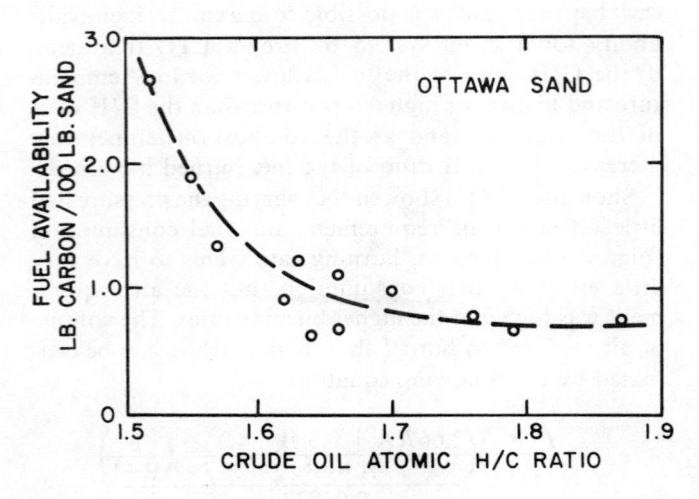

Fig. 8. Correlation of fuel availability with crude-oil H/C ratio.
(after Alexander et al, 1962)

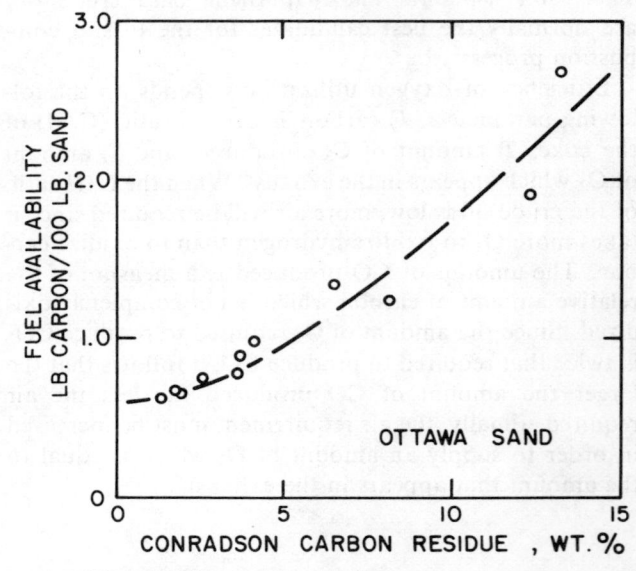

Fig. 10. Effect of crude-oil Conradson carbon on fuel availability.
(after Alexander et al, 1962)

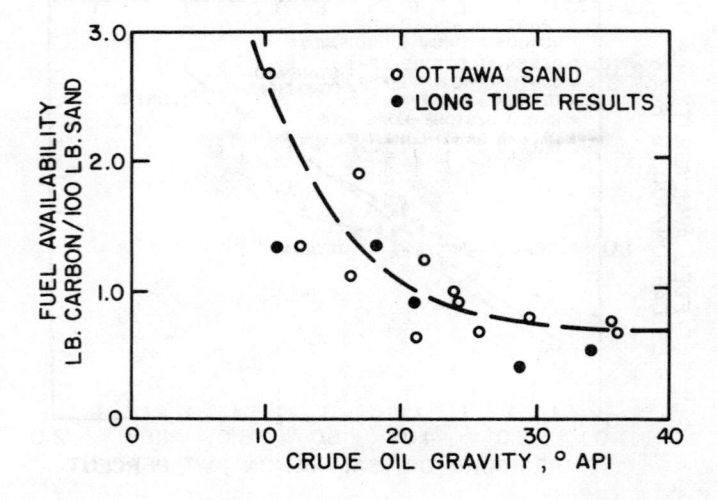

Fig. 9A. Correlation of fuel availability with crude-oil gravity.
(after Alexander et al, 1962)

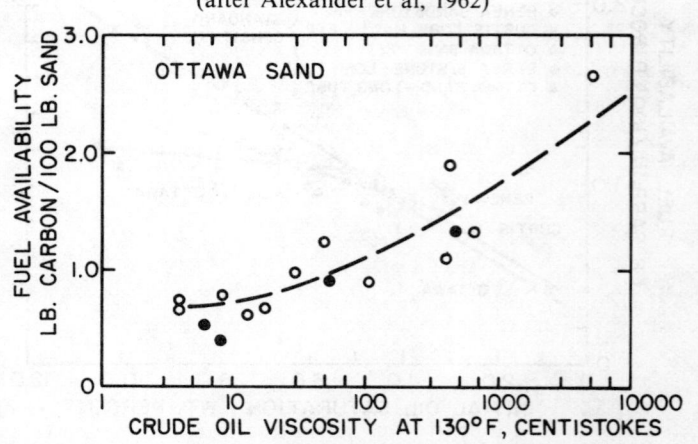

Fig. 11. Effect of crude-oil viscosity on fuel availability.
(after Alexander et al, 1962)

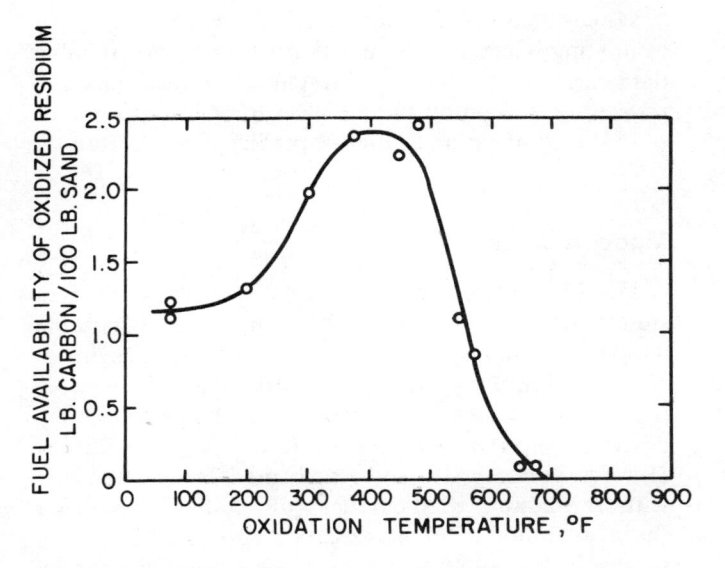

Fig. 12. Effect of low-temperature oxidation on fuel availability at 800°F.
(after Alexander et al, 1962)

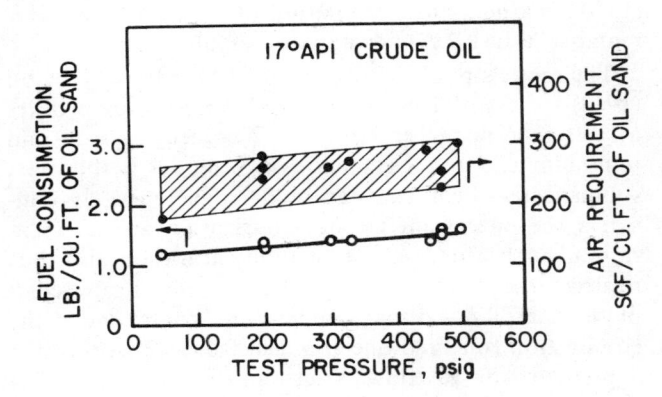

Fig. 13. Combustion-drive fuel consumption and air requirement vs test pressure.
(after Showalter, 1963)

where:

V = volume of air required to burn 1 lb of fuel (coke) (32°F, 1 atm)

A, B, C = percent CO_2, CO and O_2, respectively, in the exhaust gas corrected to an extraneous gas-free basis, i.e., any gases such as H_2 and hydrocarbon gases that that are not products of combustion must be eliminated before calculating the percentages of CO_2, CO and O_2.

R = C/H ratio of the hydrocarbon burned, and

$$R = \frac{A + B}{\frac{21 - (A+B+C)}{2.37} + \frac{B}{6}}$$

Therefore, from the API gravity of the in situ oil and with the help of Equation 4, the air requirement per cubic foot of oil sand can be estimated. It should be remembered, however, that the data of Figure 9B were obtained for an oil saturation S_o of 60% of the pore volume and water saturation S_w of 20%.

Once the volume of air required per cubic foot of fuel supply is determined, the air-oil-ratio (AOR) may be calculated by the following equation (Crawford, 1974):

$$AOR = \frac{1.9 \times 157 \times 43500 \times 0.56}{(\phi \cdot S_o \cdot 7760 - F \times 125) \times .3 + 0.37 \times \phi \times S_o \times 7760 \times .7} \quad (5)$$

where:

AOR = air-oil-ratio in standard cubic feet per barrel of oil

F = lbs of fuel (coke) per cubic foot of rock

ϕ = porosity

As an example, let the exhaust gas analysis corrected to a methane-free basis be as follows: CO_2 (17.28%); CO (1.04%); O_2 (0%). Let the fuel availability be 1.9 lb/cu ft. Therefore, C/H is calculated at 14, and V is 157 SCF of air per lb of fuel. If the porosity is 30% and the oil saturation is 75%, then substitution in equation 5 yields:

$$AOR = \frac{1.9 \times 157 \times 43500 \times 0.56}{(0.3 \times .75 \times 7760 - 1.9 \times 125).3 + 0.37 \times .3 \times .75 \times 7760 \times .7}$$

= 8031 SCF of air per barrel of oil

Experimental Studies

It is evident that proper design of an in situ combustion project largely depends on the correct evaluation of a host of parameters. While the figures presented in the previous section are helpful in determining the order of magnitude of these parameters, they are not sufficiently reliable for designing a specific project. For this reason it is essential that laboratory investigations be made for the reservoir being considered for in situ combustion, in order to measure the different parameters and formulate a strategy for field implementation.

The ultimate objectives of the laboratory tests are to determine fuel supply availability under variable LTO conditions, air requirement and AOR, as well as optimum burning rate and injection pressure. The results are not necessarily applicable to the reservoir. In the reservoir, fingering and bypassing result in lowering the sweep efficiency. Consequently, fuel availability, air requirement and AOR obtained from laboratory experiments must be adjusted in order to account for the realities of the reservoir.

The laboratory experiments can be made either on a representative reservoir core sample (Figure 14) or on a cylindrical sand section prepared by mixing reservoir oil and water with sand and kaolinite (Figure 15). The size distribution of the sand-kaolinite mixture must resemble that of the reservoir rock. Crushed cores from the reservoir rock can also be used.

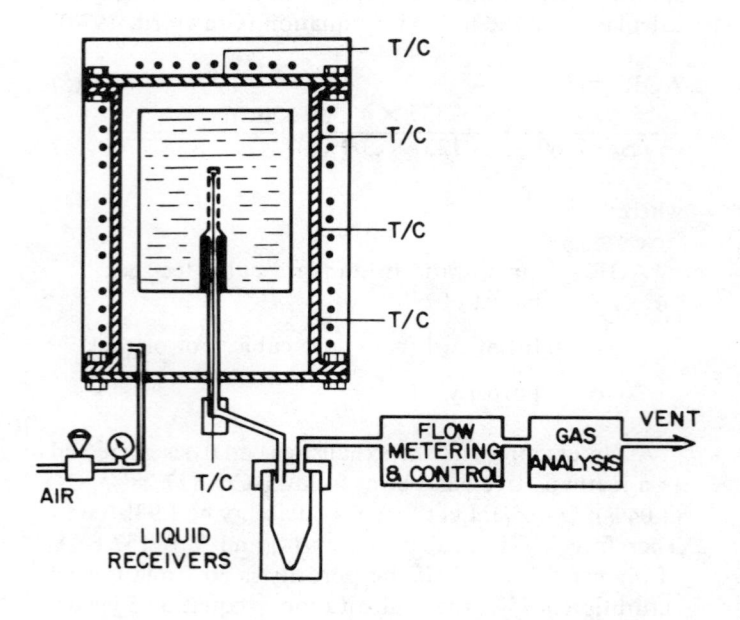

Fig. 14. Flow diagram of the fire flood-pot apparatus. (after Alexander et al, 1962)

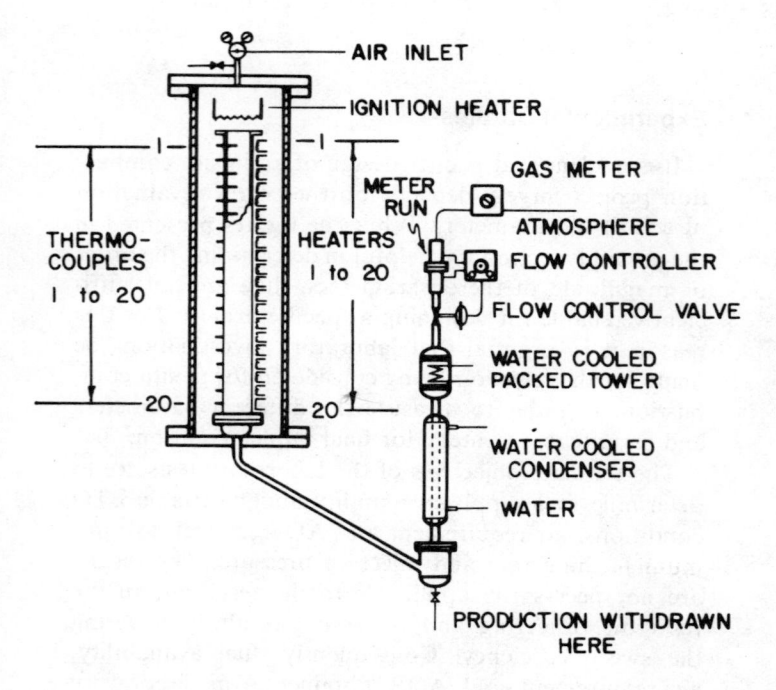

Fig. 15. Schematic of combustion-drive test cell. (after Showalter, 1963)

While laboratory tests and equations are helpful in estimating injection rates and other parameters, reliable data can only be obtained through actual field injection tests prior to combustion. However, the conformance of in situ combustion cannot be predicted with certainty.

Model Studies

The in situ combustion recovery process involves simultaneous flow of three fluid phases in a variable temperature and pressure field accompanied by chemical reactions and phase changes. Nevertheless, it is possible to formulate mathematical models which approximately describe the principal physical and chemical phenomena. However, attempts to solve such models are often met with formidable mathematical difficulties, even with the availability of high-speed computers. In addition, the usefulness of the numerical solutions is limited by the uncertainties in the values of input parameters and the simplifying assumptions that must be made in order to overcome convergence problems. According to Dietz (1970), mathematical simulation of in situ combustion remains "a task beyond man and digital computer".

The models published by Thomas (1963) and by Chu (1963) are useful in estimating peak temperature profiles at different distances from the air-injection well and minimum fuel requirements. Chu's model is the most sophisticated heat transfer model developed. It considers the energy effects of vaporization and condensation, but assumes constant fluid saturation and thus neglects the phase changes. Kuo (1968) proposed a model that allowed two temperature fronts: one at the combustion zone and one at a heat front whose position is predicted by gas flow. Smith and Farouq Ali (1971) proposed a model designed to predict sweep efficiencies in confined well patterns. In developing this model, Smith and Farouq Ali assumed fixed fuel content thoughout the reservoir and single phase gas flow. They considered heat transfer by conduction and convection, heat losses by conduction to adjacent formation and different air permeabilities on either side of the combustion zone. The model by Gottfried (1965) considered three-phase flow as well as heat flow by convection and conduction. It is considered to be the most complete model and limited only by the fact that it is a one-dimensional model. However, Gottfried's model does not include gravity and capillary effects, and no oil vaporization is allowed.

Crookston et al (1977) presented a simulator which describes the two-dimensional flow of three fluid phases and includes gravity and capillarity effects. The model accounts for heat transfer by conduction and convection in the reservoir and heat loss to base and cap rock by conduction. The model also accounts for vaporization-condensation of both oil and water, and includes four chemical reactions related to formation and oxidation of coke. Although the primary focus of the model is towards simulation of in situ combustion, it also represents water flooding, hot water flooding, steam flooding, steam stimulation and spontaneous ignition.

Economic Evaluation

Wilson and Root (1966) made an economic study in which they used a modified form of Chu's two-dimensional model. Their main objective was to compare cost of heating a reservoir either by steam injection or by forward combustion, without regard to recovery. Their sole consideration was the cost of heating to the same radial distance by either forward combustion or by steam injection. They concluded that:

1. Except for oils which yield coke in amounts less than 1 lb/cu ft and reservoir thicknesses of about 10 ft or less, heating the reservoir by steam injection is cheaper than by forward combustion as long as the cost of the fuel needed to generate the steam is low.
2. For a given thickness, pressure and rate of heat injection, either process may be cheaper, depending upon the reservoir fuel consumption and depth. However, as the price of fuel increases, the cost of heating by steam injection increases more rapidly.
3. Increased coke deposition favors steam injection.
4. Increased wellbore losses by increased depth favors combustion.
5. As the heated distance in the reservoir increases, heating by combustion becomes more favorable.
6. As the sand thickness decreases and the pressure increases, combustion is favored over steam injection.
7. As the injection rates decrease, costs of steam injection become more favorable relative to air.

Oil Recovery

Average oil recovery from in situ combustion is approximately 50%. Most of the recovery occurs before breakthrough of the combustion zone. In the case of low viscosity crudes, oil production declines very rapidly following breakthrough. In the case of viscous crudes, however, almost half of the recovery comes after breakthrough.

Breakthrough of the combustion zone is recognized by an increase in water cut of the produced oil and a rise in gas production rate and its oxygen content, followed by a sharp rise in bottom hole temperature, on the order of 100-200°F. At the same time, a noticeable drop occurs in the pH of the produced water. This increase in the acidity of the water is usually accompanied by an increase in iron and sulfate.

CONCLUSIONS

It has been demonstrated that in situ combustion is suitable in the recovery of oils of gravities in excess of 10° API. However, it is unlikely that dry combustion per se will find much wider application in the future. With continued technological improvements, it is almost certain that some form of in situ combustion such as wet and partially quenched combustion will find greater application.

The disadvantages of the in situ combustion process can be summarized as follows:

1. The in situ combustion process has a tendency to sweep only the upper part of the oil zone; therefore, vertical sweep in very thick formations is likely to be poor. The burning front produces steam both by evaporating the interstitial water and by combustion reactions. The steam mobilizes and displaces much of the heavy oil ahead of the front, but when water condenses from the steam it settles below steam vapors and combustion gases, thus causing their flow to concentrate in the upper part of the oil zone.
2. Much of the heat generated by in situ combustion is not utilized in heating the oil; rather, it heats the oil-bearing strata, interbedded shale and base and cap rock. Therefore, in situ combustion would be economically feasible when there is less rock material to be heated, i.e., when the porosity and oil saturations are high and the sand thickness is moderate.
3. Many operators feel viscous, low-gravity crudes are best suited for in situ combustion because they provide the needed fuel for combustion. However, the required air-oil ratio for viscous crudes is high while their price is generally lower than the high-gravity crudes. Some successful projects have also been reported for lower viscosity, higher gravity crudes, such as Marathon's Fry Project. This crude has an API gravity of 28.6° and a viscosity of 40 cp.
4. Installation of in situ combustion requires a large investment. However, surface installation consumes less fuel than hot water or steam-generating units.
5. Serious operational problems have been reported for in situ combustion operations. Some of these problems are: 1) formation of oil-water emulsions having consistencies of whipped cream which cause pumping problems and reduce well productivity; 2) production of low pH (acidic) hot water, rich in sulfate and iron, with attending environmental pollution and producing well corrosion problems; 3) increased sand production and cavings which cause plugging of well liner; 4) plugging of the producing wellbore due to deposition of carbon and wax as a result of thermal cracking of the oil;

5) production of environmentally hazardous gases such as carbon monoxide and hydrogen sulphide; and 6) liner and tubing failure due to excessive temperatures at the production wells.

CASE HISTORIES

In this section, three case histories are presented: the Sloss Field, Nebraska; the Miga Field, Venezuela; and the Heidleberg Field, Mississippi. In the Sloss Field, air and water were simultaneously injected in an effort to reduce AOR. The Miga Field was closed in for a two year period prior to initiation of the combustion project. Combustion began in 1964 and since that time the field has produced more than twice the primary production. The Heidleberg Field is the deepest known application of the in situ combustion process.

Sloss Field, Nebraska (USA)

This project involves a Combination Of Forward Combustion And Waterflooding (COFCAW). Air and water are injected simultaneously or alternately after a small heat bank has been formed. It is a form of wet or partially quenched combustion that was discussed earlier in this chapter. The major benefit of COFCAW is the threefold or more reduction in the amount of air required to recover a barrel of oil (Parrish and Craig, 1969).

From February 1967 to July 1971, a 960 acre COFCAW project was operated in the Sloss Field (Buxton and Pollack, 1974). Pertinent reservoir data are given in Table III. During this period, 14 MMCF of air and 11 million Bbls of water were injected. Nearly 836000 Bbls of oil were recovered, and about 340000 Bbls of hydrocarbon, vaporized by flue gas passing through the reservoir, were vented to the atmosphere. If all hydrocarbon in the vent gas had been recovered, the overall air-oil ratio would have been 11.7 MCF per Bbl.

Following termination of air injection in July 1971, five core holes were drilled in order to determine: residual oil saturation; variation in vertical sweep value with distance from the injection well; areal coverage by the burning front; maximum temperature distribution; the effective permeability of the reservoir rock; and amount and type of material deposited in the pores that could have reduced the flow capacity.

The results showed that combustion took place in the upper, more permeable part of the pay, and no permeability damage occurred, although some calcium sulphate was deposited in the heated intervals. The areal sweep was estimated at 50%, but the volumetric sweep was only 14%. Gravity separation of air and water appears to have occurred within the injection well during simultaneous injection. Thus, dry forward combustion appeared to have taken place in the upper, more permeable layers, while waterflooding took place in the lower layers. Based on these observations, the operator suggests that water and air not be injected simultaneously; rather, alternate air-water injection should be considered.

TABLE III
PERTINENT RESERVOIR DATA
SLOSS FIELD COFCAW AREA

Formation	Muddy J-1 Sandstone
Trap	Stratigraphic
Producing Mechanism	
Primary	Solution gas drive
Secondary	Waterflood
Average depth, ft	6200
Average porosity, %	19.3
Average air permeability, md	191
Oil saturation after waterflood, % PV	30 ± 10
Bottom-hole pressure, psig	
1968 survey	2274
1970 survey	3158
Bottom-hole temperature, °F	200
Crude gravity before COFCAW, °API	38.8
Oil viscosity before COFCAW, cp	0.8
Reservoir volume factor, rb/STB	1.05
Full-Scale Project	
Area, acres	960
Average net pay, ft	14.3
Oil in place (oil saturation = 30%),	
STB/acre-ft	427
MMSTB	5.9

(after Buxton and Pollock, 1974)

Miga Field, Venezuela

Depth of this reservoir ranges between 4000 and 4350 ft. The reservoir rock consists of about 25 ft of loosely consolidated sand, dipping to the north at 2°. Porosity and permeability of the sand are 22.6% and 5 darcies, respectively. Connate water saturation is 22%. The reservoir originally contained 23.2 million STB of 14° API gravity oil, of viscosity 280-630 cp at a reservoir temperature of 146°F. The field produced about 772000 Bbls of oil by primary means and was closed in from 1962-1964 when combustion began. Figure 16 shows the distribution of air injection and producing wells in the field (Terwilliger et al, 1975).

Laboratory tests indicated a fuel consumption of 2 lbs/cu ft of rock. However, estimates of fuel consumption based on observed AOR of 11.317 MCF/STB yielded an average consumption of 1.4 lbs/cu ft of rock. This amount of fuel represents 16% of the oil-in-place. Assuming 60% volumetric sweep of the reservoir, 50% of the oil-in-place is expected to be recovered. By 1975 the field had produced 2600000 Bbls of oil, or more than twice the primary production.

IN SITU COMBUSTION   51

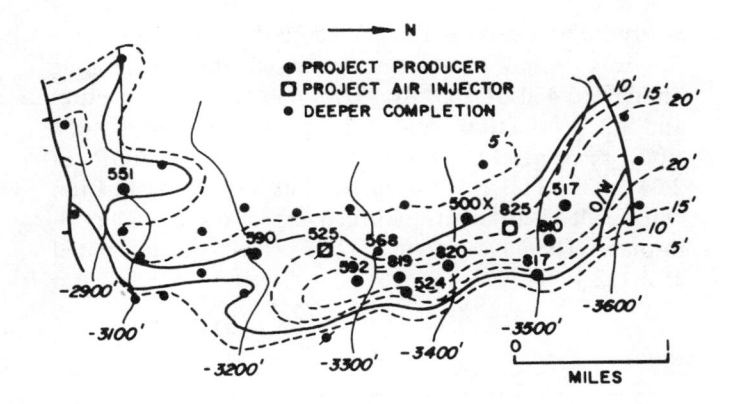

Fig. 16. MG-517 project reservoir, structure-isopach map.
(after Terwilliger et al, 1975)

Air injection is taking place at 2000 psi. (Figures 17 and 18 show the details of an air injection and a production well.) To remove condensed water from the air, a battery of glycol dehydrators were used. However, it was found that glycol tended to form highly corrosive organic acids. Now, the air is being cleaned in gravity operated centrifugal scrubbers equipped with dehumidifiers (Franco, 1974).

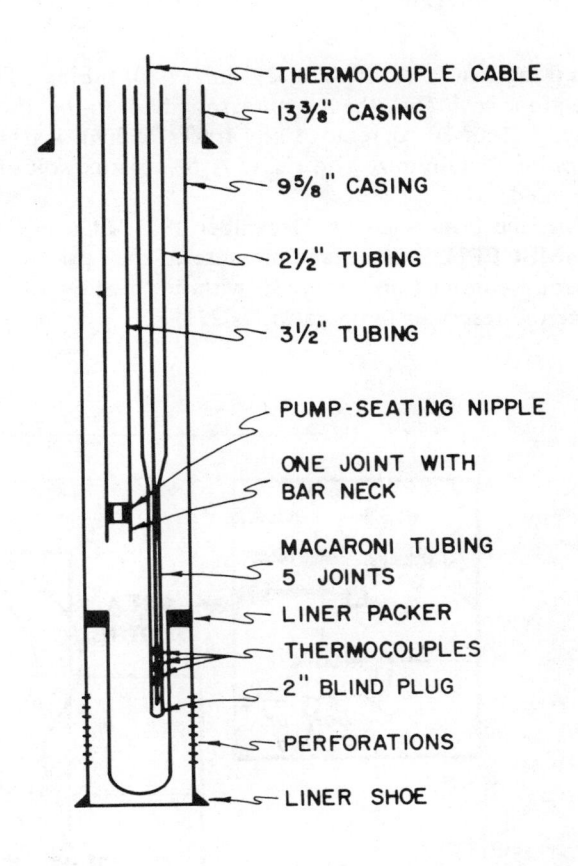

Fig. 18. Combustion producing well.
(after Franco, 1974)

Heidelberg Field, Mississippi (USA)

This project is taking place in the No. 5 Sand of the Cotton Valley Pool of Jurassic age which occurs at an average depth of 11750 ft (Mace, 1975). The No. 5 Sand covers an area of 400 acres on the western flank of the field (Figure 19).

The reservoir consists of consolidated, poorly sorted sand that was deposited in a continental environment. Average reservoir and oil properties are as follows: porosity 16.4%; permeability 39 md; connate water saturation 15%; API gravity 16-25°; average GOR 0.1 MCF/STB; oil viscosity 4 cp at the saturation pressure of 930 psia and 6 cp at the initial pressure of 5100 psia; and reservoir temperature 221°F. The thickness of the No. 5 Sand varies and attains a maximum of 40 ft.

The No. 5 Sand produced 800000 Bbls, which corresponds to a recovery factor of 6.7% of an estimated 12 million STB of oil-in-place. The pressure declined from 5100 psia to 1500 psia. This sharp decline in pressure and absence of water production indicated that the production mechanism involved is expansion of the undersaturated oil, and suggested the presence of an asphaltic mass near the oil-water contact.

The air-injection well is a converted production well A packer was set above the No. 5 Sand, and air injectio

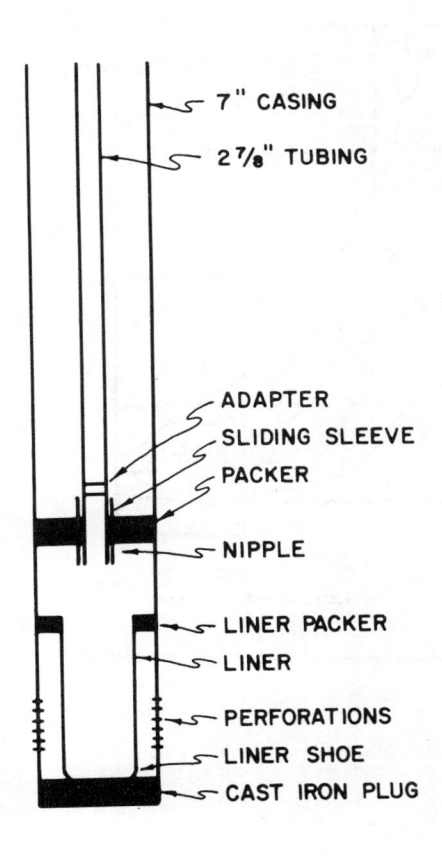

Fig. 17. Air-injection well.
(after Franco, 1974)

is taking place through a 2½-in. N-80 tubing. The injection well and the compressors had to be thoroughly cleansed to remove any traces of hydrocarbon in order to minimize fire risks. A 5% nitrox solution was used.

Air injection began in December 1971, at a rate of 1 MMSCFPD, at a surface pressure of 2500 psi. Spontaneous combustion occurred without heating of air since the reservoir temperature is 221°F.

Production response occurred 70 days after air injection was started. Production increased from a negligible amount to 450 BOPD. By 1975, cumulative air injection and oil production totaled 1007 MMCF and 381000 Bbls, respectively. This corresponds to an AOR of 2.64 MCF/STB It is estimated that in situ combustion will result in an additional recovery of 6.7% of the oil-in-place. The payout time for this project was estimated at 2 1/2 years.

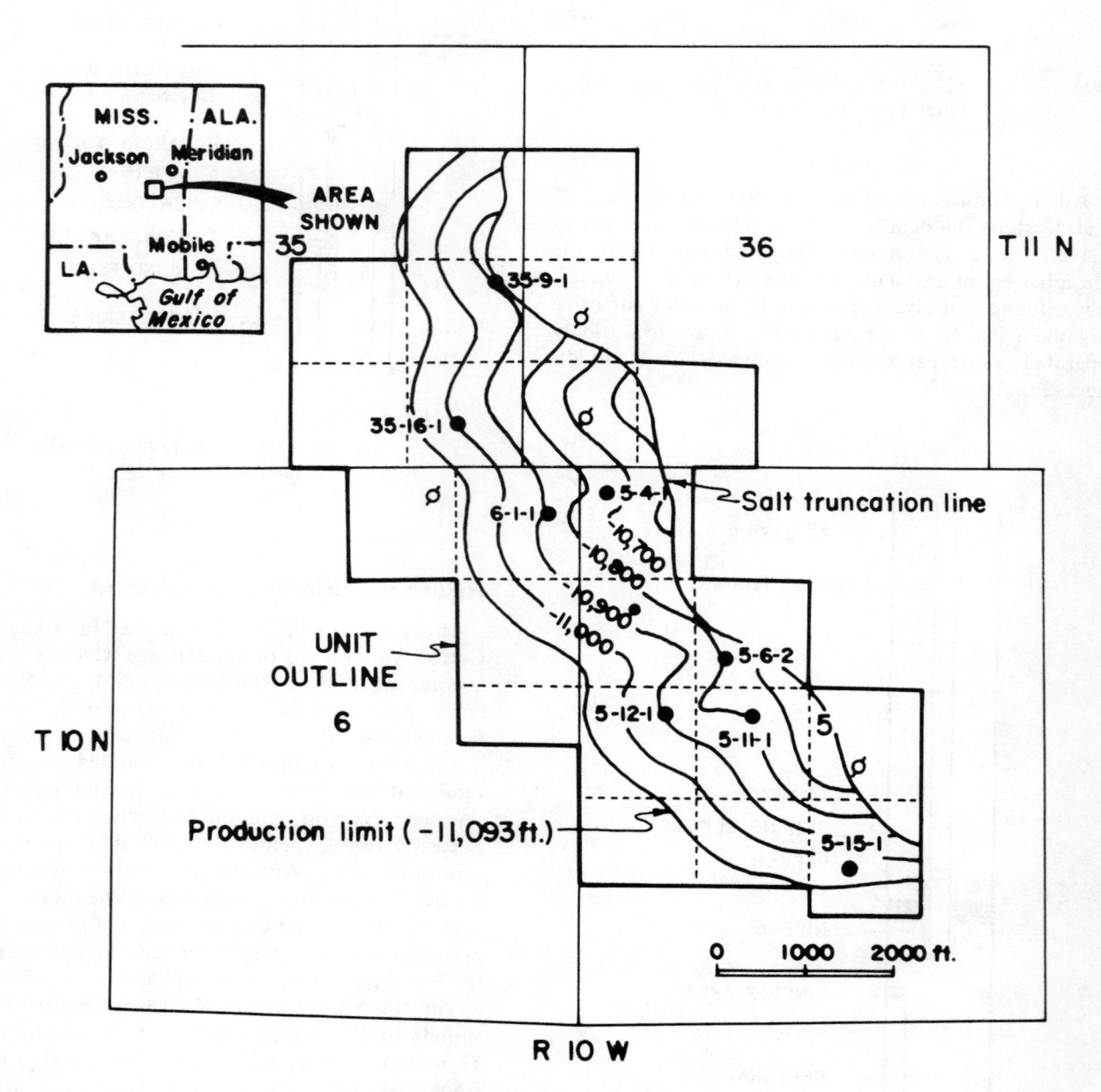

Fig. 19. Structure map of No. 5 Sand (after Mace, 1975)

BIBLIOGRAPHY

Alexander, J. D., Martin, W. L. and Dew, J. N.: "Factors Affecting Fuel Availability and Composition During In-Situ Combustion," *Jour. Pet. Tech.* (Oct., 1962).

Bae, J. H.: "Characterization of Crude Oil for Fireflooding Using Thermal Analysis Methods," *Soc. Pet. Eng. Jour.* (June, 1977).

Barnes, A. L.: "Results From a Thermal Recovery Test in a Watered-Out Reservoir," *Jour. Pet. Tech.* (Nov., 1965).

Berry, V. J. and Parrish, D. R.: "A Theoretical Analysis of Heat Flow in Reverse Combustion," *Jour. Pet. Tech.* (May, 1960).

Bousaid, I. S. and Ramey, H. J., Jr.: "Oxidation of Crude Oil in Porous Media," *Soc. Pet. Eng. Jour.* (June, 1968). *Trans.*, AIME (1968) 243.

Burger, J. and Sahuquet, B.: "Combustion 'In-Situ' Principes Et Études De Laboratoire," *Les Methodes Thermiques De Production Des Hydrocarbures.* Revue De L'Institut Francais De Pétrole (Mars-Avr., 1977).

Buxton, T. S. and Pollock, C. B.: "The Sloss COFCAW Project — Further Evaluation of Performance During and After Air Injection," *Jour. Pet. Tech.* (Dec., 1974). *Trans.*, AIME (1974) 257.

Campbell, G. G., Burwell, E. L., Sterner, T. E. and Core, L. L.: "Why a Fire Flood Project Failed," *World Oil.* (Feb. 1, 1966).

Casey, T. J., "A Field Test of the In-Situ Combustion Process in a Near-Depleted Water Drive Reservoir," *Jour. Pet. Tech.* (Feb., 1971).

Cato, B. W. and Frnka, W. A.: "Results of an In-Situ Combustion Pilot Project," *Producers Monthly.* (June, 1968).

Chu, C.: "Two-Dimensional Analysis of a Radial Heat Wave," *Jour. Pet. Tech.* (Oct., 1963).

Counihan, T. M.: "A Successful In-Situ Combustion Pilot in the Midway-Sunset Field, California," SPE 6525 prepared for the SPE-AIME 47th Annual California Regional Meeting, Bakersfield, CA. (April, 1977).

Crawford, P. B.: "In-Situ Combustion," *Secondary and Tertiary Oil Recovery Processes,* Interstate Oil Compact Commission, Oklahoma City, OK. (Sept., 1974).

Crookston, R. B., Culham, W. E. and Chen, W. H.: "Numerical Simulation Model for Thermal Recovery Processes," SPE 6724 prepared for the SPE-AIME 52nd Annual Fall Meeting, Denver, CO. (Oct., 1977).

Dietz, D. N.: "Wet Underground Combustion, State of the Art," *Jour. Pet. Tech.* (May, 1970). *Trans.*, AIME (1970) 249.

Dietz, D. N. and Weijdema, J.: "Reverse Combustion Seldom Feasible," *Producers Monthly.* (May, 1968).

Earlougher, R. C., Jr., Galloway, J. R. and Parsons, R. W.: "Performance of the Fry In-Situ Combustion Project," *Jour. Pet. Tech.* (May, 1970).

Farouq Ali, S. M.: "A Current Appraisal of In-Situ Combustion Field Tests," *Jour. Pet. Tech.* (April, 1972).

Farouq Ali, S. M.: "Forward Combustion — State of the Art," *Producers Monthly.* (Nov., 1967).

Farouq Ali, S. M.: "Reverse Combustion," *Producers Monthly.* (Dec., 1967).

Franco, A.: "How MGO Handles Heavy Crude," *O. & G. Jour.* (June 24, 1974).

Garbus, R.: "New Idea in Thermal Oil Recovery," *Pet. Eng.* (April, 1956).

Gates, C. F., Jung, K. D. and Surface, R. A.: "In-Situ Combustion in the Tulare Formation, South Belridge Field, Kern County, California," SPE 6654 prepared for the SPE-AIME 47th Annual California Regional Meeting, Bakersfield, CA. (April, 1977).

Gottfried, B. S.: "A Mathematical Model of Thermal Oil Recovery in Linear Systems," *Soc. Pet. Eng. Jour.* (Sept., 1965).

Hardy, W. C.: "Deep-Reservoir Fireflooding Economics for Independents," *O. & G. Jour.* (Jan. 18, 1971).

Hardy, W. C. et al: "In-Situ Combustion in a Thin Reservoir Containing High—Gravity Oil," *Jour. Pet. Tech.* (Feb., 1972).

Holst, P. H. and Karra, P. S.: "The Size of the Steam Zone in Wet Combustion," *Soc. Pet. Eng. Jour.* (Feb., 1975).

Ireton, E. T.: "Little Tom Thermal Recovery Demonstration Project Zavala County, Texas," *ERDA Symposium on Enhanced Oil & Gas Recovery,* Vol. 1 - Oil, Tulsa, OK. (Sept., 1976).

Joseph C.: "Bodcau In-Situ Combustion Project Bossier Parish, Louisiana," *ERDA Enhanced Oil, Gas Recovery and Improved Drilling Methods,* Vol. 1 - Oil, Tulsa, OK. (Sept., 1977).

Kuhn, C. S. and Koch, R. L.: "In-Situ Combustion ... Newest Method of Increasing Oil Recovery." *O. & G. Jour.* (Aug. 10, 1953).

Kuo, C. H.: "A Convective-Heat Transfer Model for Underground Combustion," *Soc. Pet. Eng. Jour.* (Dec., 1968).

Little, T. P.: "Successful Fireflooding of the Bellevue Field," *Pet. Eng.* (Nov., 1975).

Mace, C.: "Deepest Combustion Project Proceeding Successfully," *O. & G. Jour.* (Nov. 17, 1975).

Martin, W. L., Alexander, J. D., Dew, J. N. and Tynan, J. W.: "Thermal Recovery at North Tisdale Field, Wyoming," *Jour. Pet. Tech.* (May, 1972).

McIntyre, H.: "Oil Companies Push In-Situ Recovery," *Cdn. Chem. Proc.* (July, 1977).

Meldau, R. F. and Lumpkin, W. B.: "Phillips Tests Methods to Improve Drawdown and Producing Rates in Venezuela Fire Flood," *O. & G. Jour.* (Aug. 12, 1974).

Nelson, T. W. and McNeil, J. S.: "How to Engineer an In-Situ Combustion Project," *O. & G. Jour.* (June 5, 1961).

Parrish, D. R. and Craig, F. F., Jr.: "Laboratory Study of a Combination of Forward Combustion and Waterflooding — The COFCAW Process," *Jour. Pet. Tech.* (June, 1969). *Trans.,* AIME (1969) 246.

Patek, J. W. and Meldau, R. F.: "Wet Combustion Pilot, Paris Valley Field, California," *ERDA Enhanced Oil, Gas Recovery and Improved Drilling Methods,* Vol. 1 - Oil, Tulsa, OK. (Sept., 1977).

Showalter, W. E.: "Combustion-Drive Tests," *Soc. Pet. Eng. Jour.* (March, 1963).

Smith, C. R. and Rinehart, R. D., eds., *Thermal Recovery Techniques Short Course,* Petroleum Engineering Associates, Laramie, Wyo. (1966).

Smith, J. T. and Farouq Ali, S. M.: "Simulation of In-Situ Combustion in a Two-Dimensional System," SPE 3594 prepared for the SPE-AIME 46th Annual Fall Meeting, New Orleans, LA. (Oct., 1971).

Stoseu, J. J.: "In Situ Combustion Method for Oil Recovery State of the Art and Potential," *The Future Supply of Nature-Made Petroleum and Gas,* Pergamon Press, New York (1976).

Strange, L. K.: "Ignition: Key Phase In Combustion Recovery," *Pet. Eng.* (Nov., 1964).

Szasz, S. E. and Berry, V. J., Jr.: "Oil Recovery By Thermal Methods," #29 presented at 6th WPC in Francfort/Main. (June, 1963).

Tadema, H. J. and Weijdema, J.: "Spontaneous Ignition of Oil Sands," *O. & G. Jour.* (Dec. 14, 1970).

Terwilliger, P. L., Clay, R. R., Wilson, L. A., Jr. and Gonzalez-Gerth, E.: "Fireflood of the P_{2-3} Sand Reservoir in The Miga Field of Eastern Venezuela," *Jour. Pet. Tech.* (Jan., 1975).

Thomas, G. W.: "A Study of Forward Combustion in a Radial System Bounded by Permeable Media," *Jour. Pet. Tech.* (Oct., 1963).

van Poollen, H. K.: "Transient Tests Find Fire Front In an In Situ Combustion Project," *O. & G. Jour.* (Feb. 1, 1965).

"Wet In Situ Combustion Aids Recovery of Oil, AIChE Told," *O. & G. Jour.* (April 3, 1972).

Wilson, L. A. and Root, P. J.: "Cost Comparison of Reservoir Heating Using Steam or Air," *Jour. Pet. Tech.* (Feb., 1966).

Part II
CHEMICAL PROCESSES

Chapter 4
SURFACTANT-POLYMER INJECTION

INTRODUCTION

In 1927, L. C. Uren and E. H. Fahmy concluded: "A definite relationship exists between interfacial tension of the oil against the flooding medium and the percentage recovery obtained by flooding. The efficiency of flooding increases as the interfacial tension decreases." In that same year, a patent was issued to H. Atkinson which envisioned the use of aqueous soap solution or other aqueous solutions to decrease the "surface tension" between oil and the flooding medium and thereby increase the recovery of oil. Prior to these publications it had been reported that alkali solution flooding gave more efficient oil recovery than did ordinary waterflooding and Nutting had postulated that reactions at the solid-liquid interface led to the observed increase in oil production. In 1932, soap solutions were injected into Bradford and Venango Sand cores and it was reported that the results were inconclusive, erratic and that "further investigation is needed to determine exactly the function of the solution and to obtain a clearer insight into the phenomena involved."

During the next 25 years a major segment of the reported research on the use of surfactants to recover oil was carried out at Pennsylvania State University. The work of this group was reviewed by Calhoun et al (1951) who noted that the group recognized that both interfacial tension and wetting conditions (contact angle) were important parameters. They also recognized that surfactant adsorption was a major deterrent to economic recovery of oil by aqueous surfactant flooding. Despite these discouraging results, unanimity of opinion between the Penn State group and other research workers in the oil recovery field did not exist.

In the 1952 continuation of the Penn State work, Preston and Calhoun discussed chromatographic transport of surfactant. Ojeda et al (1954) were able to correlate residual oil saturation in cores after an aqueous surfactant flood with $\sigma/\Delta P$ and with a pore geometry parameter (k/ϕ). The data of Ojeda et al indicated that residual oil saturation could go to zero at values of $\sigma/\Delta P$ approaching zero. Identified as important parameters in determining surfactant flooding performance were: 1) pore geometry; 2) interfacial tension; 3) wettability or contact angle; 4) ΔP or $\Delta P/L$; and 5) the chromatographic transport characteristic of the surfactant in the particular system.

During the next several years, published research results (excluding the patent literature) included a number of papers concerned with the screening of surfactants for oil recovery "efficiency". Some researchers reported on wettability changes to improve oil recovery. Adsorption and chromatographic transport of surfactants were discussed in several papers and still other papers discussed continuing efforts to delineate the role of the major parameters known to affect oil recovery by aqueous surfactant flooding.

Specific contributions of note include that of Reisberg and Doscher (1956) who reported measuring interfacial tensions less than 0.01 dynes/cm in a specific crude oil-aqueous alkali system. These authors also reported achieving 100% recovery of oil from sandpacks and more than 90% recovery from Berea cores using a combination aqueous alkali-surfactant flood at flooding rates achievable in field application.

In 1959 Holm and Bernard filed for a patent in which they proposed injecting 0.1 - 3% surfactant dissolved in low viscosity hydrocarbon solvent. This procedure reduced surfactant adsorption in water-wet formations. In 1961, Csazar filed for a patent specifying the use of a mixture of an anhydrous soluble oil and a nonaqueous solvent containing up to 12% surfactant. These patents have given rise to the soluble oil flooding process.

In 1962 Gogarty and Olson filed for a patent describing the use of microemulsions in a new miscible-type recovery process known as Maraflood™. The microemulsions contain surfactant, hydrocarbon and water. Cosurfactants or an electrolyte may be added. The Gogarty and Olson patent suggests the injection of a small fraction of the pore volume of micellar solution containing a surfactant concentration greater than about 5%. Patents have been issued to Jones (1967) claiming the use of high-water-content, oil-external microemulsions and water-external micellar dispersions in oil recovery. Jones' patents, along with others, describe the use of a relatively high surfactant concentration in various aqueous systems and micellar dispersions for the recovery of oil.

The work of Wagner and Leach (1966) with model fluids in Torpedo Sandstone showed that oil recovery increased when interfacial tension was reduced to about 0.07 dynes/cm and that further small decreases in interfacial tension resulted in large increases in oil recovery. In 1968, Taber presented theoretical and experimental results which further clarified the relation between residual oil saturation and $\Delta P/L\sigma$. Taber recognized that this correlation group should include the contact angle, but in view of the well known difficulties in systematically varying or indeed in even measuring the contact angle inside rocks, he examined the effects of the other parameters only. For Berea cores, Taber found that a significant quantity of discontinuous oil (residual) was displaced when the ratio of $\Delta P/L\sigma$ reached a value of about 5 (psi/ft)/(dynes/cm). He designated this as the critical value of the ratio and noted that further increases in the value of this ratio invariably produced more residual oil. He concluded that nearly 100% of the residual oil could be displaced if very high values of $\Delta P/L\sigma$ could be obtained. He expected good recovery of the residual oil if the $\Delta P/L\sigma$ exceeded the critical value by 5 - 10 times. Using this "guideline" Taber noted that for a realistic $\Delta P/L$ value of 0.5 psi/ft, it would be necessary to have a surfactant system capable of establishing an interfacial tension of 0.01 dynes/cm before good oil recovery could be obtained from a rock having capillary characteristics like Berea. He indicated that such interfacial tensions would be difficult to achieve and maintain in a reservoir and that accordingly, attention should be given to means of achieving higher pressure gradients in field floods.

Since interfacial tensions below 0.01 dynes/cm had been measured for an aqueous alkali - crude oil system, it was believed that σ values of 0.001 or lower could be achieved with modern aqueous surfactant systems. Such a value of σ combined with a $\Delta P/L$ of 0.5 psi/ft would yield a value of 500 for the ratio of $\Delta P/L\sigma$ or about 100 times the critical value found by Taber for Berea rock.

Essentially two different concepts have developed for using surfactants to enhance oil recovery. In the first concept, a solution containing a low concentration of surfactant is injected. The surfactant is dissolved in either water or oil and is in equilibrium with aggregates of the surfactant known as micelles. Large pore volumes (about 15 - 60% or more) of the solution are injected into the reservoir to reduce interfacial tension between oil and water and, thereby, increase oil recovery. Oil may be banked with the surfactant solution process, but residual oil saturation at a given position in the reservoir will only approach zero after passage of large volumes of surfactant solution. In the second process, a relatively small pore volume (about 3 - 20%) of a higher concentration surfactant solution is injected into the reservoir. With the higher surfactant concentration, the micelles become a surfactant-stabilized dispersion of either water in hydrocarbon or hydrocarbon in water. The high surfactant concentration allows the amount of dispersed phase in the microemulsion to be high as compared with the low value in the dispersed phase of the micelles in the low concentration surfactant solutions. The injected slug is formulated with three or more components. The basic components (hydrocarbon, surfactant and water) are sufficient to form the micellar solutions. A cosurfactant fourth component (usually alcohol) can be added. Electrolytes, normally inorganic salts, form a fifth component that may be used in preparing the micellar solutions or microemulsions. The high concentration surfactant solutions displace both oil and water and rapidly displace all the oil contacted in the reservoir. As the high concentration slug moves through the reservoir, it is diluted by formation fluids and the process reverts to a low-concentration flood.

Work is under way in the laboratory and the field to select the optimum method of injecting surfactant to improve oil recovery.

Oil reservoir rocks are composed of a wide variation of minerals with corresponding variations in porosity and permeability. Cementing materials, in-place-oil, water composition, temperatures, etc., present an infinite array of variables that must be considered to design the most cost-effective surfactant system. The best process for a specific reservoir is the one which has the potential to yield the greatest profit regardless of the concentration level of the surfactant. The chemical system needs to be tailored to the reservoir.

TECHNICAL DISCUSSION
Surfactant Slug

The surfactant slug has also been referred to in the literature as micellar solutions, microemulsions, soluble oils, swollen micelles, etc. The primary purpose of these surfactant slugs is to lower interfacial tension and displace oil that cannot be displaced by water alone. The technology and mechanism by which this displacement takes place is extremely complicated and not completely understood.

The technical and patent literature contains a number of surfactant slug designs formulated to mobilize residual oil. Most of these patents describe variations in surfactant, cosurfactant, hydrocarbon-water and/or salt concentrations which yield unique slug characteristics.

Companies such as Union Oil Company and Marathon Oil Company have trademarked their processes as Uniflood and Maraflood. Several firms such as Shell,

Mobil, Texaco, Exxon, Amoco, Arco, Phillips, Conoco and others have patents in the surfactant-polymer area. Evaluations to determine which system offers the best economics for a specific field application are often complex.

External Phase

Much has been written in the literature on the oil or water-external phase of the surfactant slugs. Methods mentioned for determining which phase is continuous and which phase is disperse are the fluorescence, conductivity, phase dilution and dye solubility methods.

In determining the external phase, Dreher and Syndansk (1971) suggested that a single test may give ambiguous results. They concluded that miscibility data in combination with conductivity data give a clear indication as to whether the external phase is oil or water, even at extreme concentrations of the internal phase. It was also shown that as much as 70 weight % of the total system can exist as the internal or disperse phase in an optically clear system.

Although little data is available, the literature references lead one to believe that oil-external slugs are preferred. However, some data are available that suggest the external phase is not an important variable in surfactant-polymer flooding.

Healy and Reed (1974) conducted a series of constant-rate (1 ft/day) core floods to determine if the nature of the external phase affects displacement efficiency. In each flood a 5% PV microemulsion slug was injected into a 4-ft x 1-in x 1-in Berea core initially containing discontinuous residual oil and 1% NaCl. The slug was driven by continuous injection of 1000 ppm XC biopolymer in 1% NaCl. All slug compositions used 15% surfactant with varying amounts of oil, and adequate mobility control existed in all core flood runs. Table I illustrates the final oil saturation obtained in the floods using oil and water-external slugs.

Although the final oil saturation varied slightly, there appears to be no obvious advantage attributable to either oil-external or water-external surfactant slugs.

TABLE I
EFFECT OF EXTERNAL PHASE
ON FINAL OIL SATURATION

External Phase of Injected Slug	S_{o_f} (% PV)
Water	7.4
Water	8.0
Water	11.1
Oil	6.4
Oil	10.6

(after Healy and Reed, 1974)

Mobility Control

The concept of surfactant-polymer flooding can be described very simply. A surfactant slug, when properly designed for a specific reservoir, will essentially displace all of the oil contacted ahead of it. A stabilized oil bank will build up ahead of the slug. The saturation condition that exists in the oil bank will conform to the relative permeability characteristics of the reservoir. This oil bank will flow with a certain mobility. The surfactant slug's mobility should be less than the oil bank mobility. Likewise, for efficient displacement the mobility control polymer should be less mobile than slug and oil banks. Loss of mobility control in the fluid sequence will cause fingering and reduced displacement efficiency.

Gogarty (1967) stated that mobility control is one of the most important considerations in designing a miscible-waterflood. Two techniques were described to determine the design mobility of a miscible-waterflood.

One approach is the use of relative permeability data available for the reservoir. Miscible-waterflooding, initiated under either secondary or tertiary conditions, is characterized by a stable water-oil bank flowing ahead of the miscible zone. Flow within the stable bank conforms with the accepted theories of fractional flow for immiscible systems. The stabilized water-oil saturation conditions are dependent upon the rate of accumulation of each fluid and the water-oil relative permeability. To effect a favorable displacement, the mobility of the miscible zone must be equal to or less than the mobility of the stable bank. In the absence of detailed data, the actual mobility of the stabilized bank is difficult to predict. Imbibition water-oil relative permeability curves and field viscosity data for water and oil provide a means of estimating the minimum total relative mobility of the stable bank. Imbibition data define water-oil relationships within a stable bank.

Total flow within the stable water-oil bank, neglecting capillary pressure, is given by:

$$q_{tb} = kA \left[\frac{\Delta p}{\Delta L}\right]_b \left[\frac{k_{rw}}{\mu_w} + \frac{k_{ro}}{\mu_o}\right]_b \quad (1)$$

In terms of relative mobility, Equation 1 becomes

$$q_{tb} = kA \left[\frac{\Delta p}{\Delta L}\right]_b \left[\lambda_{rw} + \lambda_{ro}\right]_b$$

The expression, $[\lambda_{rw} + \lambda_{ro}]_b$, is the total relative mobility within the stable bank.

q_{tb} = total flow rate in stable bank

k = absolute permeability

A = area

$\left[\frac{\Delta p}{\Delta L}\right]_b$ = pressure gradient across stable bank

k_{rw} = relative permeability to water

k_{ro} = relative permeability to oil
μ_w = water viscosity
μ_o = oil viscosity
λ_{rw} = relative mobility to water
λ_{ro} = relative mobility to oil

A properly designed miscible-waterflood must have a miscible slug with relative mobility equal to or less than the total relative mobility of the stabilized bank. If the saturation conditions within the stable bank are known, the actual value for relative mobility could be calculated from field viscosities and representative imbibition water-oil relative permeability data. However, the actual value of water saturation within the stable bank is usually unknown.

Hypothetical water-oil relative permeability curves are shown in Figure 1.

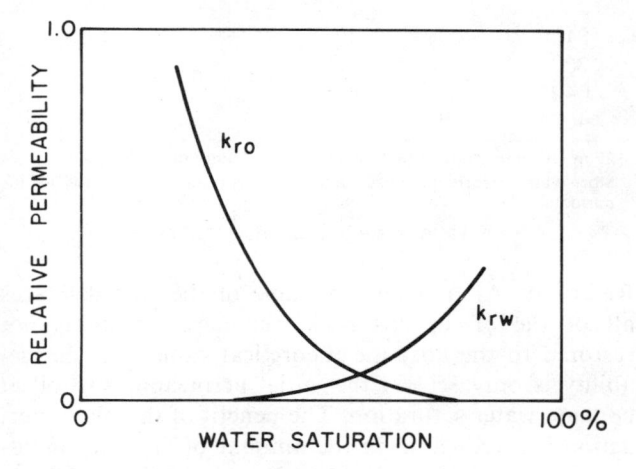

Fig. 1. Typical water-oil relative permeability curves.

Total relative mobilities from the curves in Figure 1 are shown in Figure 2.

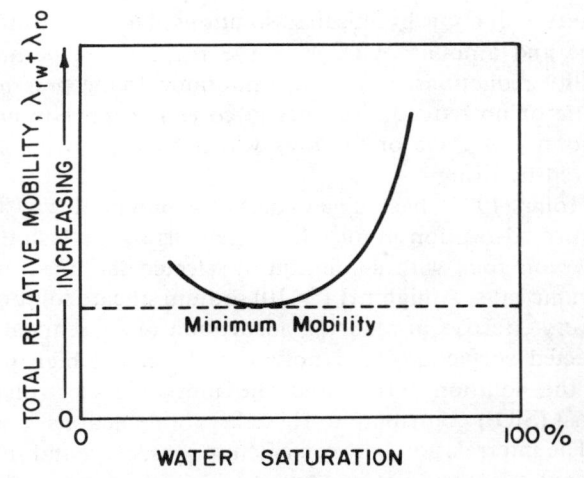

Fig. 2. Total relative mobility vs. water saturation. (after Gogarty, 1967)

In making the calculation, constant values of water and oil viscosities were assumed. Note the minimum total mobility as represented by a horizontal tangent to the curve. The actual total relative mobility within the stable bank may be equal to or greater than this minimum. The minimum total relative mobility is designated as the design mobility for the entire fluid system upstream from the stabilized water-oil bank.

Another method for determining the design mobility is based on laboratory flooding results using a miscible displacement fluid. For these floods, reservoir cores are saturated with actual reservoir fluids. Cores must be long enough to allow development of the stabilized oil-water bank ahead of the miscible fluid. Before miscible flooding, a core is: 1) saturated with water; 2) flooded with oil; and 3) waterflooded to residual oil. From these preparation floods, relative permeability values at residual saturation conditions can be obtained. These values are compared to other available relative permeability data to indicate that a representative reservoir core is being used for the miscible flood.

The differential pressure is measured across a portion of the core near the downstream end. This pressure drop is associated with the oil and water flowing in the stabilized bank ahead of the miscible fluid. From the pressure measurements, rock parameters and known flow rate, the total relative mobility of the oil-water bank can be determined by:

$$(\lambda_{rw} + \lambda_{ro})_b = \frac{q_{tb}}{kA\left[\frac{\Delta p}{\Delta L}\right]_b} \quad (2)$$

Once the mobility of the oil-water bank is determined, a selection of the slug and mobility buffer is made based on core flow experiments. Figure 3 shows the flow of two surfactant slugs at different flow rates through reservoir rock.

In order to maintain mobility control, slug B would be less mobile than the design mobility based on the flowing oil-water bank. Similar flow data would be obtained for various polymers at different concentrations. These data would be used to determine the most economical polymer that effectively displaces the slug.

A schematic diagram of the surfactant-polymer process is shown in Figure 4.

Surfactant Adsorption

Adsorption of surfactants on minerals is an important consideration in surfactant flooding. It is a major cause of surfactant retention and slug breakdown.

It has been reported that high equivalent weight sulfonates are adsorbed preferentially while the lower equivalent weight sulfonates show very little adsorption. Since the higher equivalent weight sulfonates are

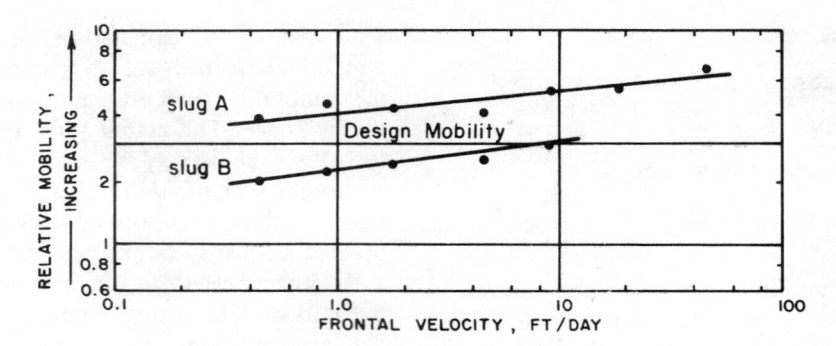

Fig. 3. Effect of slugs on mobility control design.
(after Gogarty, 1967)

Fig. 4. Schematic diagram of the surfactant-polymer process.

TABLE II
ADSORPTION DEPENDENCE UPON CLAY TYPE

Equivalent Weight	Ca Montmorillonite (mg/gm)	Kaolinite (mg/gm)
233	−2.0*	3.7
310	−1.0	4.6
333	1.7	6.0
342	8.2	9.2
400	13.3	10.8

25 ml of 0.5 percent surfactant solution in 2 percent Na_2SO_4; 7 gm Clay.
*More water adsorbed than surfactant. Equilibrium solution enriched in surfactant.

(after Gale and Sandvik, 1973)

responsible for the lowering of interfacial tension, their loss rapidly decreases the slugs ability to displace residual oil. Gale and Sandvik (1973) have shown in Figure 5 the relationship between equivalent weight and surfactant adsorbed on calcium montmorillonite.

Gale and Sandvik (1973) have noted that not all clay minerals have the same relative adsorption equivalent weight relationships. Kaolinite adsorbs more low equivalent weight surfactant than does calcium montmorillonite as indicated in Table II.

Trushenski et al (1974) discovered a surprising and beneficial effect from the interaction of the micellar slug with clay in the Second Wall Creek Sandstone Reservoir. After a large volume of the slug displaces all of the oil in this rock, the permeability is not restored to the absolute theoretical value. The permeability is only 35% of the initial permeability to oil at connate water saturation. The benefit of this clay interaction is a reduction in the amount of viscosity building material required to achieve mobility control. Nearly a three fold reduction in the micellar fluid and polymer viscosity is possible by incorporating the permeability reduction in the mobility control design. The result is a savings in polymer costs.

They further reported that the magnitude of permeability reduction by micellar solutions is related to the type and amount of clays in the rock. Large permeability reductions occur with montmorillonite present. Little or no reduction occurs in cores that have minor amounts of clays or in cores which have been desensitized by firing.

Holm (1972) has suggested that a potential way to reduce adsorption in micellar systems is to preflush the reservoir rock with a solution of selected sacrificial inorganic salts. A high pH ($>$ 10) sodium silicate solution is very effective in reducing adsorption of subsequently injected surfactants and polymers. Both the high pH of the solution (OH^-) and the inorganic electrolyte (Na^+) ($SiO_2^=$) contribute to these favorable actions.

The interrelation between calcium ions, clays and surfactant adsorption is important in selecting and applying a preflush chemical. For example, a large portion

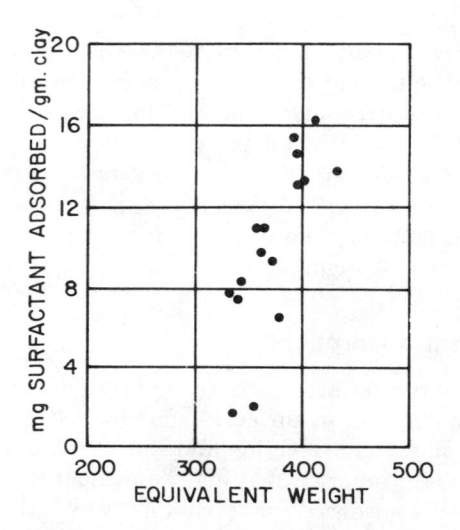

Fig. 5. Surfactant adsorption on Calcium montmorillonite.
(after Gale and Sandvik, 1973)

(> 30%) of the injected sulfonate was produced during each of the soluble oil floods shown in Table III for the Berea and Dundee sandstone cores.

A similar soluble oil flood in a Berea core which was equilibrated with a high calcium (200 ppm) water but not preflushed, recovered only 50% of the oil-in-place and produced much less sulfonate during the flood. The calcium content in the produced fluids was similar in all the above floods. These results indicated that the silicate preflush minimized sulfonate adsorption and calcium interference. With the Bell Creek cores, the amounts of sulfonate produced during each of the soluble oil floods on the silicate preflushed cores were about the same. However, large amounts of calcium (and iron) were produced during the soluble oil floods in the lower permeability (k_A < 200 md) Bell Creek core plugs. Although sulfonate adsorption was low, the excessive amounts of multivalent cations in these cores reduced oil recovery efficiency. Increased amounts of silicate preflush apparently are needed for the lower permeability rock.

Other examples of the effects of clay, calcium and preflush solutions on the oil recovery efficiency of micellar flooding are presented in Table IV.

Soluble oil-polymer floods were conducted in sand packs containing various clays with and without high pH silicate preflush. The addition of both sodium and calcium montmorillonite clays to the sandpacks caused a substantial decrease in the oil recovered by subsequent soluble oil floods, even though a polymer solution preflush was used to displace the resident brine. The high pH silicate preflush was effective on both clays and permitted the oil recovery efficiency of subsequent soluble oil floods to be at least equal to that achieved in sand without added clay.

Bae et al of Gulf (1976) have noted that sulfonate adsorption was dependent on flow rate through Berea rock. Realistic flow rates were recommended. They also noted that a sacrificial chemical such as sodium carbonate used as a preflush was effective in reducing sulfonate adsorption.

Petroleum Sulfonates

The chemistry of petroleum sulfonates is extremely complex. Since a petroleum feedstock contains a vast number of chemical structures, none of which characterize that feedstock, the sulfonation of this feedstock yields numerous sulfonates. It has been observed that laboratory oil recovery values correlate well with the equivalent weight and equivalent weight distribution of the sulfonate. The equivalent weight of a sulfonate is the molecular weight divided by the number of sulfonate groups present in the molecule. An average equivalent weight can be determined for a mixture of substances. Figure 6 shows two sodium sulfonates having identical equivalent weights but different structures.

Hill et al of Shell (1973), in developing surfactant systems, selected petroleum sulfonates because of the wide range of properties available within a common surfactant type, their relatively low cost and potential availability in large supply. Commercially available petroleum sulfonates are generally characterized as oil soluble "mahogany" sulfonates or as water soluble

TABLE III
SUMMARY OF RESULTS OF
SOLUBLE OIL-POLYMER TERTIARY FLOODS IN SANDSTONE CORES
(Starting S_{or} = 20-35% PV)

Sandstone	Brine(b) Salinity wt%	Clay Content wt%	$K_{w_{ro}}$ md.	K_{air} md.	Oil Recovered %OIP (a) (c)	Pounds Active Sulfonate & Solvent Injected Per Barrel Oil Recovered (c)
Berea	0.5 (Low Calcium)	6	50	450	87	3
	0.5 (High Calcium)	6	46	500	85	3
	10 (High Calcium)	6	50	450	80	4
Dundee	0.5 (Low Calcium)	3-4	140	2300	88	3.5
Admire (El Dorado Ka.)	10 (High Calcium)	15-20	50	500	61	6
Bell Creek (Montana)	0.5 (Low Calcium)	3	240	3000	85	3.5
	0.5 (High Calcium)	7	25	190	60	5
	0.5 (High Calcium)	5	90	800	77	4

a Micellar-Polymer Floods using 10-15% PV high pH silicate preflush in 1 ft. long linear systems (Continuous or butted core plugs). Reservoir core plugs used as received (not extracted).
b Cores equilibrated with brine prior to flood.
c Results normalized to a residual oil saturation prior to micellar floods of S_{or} = 30% PV.

(after Holm, 1972)

TABLE IV
EFFECT OF ALKALINE SILICATE PREFLUSH ON SOLUBLE OIL FLOODING OF SANDPACKS CONTAINING CLAYS
(6 ft long sandpacks containing crude oil and 9.4% brine (c))

Sandstone Type	K_w md.	Residual Oil-In-Place, %PV	Preslug Fluid	Conc. in Water-%	Slug Size %PV	Oil Recovered By Soluble Oil Polymer Flood(a) %OIP
#16 American Graded Sand Co.	3215	28.8	Pusher 700 solution	0.05	3	84.9
#16 Sand + 1% Calcium Montmorillonite clay	2577	26.3	Pusher 700 solution	0.05	3	61.4
#16 Sand + 1% Calcium Montmorillonite clay	2463	27.3	Na_2SiO_3-(b) NaOH	0.8	10	91.3
#16 Sand + 1% Sodium Montmorillonite clay	3099	25.5	Pusher 700 solution	0.05	3	70.4
#16 Sand + 1% Sodium Montmorillonite clay	2516	27.6	Na_2SiO_3-(b) NaOH	0.8	10	84.4

(a) Soluble Oil 3%PV: 46.8% crude oil, 11.6% Witco TRS-12B, 1.6% Butyl cellosolve, 40.0% water containing 6000 ppm NaCl
(b) Ratio of Na_2O to SiO_2 = 2.1
(c) Brine Composition 7.4 wt% NaCl, 1.5 wt% $CaCl_2$, .5 wt% $MgCl_2$

(after Holm, 1972)

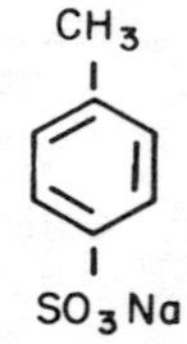

Fig. 6. Sodium sulfonates with identical equivalent weights but different structures. (Hill et al, 1973)

a. EQUIVALENT WEIGHT = 194/1 = 194

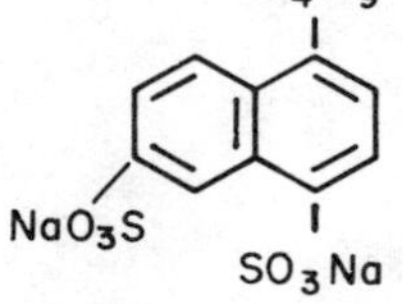

b. EQUIVALENT WEIGHT = 388/2 = 194

"sludge" or "green acid" sulfonates. The latter group generally have an average molecular weight in the range of 350 and may contain polysulfonates. The "mahogany" sulfonates are generally monosulfonates and are available as products of commerce having average molecular weights varying from approximately 400 to about 550. Frequently, intermediate average molecular weight sulfonates are prepared by blending two process products from opposite ends of this range. Petroleum sulfonates currently are used as lube oil additives, soluble oils, cutting oils and in various other applications requiring oil soluble surfactants. They possess structural and molecular weight distributions corresponding to those of the aromatic fractions of the sulfonation feedstocks and consequently can have a wide carbon number spread. It is possible to obtain a wide range of surfactant-amphiphile systems from a single basic source because the water soluble "green acid" sulfonates and the lower molecular weight fractions of the "mahogany" sulfonates solubilize high molecular weight (less water soluble) materials.

Gale and Sandvik (1973) have listed four major criteria for selecting a surfactant for an enhanced recovery process:

1) low oil-water interfacial tension
2) low adsorption
3) compatibility with reservoir fluids
4) low cost.

Low interfacial tensions reduce capillary forces that trap residual oil in porous media. Reduction of these forces frees the oil and allows it to be recovered. Attraction of surfactant to oil-water interfaces permits reduction of interfacial tensions; however, attraction to rock-water interfaces can result in loss of surfactant to rock surfaces by adsorption. Surfactant losses can also arise from precipitation due to incompatibility with reservoir fluids. Low adsorption and low cost are primarily economic considerations; whereas low interfacial tension and compatibility are necessary for workability of the process itself.

Petroleum sulfonates useful in surfactant flooding have been disclosed in several patents; however, virtually no detailed information is available in the nonpatent technical literature. One advantage of petroleum sulfonates is their potential availability in large quantities. This factor is important since approximately 100 million lb of surfactant per year would be required to flood a large reservoir.

As mentioned before, the equivalent weight distribution of a sulfonate is a function of the feedstock sulfonated. The higher equivalent weight sulfonates of about 450 or greater are generally oil soluble and not water soluble. These sulfonates are strongly related to the lowering of interfacial tension and are also strongly adsorbed on mineral surfaces. The lower equivalent weight sulfonates serve as a sacrificial adsorbate and a solubilizer of the high equivalent weight sulfonates. One characteristic of micellar surfactant solutions is that each molecule does not normally exhibit its individual solubility property, but rather the system as a whole exhibits a composite property reflecting influence from all components. Thus a mixture of high and low equivalent sulfonates will form a micellar solution exhibiting solubility properties of an intermediate equivalent weight sulfonate.

Phase Diagrams

In general, surfactant slugs are composed of three major ingredients: oil, surfactants and cosurfactants and water. The literature contains numerous ternary diagram representations for these components. These phase diagrams reveal the various phases that may form in the reservoir as residual oil is mobilized and displaced.

Healy and Reed of Exxon in 1974 defined a microemulsion to be a stable, translucent micellar solution composed of oil and water that may contain electrolytes and one or more amphiphilic compounds (surfactants, alcohols, etc.). It should be noted that according to this definition a microemulsion is not an emulsion. A microemulsion may have distinct internal and external phases, but in many cases there is no identifiable external phase.

Since a microemulsion has at least three components (oil, water and surfactant) the compositional state of the system must be specified with at least three numbers. It is therefore convenient and instructive to employ a ternary representation as shown in Figure 7.

The simple situation will involve three pure components, and the multiphase region will be bounded by a continuous binodal curve. Everywhere above the binodal curve a single phase exists that undergoes transitions among various structural states as the compositional point moves about the diagram. These transitions may be gradual, reflecting an equilibrium in which there is significant coexistence of different micellar configurations, as proposed by Winsor.

In the multiphase region, the most simple, three-component system involves only two phases throughout; one is oil-external and the other water-external. In this

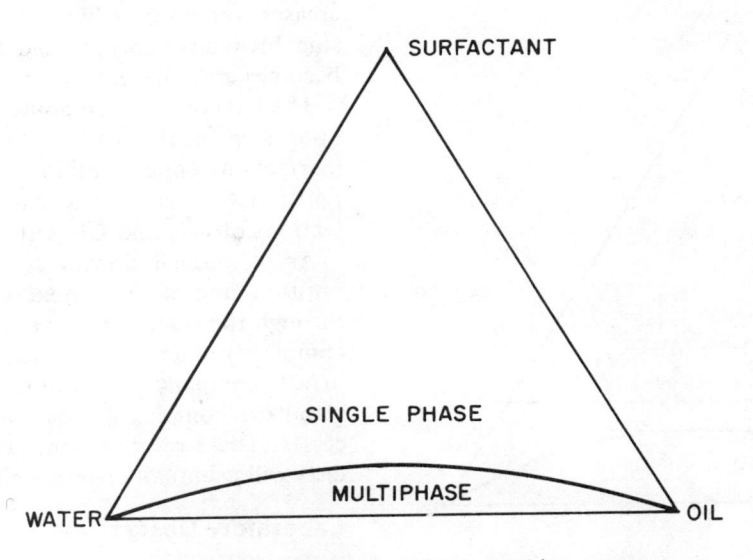

Fig. 7. Ternary representations of microemulsion system. (after Healy and Reed, 1974)

case, the two phases lie at opposite ends of a tie line and disappear equally at a plait point. The plait point would also be a phase inversion point for compositions along the binodal curve.

Systems of interest for enhanced oil recovery are not simple since there are always more than three pure components and three or more phases in equilibrium in certain portions of the multiphase region. Furthermore, these phases may contain combinations of spherical and lamellar micelles, cylindrical micelles, or simply consist of surfactant dispersed in excess water or oil. In these cases identification and construction of tie lines may be quite difficult.

In spite of these complexities, the plait point concept is often applicable. There is usually a region near the binodal curve where only two immiscible phases appear; hence, a plait point can be identified and is generally associated with a phase transition. However, this phase transition may be more complex than an abrupt transition between oil-external and water-external phases. In fact, there are instances when a unique plait point cannot be determined.

Healy et al (1975) discussed some concepts and definitions of microemulsion displacements by use of ternary diagrams. Microemulsion displacements are referred to as miscible or of a miscible-type with some misgivings, possibly in view of the micellar structures always present. For some, it is difficult to think of a displacement as miscible when just below the threshold of visibility there are discontinuities having dimensions on the order of hundreds of angstroms. Thus it appears that the concept of miscibility refers to some scale and has to do with the presence or absence of interfaces. However, the scale is arbitrary and it is convenient to use visible white light as the measure.

Figure 8 is a representation of an oil-water-surfactant system as previously described.

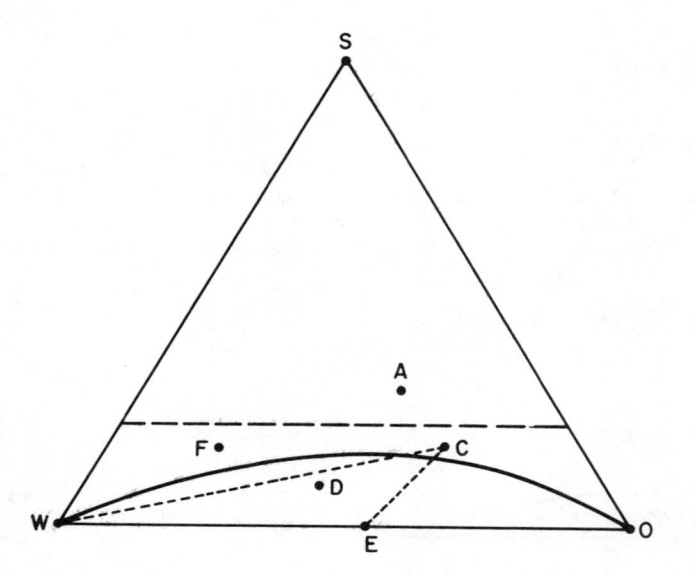

Fig. 8. Ternary representation of water-oil surfactant system. (after Healy et al, 1975)

On this diagram compositions A and C, for example, are in a miscible region because it has been determined experimentally that corresponding samples will transmit light from a white source without evidence of interfaces being present. Using the same test, composition D is not in a miscible region and therefore is in an immiscible region. Miscibility is a property of pairs of compositional points. Thus two compositions are mutually miscible, partially miscible or immiscible according to whether a straight line joining them lies, respectively, entirely interior, partially interior or exterior to a miscible region. For example, compositions A and C are miscible, compositions C and W are partially miscible and compositions D and O are immiscible.

Displacement in a porous medium is locally miscible if, over a large (relative to pore sizes) region of the core, all composition pairs are miscible. A displacement is miscible if it is everywhere locally miscible. If a displacement is not locally miscible, it is immiscible. Using this terminology, composition F can be miscibly displaced by composition C, assuming the core is initially saturated with only composition F, and ion exchange and surfactant adsorption capacities of the rock are satisfied.

Injection compositions for practical microemulsion displacements are bounded from below by the binodal curve and from above by some economic criterion that depends on bank size (dashed line). Once an oil-water bank is formed, the composition being displaced is, for example, E. For this reason a microemulsion displacement having an initial composition bounded as previously indicated is never a miscible displacement. Instead, there are generally transition regions between the microemulsion slug and the stabilized oil-water bank at the front and the drive water at the back, wherein immiscible phases grow or diminish in number as well as in saturation. As mixing progresses, the maximum surfactant concentration in the microemulsion slug decreases and eventually exits the miscible region; i.e., slug breakdown occurs and the locally miscible flood becomes an immiscible displacement.

The last composition point to leave the miscible region upon slug breakdown will be referred to as the critical (surfactant) concentration, and for the injection composition C it is approximately bounded by the intersections of $\overline{CW}$ and $\overline{CE}$ with the binodal curve.

As mentioned previously, the surfactant slug composition and its continual change as it is displaced through the reservoir is extremely complicated and not completely understood. Research is taking place in many private company laboratories and universities to understand the complexities of the surfactant polymer processes. These results, along with many additional field tests will eventually commercialize the process.

Laboratory Design

The literature contains a variety of techniques that are used to screen surfactant slugs for use as potential oil recovery agents. Some of these methods will be discussed.

Hill et al of Shell (1973) have described three different screening procedures:

1) *Measurement of interfacial tension at a crude oil-aqueous sulfonate interface.* By use of a micropendent drop apparatus values of interfacial tension against crude oil of less than 2×10^{-4} dynes/cm were measured.

2) *Microscope screening.* This procedure has proven useful for the rapid evaluation of the oil displacing ability of small quantities of surfactant solutions. It is a qualitative, somewhat subjective method which detects large decreases in interfacial tension between the aqueous solution and oil. It depends upon microscopic observation of an oil droplet adhering to a glass slide as it deforms in a slowly moving stream of surfactant solution.

A small droplet of oil, approximately 100 microns in diameter, is deposited with a fine capillary on a microscope slide and covered with a glass cover slip. It is advantageous to insure adequate separation between the two glass surfaces by inserting small strips of Saran Wrap or some similar material beneath opposite edges of the cover glass. A drop of water is placed at one edge of the cover glass and allowed, by capillary imbibition, to flow into the space separating the two surfaces and to envelop the oil droplet. Excess water which may tend to float the cover slip is removed with filter paper. Next, a drop of the surfactant solution under study is introduced at the edge of the cover slip and the oil droplet is observed under low magnification as the surfactant solution flows past. Rate of flow can be controlled to some extent by removing fluid with the aid of a piece of filter paper at the edge of the cover glass opposite to that where it is being introduced. Temperature is controlled with a microscope hot stage.

The observable effects upon the oil droplet vary from none, through a tendency toward slight elongation, to a complete breakdown into fine filaments several microns in thickness at very low interfacial tension. By application of this method, an experienced investigator can screen in excess of 50 systems per day.

3) *Displacement tests.* Conventional procedures and techniques were used in these experiments. Sandpacks prepared in water or cores mounted in plastic and saturated with water were flooded with crude oil to residual water saturation then flooded with water to residual oil saturation. Following this, the chemical system under test was injected. Cores and packs were vertical. Oil was injected vertically downward and aqueous liquids were injected vertically upward at rates equivalent to a frontal advance of about one foot per day.

Froning and Treiber (1976) of Amoco described their laboratory screening procedure as follows:

Vial tests are readily performed benchtop laboratory tests providing a *qualitative* measure of the phase behavior, miscibility and interfacial tension between surfactant formulations and oils. Thousands of these screening tests can be performed in a relatively short time. Test variables are brine composition, crude oil, temperature, cosurfactant types, added salts, polymers, etc. Micellar fluids are prepared in small vials over the range of water contents from 90 - 96% with varying ratios of surfactant to selected cosurfactants. The sodium chloride salinity is varied. The phase behavior is noted at selected temperatures. Crude oil is gently added and the ease of mixing the oil and micellar fluid is judged. If large drops of oil appear in the micellar fluid, the micellar fluid is judged to be poor. If the oil appears to be rapidly dissolved into the fluid, the micellar fluid is judged to be good. This simple test provides a coarse screen which retains most of the 90-95% brine fluids that will recover significant oil in subsequent core floods.

Table V shows data obtained with "mahogany" sulfonates with isopropyl alcohol (IPA), a new gas-oil sulfonate without cosurfactant and a new polybutene sulfonate with an ethoxylated alcohol.

Interfacial tension (IFT) determinations using the spinning drop technique provide a means for further determining the ionic tolerance and optimum salinity range for formulations of micellar fluids. The IFT data show that the vial test does pinpoint the optimum salinities over which high displacement of oil is achieved with a given micellar composition. However, there is not a direct correlation between vial test judgments and IFT measurements. The vial tests correspond more closely to miscibility than to IFT values. The vial test is used as the first screen because more conditions can be checked in a shorter time than in the IFT tests. The more promising micellar fluids mixed for the vial test are then IFT tested. When formulating more dilute micellar fluids that have little oil micibility, the IFT test is especially useful.

The oil displacement efficiency of surfactant formulations showing good vial and IFT test characteristics are determined in floods using 2-in diameter, 4-ft long Berea cores. The standard performance was that obtained with "mahogany" AA sulfonate and IPA at a 5:3 ratio using a 10% PV slug of a 92% water formulation. This formulation has been extensively studied for a number of years and is considered to give good recovery. Further tests are conducted with promising systems on cores up to 24 ft long to monitor pressures, mobility control and fluid interactions. In addition, the mobility requirements of the micellar fluid are determined and tested in reservoir cores. Adsorption tests with candidate micellar fluids are also made in reservoir cores to determine fluid bank size. Finally, after a micellar fluid has

TABLE V
VIAL TEST DATA
SECOND WALL CREEK CRUDE OIL 110°F

Sulfonate: Cosurfactant	SALINITY ppm x 10₃	92% % BRINE	94%
Mahogany Sulfonate AND IPA	6 / 3 / 1		
Gas-Oil Sulfonate and Alcohol	40 / 20 / 5		
Polybutene Sulfonate and Ethoxylated Alcohol	30 / 15 / 10 / 5		

Horizontal line indicates region of stable, good micellar fluids

(after Froning and Treiber, 1976)

passed all other tests, a pressure monitored test is performed in reservoir cores as a final check. On the basis of these tests, a chemical system is chosen for field application.

Marathon has described the Maraflood^T.M. process extensively in the literature. The concept is basically similar to most of the other processes in which a surfactant slug is displaced by a mobility buffer and finally by ordinary drive water. Early field tests used commercially available petroleum sulfonates for the surfactant portion of the slug. Recent developments have reduced slug cost by sulfonation of crude oil and by manufacturing the micellar slug in one process. By altering parameters in the manufacturing process, micellar solution properties can be varied over a wide range.

Marathon described the laboratory design for the 407 acre M-1 project in Illinois, being partially funded by DOE, as follows:

Surfactants used for M-1 project fluid system design were manufactured at the Robinson Refinery and in the Denver Research Center pilot plant. Sulfonate manufactured in these units has been used in micellar-polymer systems for another reservoir in the Robinson sand. Even with the background acquired in using this sulfonate to design fluids for a similar reservoir, it was necessary to investigate a number of variables to obtain the most cost effective micellar-polymer system for the M-1 project. Variables investigated during slug optimization work include feedstock, chemical additives, surfactant concentration, pH, sulfonate molecule cation and cosurfactant type and level. Restraints placed on the design were: 1) slug would be manufactured in the Robinson Refinery unit; 2) slug viscosity would not exceed 40 cp; and 3) the mobility buffer would utilize Dow polyacrylamide.

Laboratory fluid design tests were conducted under conditions simulating the reservoir. All flooding tests utilized samples of reservoir rock taken from a series of 6-in diameter cores cut from the M-1 project reservoir. From the time the cores were cut until they were prepared for flooding, they were handled so as to minimize alteration of their properties. Cores were shipped and stored in containers under reservoir crude or water. They were cut to a thickness of 2-in and trimmed for laboratory use with minimal exposure. Discs were cleaned by cold solvent flushing with isopropyl alcohol, toluene and hexane before drying in a vacuum oven at temperatures not exceeding 120°F. They were then re-saturated with synthetic reservoir brine and actual reservoir crude oil and waterflooded to the tertiary condition prior to laboratory experiments. These handling and cleaning procedures closely duplicated those for discs that had not been cleaned, but were simply restored to the tertiary condition using fresh crude oil and synthetic reservoir brine. The synthetic reservoir brine used in this work was based on an average analysis of waters taken from four producing wells within the project area.

Water used to mix polymers for this work is a synthetic fresh water based on analysis of an available supply water that was 80% zeolite treated to reduce hardness and polymer requirement.

Crude oil used in laboratory experiments was taken from four different 10-gal samples gathered from the project area. This was a sweet Illinois crude having a gravity of 36° API and a viscosity of 5-6 cp at the reservoir temperature of 72°F.

All laboratory experiments were conducted at a controlled temperature of 72°F. Micellar-polymer fluids were injected into a 1/8-in diameter wellbore drilled in the center of the discs. Injection was at a constant rate to give a median frontal velocity of 0.3 ft/day.

On the basis of numerous core floods run to obtain oil recovery performance, a surfactant slug was formulated for this project and Dow Pusher 700 was to be used for the mobility buffer.

Other companies such as Phillips, Texaco, Conoco, etc. discuss chemical system designs but do not go into specific detail on techniques used.

In summary, most companies use visual techniques to determine the solubility of water and oil in the surfactant slug. Interfacial tension measurements are also determined and used to screen various slugs. All companies use core flooding to determine the effectiveness of the chemical system to recover oil. The core material used in many cases is the Berea rock. Final design work is generally done in the actual reservoir rock. After the best system is designed the field test is the next step.

Description of Various Surfactant-Polymer Processes

Foster of Mobil in 1972 described a low tension waterflood process. It was assumed the process would apply to a nearly or completely watered out sandstone reservoir. The water phase present in the reservoir at this stage is assumed to be a typical oil field brine, high in total dissolved solids and in divalent cations, particularly calcium and magnesium. A regular pattern from the existing injectors and producers is chosen with high areal sweep as an important design criterion.

The process consists of injecting three slugs of water with different chemical compositions. These will be designated as the protective slug, surfactant slug and mobility control slug, or as slugs I, II and III, respectively.

The protective slug is an aqueous solution of sodium chloride. Within limits, its volume is somewhat arbitrary, in the range of one-tenth of the pattern pore volume. For most applications the concentration of sodium chloride in this slug would be between 1.0-2.0 weight %. The primary purpose of this slug is to screen the low tension surfactant from the reservoir brine and to base exchange the reservoir solids, replacing magnesium and calcium with sodium ions. In certain reservoirs, those containing brines with relatively low total dissolved solids and low amounts of divalent cations, this slug can be much less than one-tenth of a pore volume as these waters are much more compatible with low tension surfactants. The rear portion of this slug will usually contain other inorganic salts such as sodium tripolyphosphate and/or sodium carbonate whose presence reduces surfactant adsorption and improves the water-wetness of the reservoir rock surfaces.

The surfactant slug contains the same sodium chloride content as slug I and the same inorganic sacrificial chemicals mentioned above. This slug also contains a selected petroleum sulfonate in concentrations ranging from about 1.0-3.0 weight %. The size of this slug is arbitrary (in the range of one-tenth of a pore volume) but the amount of surfactant in it is carefully set by adsorption criteria. It is sometimes necessary to increase the viscosity of the rear part of this slug with an anionic biopolymer. The primary purpose of this slug is to reduce the interfacial tension between the oil and the water to the order of 0.001 dyne/cm. Additionally, the slug material (probably through the formation of an emulsion) can affect significant improvement in the viscosity ratio. Laboratory results suggest that a water-to-oil viscosity ratio of four or more is often provided by an oil-in-water emulsion which spontaneously forms in situ. For such systems, the biopolymer might not be needed in this slug during a field application. When these conditions are realized under typical flooding rates (0.5 ft/day) residual oil is reduced to zero and most of the oil phase moves faster than the water phase. This set of circumstances produces a tertiary oil bank.

The sodium chloride content of the mobility control slug is considerably less than that of the previous two, ranging from 0.2-0.6 weight %. Although the salt concentration level is somewhat arbitrary, it should be less than the minimum salt concentration which produces very low tension. This slug also contains a water soluble biopolymer in concentration sufficient to provide a locally stable mobility condition at the rear of the oil bank. The exact level of water viscosity necessary to achieve this condition depends on a number of factors outside the scope of this summary. However, a not unusual value would be in the order of ten times the oil viscosity. As it is not economically feasible to maintain such a water viscosity over an appreciable fraction of the reservoir volume, the concentration of biopolymer must be decreased with distance from the rear of the oil bank, until it vanishes. Such a graded viscosity zone provides a mildly unstable hydrodynamic condition. The pore volume of this mobility control slug should be around 15% PV for reservoirs containing typical medium gravity crude oils.

This sequence of three slugs can be driven by the original reservoir brine. An ideal situation results if the combined volume of these slugs just equals that of the oil mobilized and subsequently produced. Such an overall material balance would provide a disposal mechanism for the greater part of the reservoir brine, given a well-designed schedule of multi-pattern development.

Initial development of Union Oil's soluble oil flooding process began in 1957. Known as Uniflood[T.M.] it has been described as follows:

1. Injection of a 10% PV slug of a preflush chemical solution.
2. Injection of a 3-4% PV slug of soluble oil; this slug of micellar solution is developed in the reservoir by alternate injection of a substantially anhydrous soluble oil formulation and fresh water.
3. Injection of a 60% PV slug of polymer solution.
4. Injection of approximately 70% PV of any fresh water that is readily available.

Soluble oils are oleic compositions that spontaneously solubilize water, but not oil. They consist essentially of hydrocarbons, surfactants and a stabilizing agent. The stabilizing agent can be a solvent, an electrolyte or both. The soluble oil can also contain water ranging from a few percent to more than half of the soluble oil blend.

The amount of oil recovered by soluble oil flooding appears to be a linear function of concentration of the sulfonate in the soluble oil until the concentration gets below 5%. The results in Figure 9 illustrate this relationship in which only the concentration of sulfonates and solvent in the soluble oil was changed.

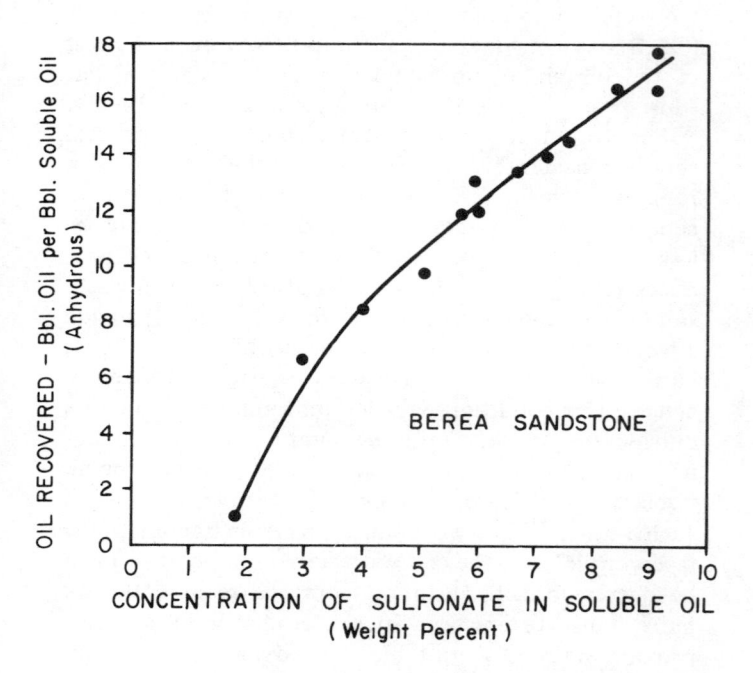

Fig. 9. Effect of concentration of sulfonate on recovery by soluble oil polymer floods. (after Holm, 1972)

In 1967, Gogarty of Marathon described the Maraflood Oil Recovery Process. It was described as a new oil recovery method for producing oil under both secondary and tertiary reservoir conditions. The concept of the process is that a miceller solution slug is injected to displace oil and water. Following the slug a mobility buffer is injected to protect the slug. Finally, drive water is injected to propel the slug and mobility buffer through the reservoir.

Micellar solution slugs used in the process contain primarily hydrocarbon, surfactant, cosurfactant, electrolyte and water. Table VI shows the compositional ranges of three categories of micellar solutions which have been developed for field use.

TABLE VI
MICELLAR SOLUTION COMPOSITIONS

Component	Low Water Oil External	High Water Oil External	High Water Water External
Surfactant	>5	>4	>4
Hydrocarbon	35-80	4-40	2-50
Water	10-55	55-90	40-95
Cosurfactant	<4	0.01-20	0.01-20
Electrolyte	.001-5*	0.001-4*	0.001-4*

*Wt. % Based on Water.

(after Gogarty et al, 1968)

The micellar solution is displaced through the reservoir with either water-external emulsions or water solutions containing polymers. Because of economics, polymers are preferred. At the interface between the buffer and water an unfavorable mobility ratio exists. Under these conditions excessive mixing takes place as water moves faster than the buffer. This mixing can be reduced by grading the concentration of the mobility buffer. After injection of the mobility buffer regular water injection continues, and field operations become identical to a regular waterflood.

Trushenski et al (1974) of Amoco described a process in which a 92% water micellar solution displaced 85% of the tertiary oil when a 10% PV slug was injected into laboratory cores. The micellar fluids developed were high water content (85 - 95 weight %) microemulsions. These fluids generally were prepared with from 4-10% oil soluble hydrocarbon sulfonate with an equivalent weight of from 350 to 475. Also, an oil or water soluble alcohol cosurfactant was used.

These micellar slugs were driven through the reservoir by a mobility buffer bank containing either biopolymers or polyacrylamides. The mobility buffer was then followed by the drive water.

Hill et al of Shell in 1973 described an aqueous surfactant system for oil recovery. An aqueous surfactant system containing 2 % petroleum sulfonate was combined with polymer to obtain appropriate mobility characteristics for flooding the Tar Springs sand in the Benton Field in Illinois. Sodium tripolyphosphate was incorporated to increase the system's tolerance for multivalent cations found in field waters. A slug of this chemical system driven by a polymer solution through previously waterflooded cores recovered 85% of the residual oil.

It has been mentioned previously that the various combinations of the ingredients that can be incorporated into a surfactant slug are essentially infinite. A laboratory screening program should be undertaken to determine the system with the most economic potential. Some screening methods mentioned in the literature are interfacial tension measurements, phase diagrams, phase stability tests, adsorption determinations and oil recovery experiments. However, the ultimate determination is either a small pilot test or a field-wide commercial project.

FIELD HANDLING SYSTEMS

Water Treating System

Water treatment facilities will vary considerably according to the condition of the water supply and the requirements of the injection program. In some cases the minimum treatment requires filtration of the water through a diatomaceous earth pressure filter.

If the water is being used for a miscible slug or polymer formulation, it may be necessary to process the filtered water through an ion exchange water softener. This step is used to replace various detrimental cations with sodium ions from the resin in the water softener. (See Figure 10.)

Surfactant Slug Blending System

The individual components of a miscible slug vary in composition, with many patented micellar formulas. Most slugs consist of at least four distinct components: a petroleum sulfonate, an aqueous phase, a hydrocarbon phase and a cosurfactant. All components except the cosurfactant are metered into a large blending tank where they are mixed until homogeneous. (See Figure 11.) If filtration is necessary the slug is generally heated prior to pumping through the filter. The preheating serves several purposes. It stabilizes the slug, improves filterability because of reduced slug viscosity and reduces possible paraffin precipitation in the injection well. After filtration, the cosurfactant, which is almost always an alcohol, is metered into the slug. The cosurfactant increases the stability of the micellar system and simultaneously changes the viscosity to meet mobility requirements in the reservoir. The slug is generally placed in a preinjection hold tank prior to injection. A positive displacement pump is used to inject the slug at a predetermined flow rate.

Polymer Blending System

Polymer mixing is generally accomplished in a facility of the type shown in Figure 12. The heart of this system is the dry polymer mixer which meters the powdered or granulated polymer into a controlled water stream to give a uniform dispersion. The typical arrangement allows the polymer to contact a tangentially swirling stream of water in a funnel-shaped device. The GACO and Dow mixers are of this type. The polymer feed rate to the mixer is controlled by a variable speed feed auger. The water rate is adjusted as necessary to provide the needed mixing in the funnel. The remainder of the water needed to meet the target polymer concentration is introduced as a by-pass stream which mixes

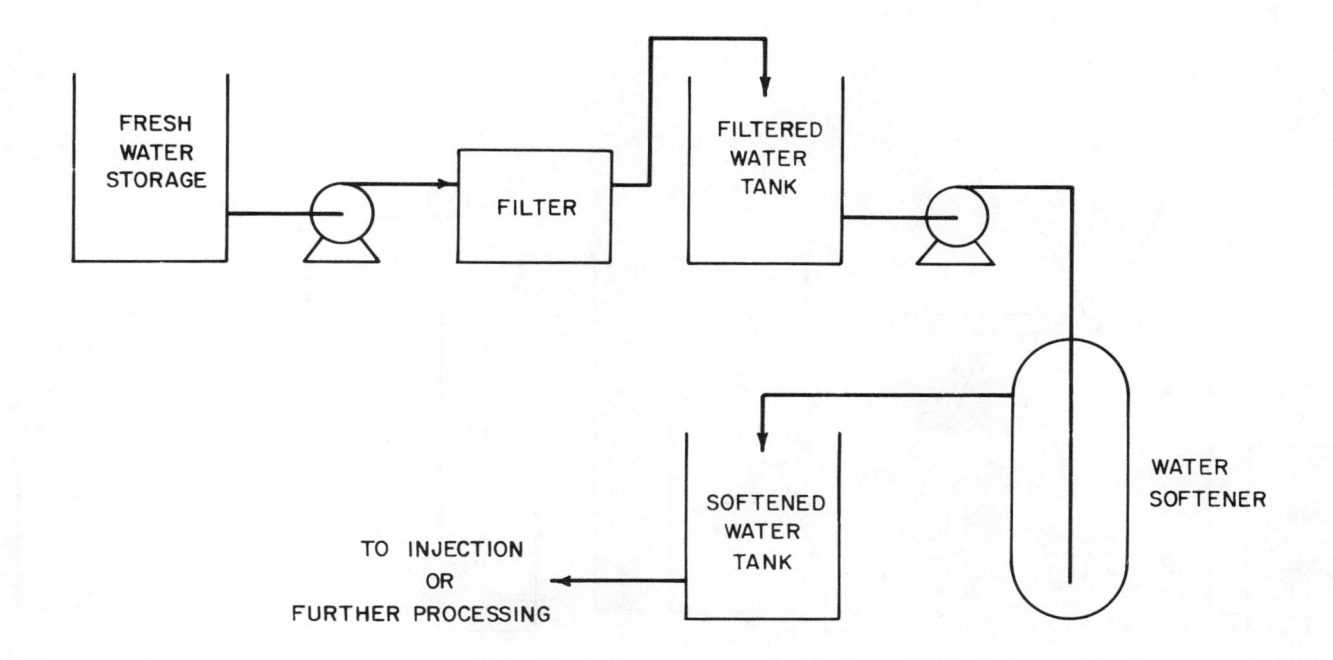

Fig. 10. Water treatment system diagram.

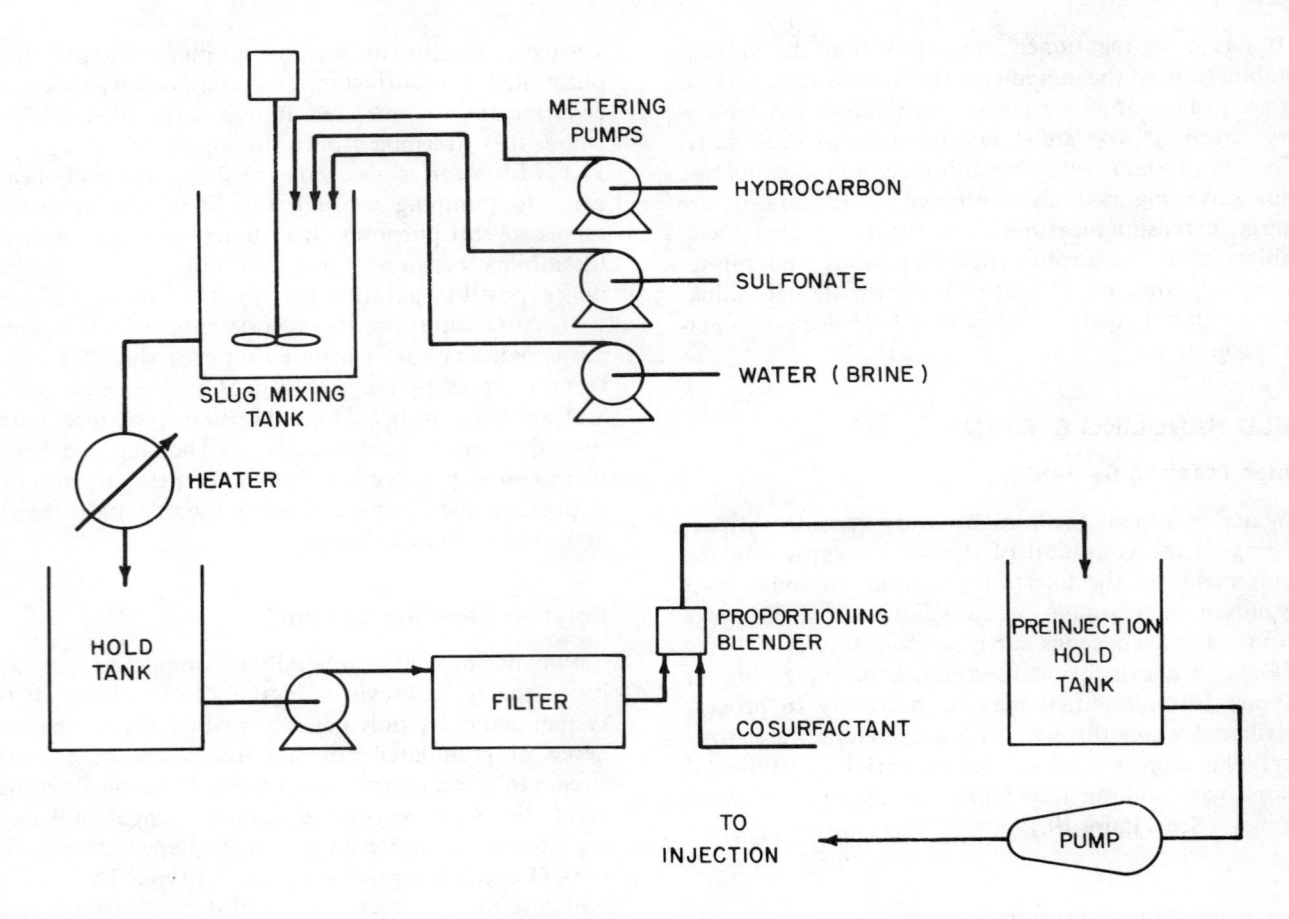

Fig. 11. Surfactant slug blending system diagram.

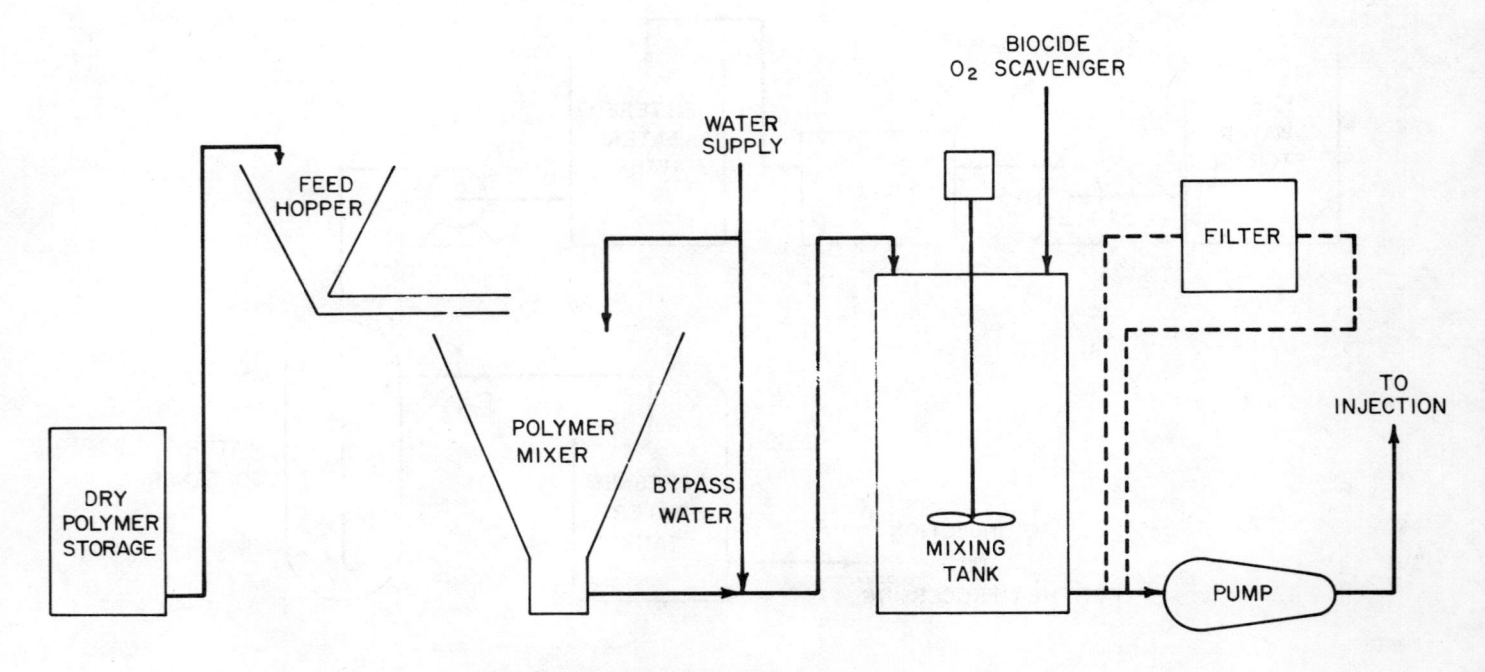

Fig. 12. Dry polymer blending system diagram.

with the polymer dispersion immediately down stream of the mixer.

Maintenance of a sufficient dry polymer inventory in the feed hopper is generally accomplished in one of two ways. In small scale operations, fifty pound bags of polymer are emptied into the feed hopper or into a storage bin which is connected to the feed hopper. In large scale operations, the polymer is stored in bulk and transported to the feed hopper pneumatically. Care must be exercised to control the polymer dust in these areas and personnel are required to wear masks and possibly additional protective gear.

Because the rate of dilution of high concentration polymer is quite slow, relatively large mixing tanks with gentle agitation are needed downstream of the polymer mixers. These tanks are generally blanketed with nitrogen to exclude atmospheric oxygen. This is also a common place to inject an oxygen scavenger or biocide as necessary.

The thoroughly mixed polymer from the tank is generally injected directly with a piston-type positive displacement pump. In some cases, when face plugging of injection wells is feared, a wellhead cartridge filter may be used to ensure the injected polymer contains no gelled agglomeration of high concentration polymer.

Preparation of polymer solution from emulsion polymer supply is considerably less complicated. Only the metered dilution water and chemical additions are required. Polymer dilution can frequently be accomplished with static or in-line mixers eliminating the large mixing tank. High concentration liquid polymer is stored in a tank with a metering pump used to control polymer rate to the mixer.

Fluid Injection System

Fluid injection into a reservoir through numerous wells is generally accomplished by means of a manifold system. Figure 13 illustrates a simplified case. Because variable speed positive displacement pumps are generally used to inject fluids into a reservoir, the total volumetric flow rate can be controlled to meet the overall injection program. However, without flow control devices at each well the relative flow is determined by the flow resistance in each injection well. To compensate for this uncontrolled injection, some type of flow controller is needed at each well. In many cases when the fluid being injected is water or a miscible slug, a simple throttling valve is sufficient to regulate the flow. If the number of wells accepting fluid from a single pump is great, the control can be somewhat unstable because the entire system is interactive. Any change in a throttling device at one well causes a change in flow at all the remaining wells because the total flow rate remains constant. Still this system is workable in most cases with sufficient monitoring of injection rates at each well.

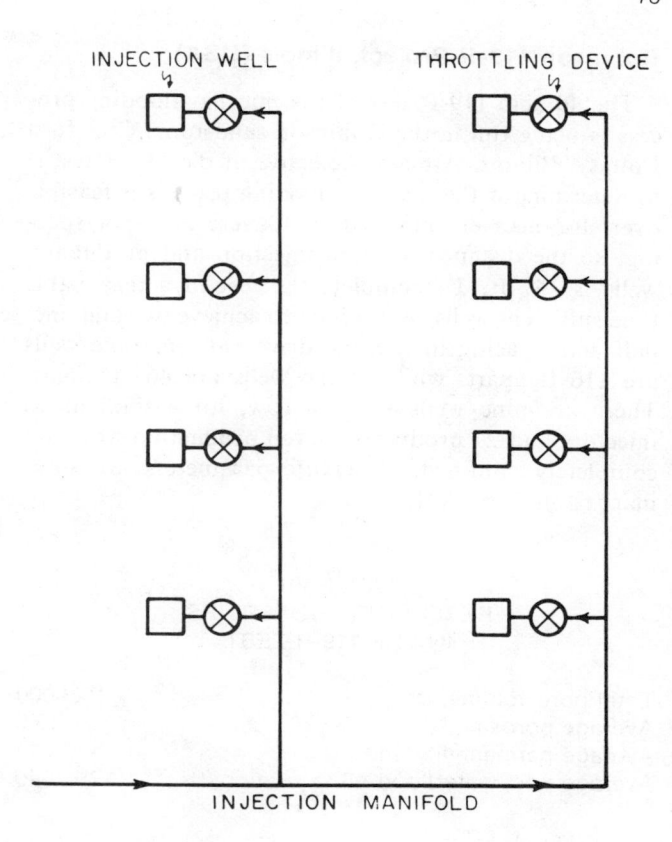

Fig. 13. Fluid injection distribution manifold system diagram.

The injection of polyacrylamide polymer requires a special solution to the problem of controlling injection rates. These polymers are susceptible to shear degradation when they pass through a throttling valve. The method commonly used for rate control is the insertion of a long coil of tubing of relatively small diameter. Because polymers are less sensitive to the viscous shear in a pipe than the viscoelastic shear through an orifice or similar device, these coils accomplish the goal of flow control without polymer degradation. The diameter of the tube coil is calculated based on the shear rate for the desired flow rate, while the length of the coil is calculated based on the pressure which must be dissipated prior to entering the wellhead.

CASE HISTORIES

There are few large field projects utilizing the surfactant polymer process. Four projects are briefly described herein; they are as follows:

Field	Operator	Start Date	Acres
119-R	Marathon	1968	40
219-R	Marathon	1975	113
M-1	Marathon	1977	407
N. Burbank	Phillips	1975	90

Robinson 119-R Project, Illinois (USA)

The 40 acre 119-R test of the micellar flooding process is being run in the Robinson sandstone, Crawford County, Illinois. A major objective of the 119-R test is to determine if the micellar flooding process is feasible over distances encountered in 10 acre five-spot spacing, so the distance between injection and production wells is 467 ft. To complete the test in a reasonable time sufficient wells were added to achieve two and one-half acre spacing in the line drive pattern. Like wells are 116 ft apart, while unlike wells are 467 ft apart. There are nine wells in each row, for a total of 18 injectors and 27 producers. Seven production wells are completely confined. Reservoir parameters are summarized in Table VII.

TABLE VII
RESERVOIR PROPERTIES
IN THE 119-R TEST

Total pore volume, Bbls	1635000
Average porosity, %	19.3
Average permeability, md	211
Average post waterflood oil saturation, %	40

Micellar solution injection started in September 1968. The micellar slug contained: a mixture of Marathon manufactured crude petroleum sulfonate (made using a sulfuric acid process in the Detroit Refinery) and a commercially available sulfonate as the surfactant; lease crude as the hydrocarbon; primary amyl alcohol as the cosurfactant; and fresh water. Slug viscosity was about 30 cp at 72°F. After injection of 114000 Bbls of micellar solution (about 7% PV), mobility buffer injection began in January 1969. A polymer solution, Dow Pusher 700 in water, was used as the mobility buffer in this test. Initial concentration was approximately 1200 ppm; concentration was decreased throughout the flooding life until February 1973, when injection was switched to fresh water alone. Approximately 100% PV of polymer solution at an average concentration of about 470 ppm was injected. Plans call for injecting about 45% PV of water.

Production increased from approximately 25 BOPD to 240 BOPD and oil cut increased from 1.5-22%. Cumulative oil production to September 1974 was approximately 243200 Bbls from the total pattern. This represents 39% of the estimated 620000 Bbls of oil-in-place at the end of waterflood. One objective of the test is to determine the recovery factor for the confined portion of the test, approximately 20 acres with a 670000 Bbl pore volume. The September 1, 1974 cumulative production from this area was 120400 Bbls; this represents a 45% recovery or about 270 Bbls per acre ft. Estimated ultimate recovery from this confined area is about 129000 Bbls or 48% of the oil-in-place.

Table VIII summarizes the various production phases of this reservoir:

TABLE VIII
RECOVERY HISTORY
IN THE 119-R AREA

	Years	Recovery
Primary (Dissolved Gas)	1906-1957	300 STB/acre-ft
Secondary (Waterflooding)	1957-1968	180 STB/acre-ft
Tertiary (Maraflood) (estimated ultimate confined pattern)	1968-1974	291 STB/acre-ft
(achieved to 9/1/74, total pattern)		225 STB/acre-ft

Robinson 219-R Project, Illinois (USA)

A large-scale application of the Maraflood$^{T.M.}$ oil recovery process is being operated by Marathon Oil Company on the Henry Unit in Crawford County, Illinois. The 219-R project is being conducted on a 113 acre tract in the Robinson sand at a depth of about 1000 ft. The 219-R project is contiguous to the 119-R project.

The 219-R project was developed using a five-spot pattern on a 3 acre spacing. It uses 39 injection wells and 55 production wells. Injection of 312266 Bbls of surfactant slug was initiated in October 1975 and was completed in April 1976. Polymer (Dow Pusher 700) was injected behind the slug at a spike concentration of 1156 ppm. Approximately 105% PV of mobility buffer at an average concentration of about 566 ppm is planned. Cumulative fluid injection into the project as of January 1, 1978 totals 82% PV. A production response occurred after about 18% PV injection. The production rate increased from approximately 40 BOPD at the beginning of the project to an August 1977 peak production of 536 BOPD. Cumulative oil recovery to January 1, 1978 is 219419 Bbls or about 18.1% of the oil-in-place. At the present time it is too early to extrapolate the production curve to estimate ultimate recovery.

Robinson M-1 Project, Illinois (USA)

The M-1 project consists of 407 acres located in Crawford County, Illinois. The Robinson Sand Reservoir is similar to the reservoir being flooded in the 119-R and 219-R projects. This project is being partially funded by DOE.

The average sand thickness is 28 ft with a 19% porosity and 103 md arithmetic mean permeability. The oil saturation is estimated to be 40%. This consolidated sand is a point-bar type river deposit. Permeability is reduced by the presence of kaolinite and calcite in the sand.

Fluid system design for the M-1 was completed in 1976 based on numerous core flood studies. The micellar solution chosen was a crude oil sulfonate slug with 10 weight % active sulfonate. Primary hexanol was used as a cosurfactant to reduce slug viscosity below 40 cp. The polymer chosen was Dow 700, a partially hydrolyzed polyacrylamide. The injection schedule was as follows:

1) 10% PV micellar solution
2) 105% PV polymer, starting at 1150 ppm and decreasing in six unequal increments
3) 35% PV produced water

Through September 30, 1978 about 9.2% PV slug had been injected into a project pore volume of 16.6 million Bbls. Cumulative oil production through this time was 95 MSTB oil at a cumulative oil cut of 4.6%. It is too early for the oil production to respond to the surfactant slug. The 219-R project in a similar reservoir started responding after a cumulative injection of about 18% PV.

North Burbank Project, Oklahoma (USA)

A Phillips-DOE tertiary pilot test was initiated in May 1975 on a 90 acre tract of the North Burbank Field using a surfactant-polymer system developed specifically for this oil-wet reservoir.

The North Burbank Unit was selected as a highly desirable reservoir for a surfactant-polymer tertiary recovery pilot test because it had been successfully waterflooded for over 25 years and production had declined to the stripper level. However, the reservoir still contains about 400 million Bbls of sweet 39° API oil and thus offers a worthwhile opportunity to increase recoverable reserves. The site selected was NBU Tract 97, a 160 acre area having a history of good performance under waterflood. It is characterized by an average net pay of 43 ft, average permeability of 52 md and a current oil saturation estimated at 35%. The sand is rather oil-wet, and laboratory tests showed that a specially designed surfactant formulation would be required for effective oil recovery. The test pattern of 25 wells consists of nine inverted ten acre five-spots. The pattern PV is 5.3 million Bbls and contains about 1.8 million Bbls of oil. The predicted recovery by the surfactant flood was 0.6 million Bbls.

The fluid system for this project consisted of a preflush of fresh and salt water of 2.4 million Bbls. The surfactant slug contained: 6 weight % Witco TRS10-410; 3 weight % isobutyl alcohol; and 91% Ark-Burbank water containing 1.5 weight % sodium chloride. Approximately 295000 Bbls of slug were injected. The mobility buffer of about 2.67 million Bbls contained 790000 pounds of Betz Hi-Vis polyacrylamide. A fresh water buffer of about 2.7 million has been injected as of January 1, 1979.

Cumulative enhanced oil production through January 1, 1979 has been about 118000 Bbls. Present enhanced oil production is about 190 BOPD and is stabilized and possibly declining. No forecast has been made for the ultimate production, but recovery to date is about 7% of the oil-in-place.

BIBLIOGRAPHY

Abrams, A.: "The Influence of Fluid Viscosity, Interfacial Tension and Flow Velocity on Residual Oil Saturation Left by Waterflood, *Soc. Pet. Eng. Jour.* (Oct., 1975).

Anderson, D. R. et al: "Interfacial Tension and Phase Behavior in Surfactant-Brine-Oil Systems," SPE 5811 presented at the SPE-AIME 4th Symposium on Improved Methods for Oil Recovery, Tulsa, OK. (March, 1976).

Ayers, R. C., Jr.: "Two-Bank Miscible Tertiary Oil Recovery Process," SPE 3801 prepared for the SPE-AIME Improved Oil Recovery Symposium, Tulsa, OK. (April, 1972).

Bae, J. H. and Petrick, C. B.: "Adsorption/Retention of Petroleum Sulfonates in Berea Cores," *Soc. Pet. Eng. Jour.* (Oct., 1977).

Baker, L. E.: "Effects of Dispersion and Dead-End Pore Volume in Miscible Flooding," SPE 5632 prepared for the SPE-AIME 50th Annual Fall Meeting, Dallas, TX. (Sept., 1975).

Baviere, M.: "Phase Diagram Optimization in Micellar Systems," SPE 6000 presented at the SPE-AIME 51st Annual Fall Meeting, New Orleans, LA. (Oct., 1976).

Benham, A. L., Dauden, W. E. and Kunzman, W. J.: "Miscible Fluid Displacement - Prediction of Miscibility," *Trans.*, AIME (1960) 219.

Benton, W. J., Hwan, R., Miller, C. A. and Fort, T., Jr.: "Structure of Solutions of Synthetic Petroleum Sulfonates Before and After Addition of Oil," *ERDA Enhanced Oil, Gas Recovery and Improved Drilling Methods*, Vol. 1 - Oil, Tulsa, OK. (Sept., 1977).

Bentsen, R. G. and Nielsen, R. F.: "A Study of Plane Radial Miscible Displacement in a Consolidated Porous Medium," *Soc. Pet. Eng. Jour.* (March, 1965).

Berg, R. L., Noll, L. A. and Good, W. D.: "ERDA In-House Research on Thermodynamics of Oil Recovery, Micellar Systems," *ERDA Enhanced Oil, Gas Recovery and Improved Drilling Methods*, Vol. 1 - Oil, Tulsa, OK. (Sept., 1977).

Bernard, G. G.: "Effects of Clays, Limestone, and Gypsum on Soluble Oil Flooding," *Jour. Pet. Tech.* (Feb., 1975).

Blackwell, R. J., Rayne, J. R. and Terry, W. M.: "Factors Influencing the Efficiency of Miscible Displacement," *Trans.,* AIME (1959) 216.

Blair, C. M., Jr., and Lehmann, S., Jr.: "Process for Increasing Productivity of Subterranean Oil-Bearing Strata," U.S. Patent No. 2,356,205. (1942).

Bleakley, W. B.: "Journal Survey Shows Recovery Projects Up," *O. & G. Jour.* (March 25, 1974).

Boneau, D. F. and Clampitt, R. L.: "A Surfactant System for the Oil Wet Sandstone of the North Burbank Unit," *Jour. Pet. Tech.* (May, 1977).

Boneau, D. F., Trantham, J. C., Jackson, K. M. and Threlkeld, C. B.: "Performance, Monitoring and Control of Phillips Surfactant Flood in the North Burbank Unit - First 18 Months," 2nd Tertiary Oil Recovery Conference, Wichita, KS. (April, 1977).

Bowcatt, J. E. and Schulman, J. H.: "Emulsions," Zuts für Elect. Vol. 59. (1955).

Buckley, S. E. and Leverett, M. C.: "Mechanism of Fluid Displacement in Sands," *Trans.,* AIME (1942) 146.

Calhoun, J. C. Jr., Stahl, C. D., Preston, F. W. and Nielson, R. F.: "A Review of Laboratory Experiments on Wetting Agents for Waterflooding," *Producers Monthly.* (Jan., 1951).

Carpenter, D. H.: "Micellar Fluid Termed Promising Oil-Recovery Tool," *O. & G. Jour.* (Dec. 4, 1972).

Carpenter, D. H. and Davies, J. B.: "JPT Forum: The Energy Yield of Micellar Flooding," *Jour. Pet. Tech.* (Jan., 1976).

Cash, L. et al: "The Application of Low Interfacial Tension Sealing Rules to Binary Hydrocarbon Mixtures," *Journal of Colloid and Interfacial Science.* (March 15, 1977).

Cheesman, D. F. and King, A.: "The Properties of Dual Emulsions," *Trans. Fara. Soc.* Vol. 34 (1938).

Claridge, E. L.: "A Method of Design of Graded Viscosity Banks," SPE 6848 presented at the SPE-AIME 52nd Annual Fall Meeting, Denver, CO. (Oct., 1977).

Claridge, E. L. and Bondor, P. L.: "A Graphical Method for Calculating Linear Displacements with Mass Transfer and Continuously Changing Mobilities," *Soc. Pet. Eng. Jour.* (Dec., 1974).

Cooke, C. E., Jr.: "Microemulsion Oil Recovery Process," U.S. Patent No. 3,373,809.(1965).

Craig, F. F., Jr. and Owens, W. W.: "Miscible Slug Flooding - A Review," *Jour. Pet. Tech.* (April 1960).

Csaszar, A. K.: "Solvent - Waterflood Oil Recovery Process," U.S. Patent No. 3,163,214. (1961).

Dabbous, M. K.: "Displacement of Polymers in Waterflooded Porous Media and Its Effects on a Subsequent Micellar Flood," *Soc. Pet. Eng. Jour.* (Oct., 1977).

Dabbous, M. K. and Elkins, L. E.: "Preinjection of Polymers to Increase Reservoir Flooding Efficiency," SPE 5836 presented at the SPE-AIME 4th Symposium on Improved Methods for Oil Recovery, Tulsa, OK. (March, 1976).

Danielson, H. H. and Paynter, W. T.: "Bradford Sand Micellar-Polymer Flood, Bradford, Pennsylvania," *ERDA Enhanced Oil, Gas Recovery and Improved Drilling Methods,* Vol. 1-Oil, Tulsa, OK. (Sept., 1977).

Danielson, H. H., Paynter, W. T. and Milton, H. W., Jr.: "Tertiary Recovery by the Maraflood Process in the Bradford Field," SPE 4753 presented at the SPE-AIME Improved Oil Recovery Symposium, Tulsa, OK. (April, 1974).

Davis, H. T. and Scriven, L. E.: "Ultralow Interfacial Tension, Phase Behavior and Chemical Flooding Processes for Improved Petroleum Recovery," *ERDA Enhanced Oil, Gas Recovery and Improved Drilling Methods,* Vol. 1 - Oil, Tulsa, OK. (Sept., 1977).

Davis, J. A., Jr.: "Maraflood Process — A New Oil Recovery Method," *Producers Monthly.* (Feb., 1968).

Davis, J. A., Jr.: "Microscopic Displacement Mechanisms of Micellar Solutions," SPE 1847 - D prepared for the SPE-AIME 42nd Annual Fall Meeting, Houston, TX. (Oct., 1967).

Davis, J. A., Jr. and Jones, S. C.: "Displacement Mechanisms of Micellar Solutions," *Jour. Pet. Tech.* (Dec., 1968). *Trans.,* AIME (1968) 243.

DeGroot, M.: "Flooding Process for Recovering Oil From Subterranean Oil-Bearing Strata," U.S. Patent No. 1,823,439. (1929).

DeGroot, M.: "Flooding Process for Recovering Fixed Oil From Subterranean Oil-Bearing Strata," U.S. Patent No. 1,823,440.

Des Brisay, C. L., Gray, J. W. and Spivak, A.: "Miscible Flood Performance of the Intisar "D" Field, Libyan Arab Republic," *Jour. Pet. Tech.* (Aug., 1975).

DeVault, D. J.: "The Theory of Chromatography," *Journal American Chemical Society*, Vol. 65. (1943).

Dey, N. C. and Mony, G. S.: "JPT Forum: Discussion of Core Analysis for the El Dorado Micellar - Polymer Project," *Jour. Pet. Tech.* (Aug., 1977).

Doe, P. H., Wade, W. H. and Schechter, R. S.: "Alkyl Benzene Sulfonates for Producing Low Interfacial Tensions Between Hydrocarbons and Water," *Journal of Colloid and Interfacial Science.* (May, 1977).

Dombrowski, H. S. and Brownwell, L. E.: "Residual Equilibrium Saturation of Porous Media," *Ind. Eng. Chem.*, Vol. 46. (1954).

Dreher, K. D. and Syndansk, R. D.: "Observation of Oil-Bank Formation During Micellar Flooding," SPE 5838 presented at the SPE-AIME 4th Symposium on Improved Methods for Oil Recovery, Tulsa, OK. (March, 1976).

Dreher, K. D. and Syndansk, R. D.: "On Determining the Continuous Phase in Microemulsions," *Jour. Pet. Tech.* (Dec., 1971).

Dunlap - Wilson, P. M. and Brandner, C. F.: "Aqueous Surfactant Solutions Which Exhibit Ultralow Tensions at the Oil-Water Interface," *Journal of Colloid and Interfacial Science.* (July, 1977).

Dyes, A. B., Caudle, B. H. and Erickson, R. A.: "Oil Production After Breakthrough - As Influenced by Mobility Ratio," *Trans.*, AIME (1954) 201.

Earlougher, R. C., O'Neal, J. E. and Surkalo, H.: "Micellar Solution Flooding: Field Test Results and Process Improvements," SPE 5337 prepared for the SPE-AIME Rocky Mountain Regional Meeting, Denver, CO. (April, 1975).

Eicke, H. F.: "Thermodynamical and Statistical Considerations on Some Microemulsion Phenomenon," *Journal of Colloid and Interfacial Science.* (April, 1977).

El-Emary, M. M. et al: "The Low Interfacial Tension Behavior of Pure Sodium Alkylbenzene Sulfonates," *ERDA Enhanced, Oil Gas Recovery and Improved Drilling Methods,* Vol. 1 - Oil, Tulsa, OK. (Sept., 1977).

El-Saleh, M. M. and Farouq Ali, S. M.: "Oil Recovery and Slug Breakdown Behavior in the Alcohol Slug Process," *Producers Monthly.* (Feb., 1967).

Flumerfelt, R. W.: "Measurement of Dynamic Interfacial Properties Using a Prop Deformation Method: Theoretical Extensions," *ERDA Enhanced Oil, Gas Recovery and Improved Drilling Methods,* Vol. 1 - Oil, Tulsa, OK. (Sept., 1977).

Foster, W. R.: "A Low-Tension Waterflooding Process," *Jour. Pet. Tech.* (Feb., 1973) *Trans.*, AIME (1973) 255.

Foster, W. R.: "A Low Tension Waterflooding Process Employing a Petroleum Sulfonate, Inorganic Salts and a Biopolymer," SPE 3803 prepared for the SPE-AIME Improved Oil Recovery Symposium, Tulsa, OK. (April, 1972).

French, M. S. et al: "Field Test of an Aqueous Surfactant System for Oil Recovery, Benton Field, Illinois," SPE 3799 prepared for the SPE-AIME Improved Oil Recovery Symposium, Tulsa, OK. (April, 1972).

Friedman, F. and Ramirez, W. F.: "A Single Phase Model of Mechanisms Effecting Miscible Surfactant Oil Recovery," *Chemical Engineering Science.* (1977).

Froning, H. R. and Treiber, L. E.: "Development and Selection of Chemical Systems for Miscible Waterflooding," SPE 5816 presented at the SPE-AIME 4th Symposium on Improved Methods for Oil Recovery, Tulsa, OK. (March, 1976).

Gale, W. W. and Sandvik, E. I.: "Tertiary Surfactant Flooding: Petroleum Sulfonate Composition-Efficacy Studies," *Soc. Pet. Eng. Jour.* (Aug., 1973).

Gardner, G. H. F., Downie, J. and Kendall, H. A.: "Gravity Separation of Miscible Fluids in Linear Models," *Soc. Pet. Eng. Jour.* (June, 1962).

Gatlin, C. and Slobod, R. L.: "The Alcohol Slug Process for Increasing Oil Recovery," *Trans.*, AIME (1960) 219.

Geffen, T. M.: "Here's What's Needed to Get Tertiary Recovery Going," *World Oil.* (March, 1975).

Gibbons, L. C.: "Petroleum Resources in Illinois," Fifth Annual Illinois Energy Conference. (Sept., 1977).

Gidley, J. L.: "Stimulation of Sandstone Formations with Acid - Mutual Solvent Method," *Jour. Pet. Tech.* (May, 1971).

Gogarty, W. B.: "Miscible - Type Waterflooding: The Maraflood Oil Recovery Process," SPE 1847-A prepared for the SPE-AIME 42nd Annual Fall Meeting, Houston, TX. (Oct., 1967).

Gogarty, W. B.: "Status of Surfactant or Micellar Methods," *Jour. Pet. Tech.* (Jan., 1976).

Gogarty, W. B. and Davis, J. A., Jr.: "Field Experience with the Maraflood Process," SPE 3806 prepared for the SPE-AIME Improved Oil Recovery Symposium, Tulsa, OK. (April, 1972).

Gogarty, W. B., Kinney, W. L. and Kirk, W. B.: "Injection Well Stimulation with Micellar Solutions," *Jour. Pet. Tech.* (Dec., 1970).

Gogarty, W. B., Meabon, H. P. and Milton, H. W.: "Mobility Control Design for Miscible - Type Waterfloods," *Jour. Pet. Tech.* (Feb., 1970).

Gogarty, W. B. and Surkalo, H.: "A Field Test of Micellar Solution Flooding," *Jour. Pet. Tech.* (Sept., 1972).

Gogarty, W. B. and Tosch, W. C.: "Miscible - Type Waterflooding: Oil Recovery with Micellar Solutions," *Jour. Pet. Tech.* (Dec., 1968).

Goldburg, A.: "Micellar - Polymer Oil Recovery Demonstration in the Bell Creek Field, Montana," *ERDA Enhanced Oil, Gas Recovery and Improved Drilling Methods,* Vol. 1 - Oil, Tulsa, OK. (Sept., 1977).

Goldburg, A.: "Selection Methodology as Between Competing Micellar - Polymer Designs," SPE 6729 presented at the SPE-AIME 52nd Annual Fall Meeting, Denver, CO. (Oct., 1977).

Gupta, S. P. and Trushenski, S. P.: "Micellar Flooding - the Design of the Polymer Mobility Buffer Bank," *Soc. Pet. Eng. Jour.* (Feb., 1978).

Haberman, B.: "The Efficiency of Mobile Displacement as a Function of Mobility Ratio," *Trans.,* AIME (1960) 219.

Healy, R. N. and Reed, R. L.: "Immiscible Microemulsion Flooding," *Soc. Pet. Eng. Jour.* (April, 1977). *Trans.,* AIME (1977) 263.

Healy, R. N. and Reed, R. L.: "Physiochemical Aspects of Microemulsion Flooding," *Soc. Pet. Eng. Jour.* (Oct. *Trans.* AIME (1974) 257.

Healy, R. N., Reed, R. L. and Carpenter, C. W., Jr.: "A Laboratory Study of Microemulsion Flooding," *Soc. Pet. Eng. Jour.* (Feb., 1975).

Healy, R. N., Reed, R. L. and Stenmark, D. G.: "Multiphase Microemulsion Systems," *Soc. Pet. Eng. Jour.* (June, 1976).

Hesselink, F. T.: "Effect of Surfactant Phase Behavior and Interfacial Activity on the Recovery of Capillary - Trapped Residual Oil," Publication 498 presented at Symposium on Enhanced Oil Recovery by Displacement with Saline Solutions, London, England. (May, 1977).

Hill, H. J., Reisberg, J. and Stegemeier, G. L.: "Aqueous Surfactant Systems for Oil Recovery," *Jour. Pet. Tech.* (Feb., 1973). *Trans.,* AIME (1973) 255.

Hoar, T. P. and Schulman, J. H.: "Transparent Water-in-Oil Dispersions: The Oleopathic Hydro-Micelle," *Nature,* Vol. 152. (1943).

Holbrook, O. C.: "Surfactant - Water Secondary Recovery Process," U.S. Patent No. 3, 006,411. (1958).

Holm, L. W.: "Use of Soluble Oils for Oil Recovery," *Jour. Pet. Tech.* (Dec., 1971).

Holm, L. W. and Bernard, G. G.: "Secondary Recovery Waterflood Process," U.S. Patent No. 3,082,822. (1959).

Holm, L. W. and Csaszar, A. K.: "Oil Recovery by Solvents Mutually Soluble in Oil and Water," *Soc. Pet. Eng. Jour.* (June, 1962).

Holm, L. W. and Josendal, V. A.: "Reservoir Brines Influence Soluble - Oil Flooding Process," *O. & G. Jour.* (Nov. 13, 1972).

Howell, J. C., McAtee, R. W., Snyder, W. O. and Tonso, K. L.: "Large-Scale Field Application of Micellar Polymer Flooding," SPE 7089 presented at the SPE-AIME 4th Symposium on Improved Methods for Oil Recovery, Tulsa, OK. (April, 1978).

Johnson, J. S., Jr.: "Enhanced Oil Recovery by Micellar Floods: Chemicals from Wastes, Tracers and Ion-Exchange Aspects," *ERDA Enhanced Oil, Gas Recovery and Improved Drilling Methods,* Vol. 1 - Oil, Tulsa, OK. (Sept., 1977).

Jones, S. C.: "Laboratory Displacement with Micellar Solutions," SPE 1847 prepared for the SPE-AIME 42nd Annual Fall Meeting, Houston, TX. (Oct., 1967).

Jones, S. C.: "High Water Content Oil - External Micellar Dispersions," U.S. Patent No. 3,497,006. (1967).

Jones, S. C.: "Use of Water - External Micellar Dispersions in Oil Recovery," U.S. Patent No. 3,506,070. (1967).

Jones, S. C. and Dreher, K. D.: "Cosurfactants in Micellar Systems Used for Tertiary Oil Recovery," SPE 5566 presented at the SPE-AIME 50th Annual Fall Meeting, Dallas, TX. (Sept., 1975).

Jones, S. C. and McAtee, R. W.: "A Novel Single Well Field Test of a Micellar Solution Slug," *Jour. Pet. Tech.* (Nov., 1972).

Kehn, D. M., Pyndus, G. T. and Gaskell, M. H.: "Laboratory Evaluation of Prospective Enriched Gas Drive Projects," *Trans.,* AIME (1958) 213.

Kleinschmidt, R. F., Trantham, J. C., Boneau, D. F. and Patterson, H. L.: "North Burbank Tertiary Recovery Pilot Test - Two Year Status Report," *ERDA Enhanced Oil, Gas Recovery and Improved Drilling Methods,* Vol. 1 - Oil, Tulsa, OK. (Sept., 1977).

Knaggs, E. A., Nussbaum, M. L., Carlson, J. B. and Guenzani, R. C.: "Petroleum Sulfonate Utilization in Enhanced Oil Recovery Systems," SPE 6006 presented at the SPE-AIME 51st Annual Fall Meeting, New Orleans, LA. (Oct., 1976).

Knight, B. L.: "Commercial Scale Demonstration of the Maraflood Process: M-1 Project, Crawford County, Illinois," *ERDA Enhanced Oil, Gas Recovery and Improved Drilling Methods,* Vol. 1 - Oil, Tulsa, OK. (Sept., 1977).

Knight, R. K. and Baer, P. J.: "A Field Test of Soluble - Oil Flooding at Higgs Unit," *Jour. Pet. Tech.* (Jan., 1973).

Knobloch, T. S., Farouq Ali, S. M. and Trevino Diaz, M. J.: "The Role of Acid-Additive Mixtures on Asphaltene Precipitation," SPE 7627 prepared for the SPE-AIME Eastern Regional Meeting, Washington, D.C. (Nov., 1978).

Kossack, C. A. and Bilhartz, H. L., Jr.: "The Sensitivity of Micellar Flooding to Reservoir Heterogeneities," SPE 5808 presented at the SPE-AIME 4th Symposium on Improved Methods for Oil Recovery, Tulsa, OK. (March, 1976).

Kovitz, A. A. and Yannimaras, D.: "Measurement of Interfacial Tension by the Rod-In-Free Surface Meniscal Breakoff Method," *ERDA Enhanced Oil, Gas Recovery and Improved Drilling Methods,* Vol. 1 - Oil, Tulsa, OK. (Sept., 1977).

Kyle, C. R. and Perrine, R. L.: "Experimental Studies of Miscible Displacement Instability," *Soc. Pet. Eng. Jour.* (Sept., 1965).

Lake, L. W. and Helfferich, F. G.: "The Effect of Dispersion, Cation Exchange and Polymer/Surfactant Adsorption on Chemical Flood Environment," SPE 6769 prepared for the SPE-AIME 52nd Annual Fall Meeting, Denver, CO. (Oct., 1977).

Larson, R. G.: "The Influence of Phase Behavior on Surfactant Flooding," SPE 6774 presented at the SPE-AIME 52nd Annual Fall Meeting, Denver, CO. (Oct., 1977).

Larson, R. G. and Hirasaki, G. J.: "Analysis of Physical Mechanisms of Chemical Flooding," SPE 6003 presented at the SPE-AIME 51st Annual Fall Meeting, New Orleans, LA. (Oct., 1976).

Latimer, J. R., Jr.: "Perspective on Improved Oil Recovery," *Pet. Eng.* (Jan., 1965).

Lifebvre du Prey, E. J.: "Factors Affecting Liquid - Liquid Relative Permeabilities of a Consolidated Porous Medium," *Soc. Pet. Eng. Jour.* (Feb., 1973).

Leverett, M. C.: "Flow of Oil-Water Mixtures through Unconsolidated Sands," *Trans.,* AIME (1939) 132.

Lo, I. et al: "The Influence of Surfactant HLB and the Nature of the Oil Phase on the Phase Diagrams of Nonionic Surfactant-Oil-Water-Systems," *Journal of Colloid and Interfacial Science.* (April, 1977).

Melrose, J. C. and Brandner, C. F.: "Role of Capillary Forces in Determining Microscopic Displacement Efficiency for Oil Recovery by Waterflooding," *Jour. Cdn. Pet. Tech.* (1974).

Miller, R. J.: "The Eldorado Micellar-Polymer Project, El Dorado, Kansas," *ERDA Enhanced Oil, Gas Recovery and Improved Drilling Methods,* Vol. 1 - Oil, Tulsa, OK. (Sept., 1977).

Miller, R. J. and Richmond, C. N.: "El Dorado Micellar Polymer Project Facility," *Jour. Pet. Tech.* (Jan., 1978).

Mohan, V., Malviya, B. K. and Wasan, D. T.: "Interfacial Viscoelastic Properties of Adsorbed Surfactant and Polymerc Films at Fluid Interfaces," *Cdn. Jour. Chem. Eng.* (Dec., 1976).

Moore, T. F. and Slobod, R. L.: "The Effect of Viscosity and Capillarity on the Displacement of Oil by Water," *Producers Monthly,* Vol. 20, No. 10. (1956).

Mungan, N.: "Role of Wettability and Interfacial Tension in Water Flooding," *Soc. Pet. Eng. Jour.* (June, 1964).

Murphy, C. L., Thiede, D. M. and Eskew, J. O.: "A Successful Low Tension Waterflood Two-Well Test," SPE 6005 presented at the SPE-AIME 51st Annual Fall Meeting, New Orleans, LA. (Oct., 1976).

Nelson, R. C. and Pope, G. A.: "Phase Relationships in Chemical Flooding," SPE 6773 presented at the SPE-AIME 52nd Annual Fall Meeting, Denver, CO. (Oct., 1977).

Ojeda, E., Preston, F. W. and Calhoun, J. C., Jr.: "Correlations of Oil Residuals following Surfactant Floods," *Producers Monthly,* Vol. 18. (1953).

Parsons, R. W.: "Velocities in Developed Five-Spot Patterns," *Jour. Pet. Tech.* (May, 1974).

Parsons, R. W. and Jones, S. C.: "Linear Scaling in Slug Type Processes — Application to Micellar Flooding," *Soc. Pet. Eng. Jour.* (Feb., 1977).

Pease, R. W., Lohse, E. A. and Lane, R. N.: "Selection and Evaluation of West Virginia Oil Reservoirs as Candidates for Enhanced Oil Recovery Technology," SPE 7636 presented at the SPE-AIME Eastern Regional Meeting, Washington, D.C. (Nov., 1978).

Phillips Petroleum Company, *North Burbank Unit Tertiary Recovery Pilot Test,* BERC/TPR-76-2. (July, 1976).

Poettmann, F. H.: "Update on Microemulsion Flooding," *Pet. Eng.* (Nov., 1975).

Pope, G. A. and Nelson, R. C.: "A Chemical Flooding Compositional Simulator," SPE 6725 presented at the SPE-AIME 52nd Annual Fall Meeting, Denver, CO. (Oct., 1977).

Preston, F. W. and Calhoun, J. C., Jr.: "Application of Chromatography to Petroleum Production Research," *Producers Monthly,* Vol. 16, No. 5. (1952).

Puerto, M. C. and Gale, W. W.: "Estimation of Optimal Salinity and Solubilization Parameters for Alkylorthoxylene Sulfonate Mixtures," *Soc. Pet. Eng. Jour.* (June, 1977).

Pursley, S. A. and Graham, H. L.: "Borregos Field Surfactant Pilot Test," *Jour. Pet. Tech.* (June, 1975).

Pursley, S. A., Healy, R. N. and Sandvik, E. I.: "A Field Test of Surfactant Flooding," SPE 3805 prepared for the SPE-AIME Improved Oil Recovery Symposium, Tulsa, OK. (April, 1972).

Pusch, W. H.: "JPT Forum: Core Analysis Study for the El Dorado Micellar-Polymer Project," *Jour. Pet. Tech.* (Aug., 1977).

Reed, M. G.: "Formation Permeability Damage by Mica Alteration and Carbon Dissolution," *Jour. Pet. Tech.* (Sept., 1977).

Reed, R. L.: "Mechanism of Alcohol Displacement of Oil from Porous Media," *Soc. Pet. Eng. Jour.* (1961).

Reisberg, J.: "Secondary Recovery Method," U.S. Patent No. 3,330,344. (1964).

Reisberg, J.: "Surfactants for Oil Recovery by Waterfloods," U.S. Patent No. 3,348,611. (1965).

Reisberg, J. and Doscher, T. M.: "Interfacial Phenomena in Crude Oil-Water Systems," *Producers Monthly.* (Sept., 1956).

Reitzel, F. A. and Callow, G. O.: "Pool Description and Performance Analysis Leads to Understanding Golden Spike's Miscible Flood," *Jour. Pet. Tech.* (July, 1977).

Robbins, M. L.: "Theory for the Phase Behavior of Microemulsions," SPE 5839 presented at the SPE-AIME Improved Oil Recovery Symposium, Tulsa, OK. (March, 1976).

Salter, S. J.: "The Influence of Type and Amount of Alcohol on Surfactant-Oil-Brine Behavior and Properties," SPE 6843 presented at the SPE-AIME 52nd Annual Fall Meeting, Denver, CO. (Oct., 1977).

Sancevic, Z. A.: "Effect of Adverse Mobility Ratios and Graded Viscosity Zones on Viscous Fingering in Miscible Displacements," M.S. Thesis, Department of Petroleum and Natural Gas Engineering, The Pennsylvania State University. (Jan., 1961).

Sandvik, E. I., Gale, W. W. and Denekas, M. O.: "Characterization of Petroleum Sulfonates," *Soc. Pet. Eng. Jour.* (June, 1977).

Sayyouh, M. H., Farouq Ali, S. M. and Stahl, C. D.: "Effect of Frontal Advance Rate on Oil Recovery by Micellar-Polymer Displacement," *ERDA Enhanced Oil, Gas Recovery and Improved Drilling Methods,* Vol. 1 - Oil, Tulsa, OK. (Sept., 1977).

Sayyouh, M. H., Farouq Ali, S. M. and Stahl, C. D.: "Rate Effects in the Tertiary Micellar Flooding of the Bradford Crude Oil," SPE 7639 prepared for the SPE-AIME Eastern Regional Meeting, Washington, D.C. (Nov., 1978).

Schulman, J. H.: "Colloid Chemistry," *Annual Revue of Physical Chemistry*, Vol. 11. (1960).

Schulman, J. H. and Friend, J. A., "Light Scattering Investigation on the Structure of Transparent Oil-Water Disperse Systems, II," *Journal of Colloid Science*, Vol. 4. (1949).

Schulman, J. H. and McRoberts, T. S.: "On the Structure of Transparent Water and Oil Dispersions (Solubilized Oils)," *Trans. Fara. Soc.*, Vol. 42B. (1946).

Schulman, J. H. and Montagne, J. B.: "Formation of Microemulsions by Amino Alkyl Alcohols," *Ann. N.Y. Acad. of Sci.*, Vol. 92, No. 2. (1961).

Schulman, J. H. and Riley, D. P.: "X-Ray Investigation of the Structure of Transparent Oil-Water Disperse Systems, I," *Journal of Colloid Science*, Vol. 3. (1948).

Schulman, J. H., Stoeckenius, W. and Prince, L. M.: "Mechanism of Formation and Structure of Micro Emulsions by Electron Microscopy," *Journal of Physical Chemistry*, Vol. 63. (1959).

Sharp, J. M. "GURC Report No. 140-S, Final Report on a Survey of Field Tests of Enhanced Recovery Methods for Crude Oil," presented to the National Science Foundation and the Federal Energy Admin. (Dec. 27, 1974).

Sheffield, M.: "Three Phase Fluid Flow Including Gravitational, Viscous and Capillary Forces," SPE 2012 prepared for the Symposium on Numerical Simulation, Dallas, TX. (April, 1968).

Slattery, J. C.: "Effect of Surface Viscosities in the Recovery of Residual Oil by Displacement," *ERDA Enhanced Oil, Gas Recovery and Improved Drilling Methods*, Vol. 1 - Oil, Tulsa, OK. (Sept., 1977).

Slobod, R. L. and Howlett, W. E.: "The Effects of Gravity Segregation in Laboratory Studies of Miscible Displacement in Vertical Unconsolidated Porous Media," *Soc. Pet. Eng. Jour.* (March, 1964).

Smith, G. D., Donelan, C. E. and Borden, R. E.: "Oil-Continuous Microemulsions Composed of Hexane, Water and Z-Propanol," *Journal of Colloid and Interfacial Science*. (July, 1977).

Snyder, L. J.: "Two-Phase Reservoir Flow Calculations," SPE 2014 prepared for the Symposium on Numerical Simulation of Reservoir Engineering, Dallas, TX. (April, 1968).

Stahl, C. D.: "Miscible Phase Displacement — A Survey," Parts 1, 2, 3, 4 and 5. *Producers Monthly*, Vol. 29, Nos. 1, 2, 3, 4 and 5. (1965).

Stegemeier, G. L.: "Mechanisms of Entrapment and Mobilization of Oil in Porous Media," presented at the 81st National Meeting of the American Institute of Chemical Engineers, Kansas City, KS. (April, 1976).

Strange, L. K. and Talash, A. W.: "Analysis of Salem Low-Tension Waterflood Test," *Jour. Pet. Tech.* (Nov., 1977).

Szabo, M. T.: "Micellar Shear Degradation, Formation Plugging and Inaccessible Pore Volume," SPE 6772 presented at the SPE-AIME 52nd Annual Fall Meeting, Denver, CO. (Oct., 1977).

Taber, J. J.: "Dynamic and Static Forces Required to Remove a Discontinuous Oil Phase from Porous Media Containing Both Oil and Water," *Soc. Pet. Eng. Jour.* (March, 1969).

Taber, J. J.: "The Injection of Detergent Slugs in Water Floods," *Trans.*, AIME (1958) 213.

Taber, J. J., Kirby, J. C. and Schroeder, F. U.: "Studies on the Displacement of Residual Oil: Viscosity and Permeability Effects," Paper 47b presented at the AIChE Symposium on Transport Phenomena in Porous Media, 71st National Meeting, Dallas, TX. (Feb., 1972).

Talash, A. W.: "Experimental and Calculated Relative Permeability Data for Systems Containing Tension Additives," SPE 5810 presented at the SPE-AIME Improved Oil Recovery Symposium, Tulsa, OK. (March, 1976).

Thomas, C. P., Winter, W. K. and Fleming, P. D.: "Application of a General Multiphase, Multicomponent Chemical Flood Model to Ternary, Two-Phase Surfactant Systems," SPE 6727 prepared for the SPE-AIME 52nd Annual Fall Meeting, Denver, CO. (Oct., 1977).

Thomas, R. D., Walker, C. J. and Burtch, F. W.: "ERDA's Micellar-Polymer Flood Project in Northeast Oklahoma," *ERDA Enhanced Oil, Gas Recovery and Improved Drilling Methods*, Vol. 1 - Oil. (Sept., 1977).

Tornberg, E.: "A Surface Tension Apparatus According to the Drop Volume Principal," *Journal of Colloid and Interfacial Science*. (June, 1977).

Tosch, W. C.: "Technology of Micellar Solutions," SPE 1847-B prepared for the SPE-AIME 42nd Annual Fall Meeting, Houston, TX. (Oct., 1967).

Tosch, W. C. and Burdge, D. N.: "Viscosity Control in Miscible Floods," U.S. Patent No. 3,330,343 (1967).

Trantham, J. C. and Clampitt, R. L.: "Determination of Oil Saturation after Waterflooding in an Oil-Wet Reservoir — The North Burbank Unit, Tract 97 Project," *Jour. Pet. Tech.* (May, 1977).

Trantham, J. C., Patterson, H. L. and Boneau, D. F.: "The North Burbank Unit Tract 97 Surfactant/Polymer Pilot-Operation and Control," SPE 6746 presented at the SPE-AIME 52nd Annual Fall Meeting, Denver, CO. (Oct., 1977).

Trogus, F. J., Schechter, R. S., Pope, G. A. and Wade, W. H.: "Adsorption of Mixed Surfactant Systems," SPE 6845 prepared for the 52nd Annual Fall Meeting, Denver, CO. (Oct., 1977).

Trushenski, S. P., Dauben, D. L. and Parrish, D. R.: "Micellar Flooding — Fluid Propagation, Interaction and Mobility," *Soc. Pet. Eng. Jour.* (Dec., 1974). *Trans.,* AIME (1974) 257.

Uren, L. C. and Fahmy, E. H., "Factors Influencing the Recovery of Petroleum from Unconsolidated Sands by Waterflooding," *Trans.,* AIME (1927) 77.

Wade, J. E.: "Micellar-Polymer Joint Demonstration Project, Wilmington Field, California," *ERDA Enhanced Oil, Gas Recovery and Improved Drilling Methods,* Vol. 1 - Oil, Tulsa, OK. (Sept., 1977).

Wade, W. H. et al: "Interfacial Tension and Phase Behavior of Surfactant Systems," SPE 6844 prepared for the SPE-AIME 52nd Annual Fall Meeting, Denver, CO. (Oct., 1977).

Wagner, O. R. and Leach, R. O.: "Effect of Interfacial Tension on Displacement Efficiency," *Soc. Pet. Eng. Jour.* (Dec., 1966).

Wasan, D. T. et al: "The Mechanism of Oil Bank Formation and Coalescence in Porous Media," *ERDA Enhanced Oil, Gas Recovery and Improved Drilling Methods,* Vol. 1 - Oil, Tulsa, OK. (Sept., 1977).

Wasan, D. T. et al: "Observations on the Coalescence Behavior of Oil Droplets and Emulsion Stability in Enhanced Oil Recovery," SPE 6846 prepared for the SPE-AIME 52nd Annual Fall Meeting, Denver, CO. (Oct., 1977).

Weinaug, C. F. and Ling, D.: "Production of Hydrocarbon Material," U.S. Patent No. 2, 876, 277. (1959).

Whiteley, R. C. and Ware, J. W.: "Low Tension Waterflood Pilot at the Salem Unit, Marion County, Illinois — Part 1: Field Implementation and Results," *Jour. Pet. Tech.* (Aug., 1977).

Whorton, L. P. and Kieschnick, W. F., Jr.: "Oil Recovery by High Pressure Gas Injection," *O. & G. Jour.* (April, 6, 1950).

Widmyer, R. H., Satter, A., Frazier, G. D. and Graves, R. H.: "Low-Tension Waterflood Pilot at the Salem Unit, Marion County, Illinois — Part 2: Performance Evaluation," *Jour. Pet. Tech.* (Aug., 1977).

Wilson, J. F.: "Miscible Displacement — Flow Behavior and Phase Relationships for a Partially Depleted Reservoir," *Trans.,* AIME (1960) 219.

Wilson, P. M., Murphy, C. L. and Foster, W. R.: "The Effects of Sulfonate Molecular Weight and Salt Concentration on the Interfacial Tension of Oil-Brine-Surfactant Systems," SPE 5812 presented at the SPE-AIME 4th Symposium on Improved Methods for Oil Recovery, Tulsa, OK. (March, 1976).

Winsor, P. A.: "Solvent Properties of Amphiphillic Compounds," Butterworths, London, England. (1954).

Zetik, D. F.: "Reservoir Simulation for the El Dorado Micellar-Polymer Project," SPE 5807 presented at the SPE-AIME 4th Symposium on Improved Methods for Oil Recovery, Tulsa, OK. (March, 1976).

Chapter 5
POLYMER FLOODING

INTRODUCTION

Polymer flooding can yield significant increases in percentage recovery when compared to conventional waterflood projects in certain reservoirs. The mechanisms by which polymers accomplish this enhanced oil recovery are complex and not fully understood. However, several principles have been clearly demonstrated as primary factors in the determination of polymer effectiveness compared to waterflooding. Two of these are reservoir heterogeneity and mobility ratio of reservoir fluids.

TECHNICAL DISCUSSION

Reservoir Heterogeneity

Although oil reservoirs are characterized as porous media with certain porosities and permeabilities, they are almost never homogeneous beds with constant properties. Generally, there are numerous strata with wide-ranging properties. In terms of enhanced oil recovery, the divergence of reservoir permeability is a significant factor. There may be numerous fractures also. Together, the permeability variation and fractures can have a profound effect on the flow of fluids in a reservoir and thereby influence oil recovery.

When water or other fluids are injected under pressure, they will seek the path of least resistance to the point of lowest pressure, which is generally the producing well. Because the high permeability zones and the fractures offer the least resistance to flow, most of the injected fluid follows this path. In doing so, most of the oil remaining in the lower permeability zones is by-passed. The oil which is displaced from the high permeability zones and produced is replaced with injected fluid, lowering the residual oil saturation in these regions. As the oil saturation decreases, the permeability to water increases, further exaggerating the inequality in relative flow rates between the high and low permeability zones. The result is ever-increasing water-to-oil ratios in the producing wells and low ultimate recovery of oil-in-place (Garland, 1966).

Variations of permeability in the vertical plane cause the injected fluid to advance from the injection point as an irregular front. A measure of this variation is the vertical sweep efficiency or invasion efficiency. It is defined as the cross-sectional area contacted by the injected fluid divided by the cross-sectional area enclosed in all layers behind the injected fluid front.

Similarly, the variations in the horizontal plane cause injected fluid to advance at an uneven rate. Thus injected fluid will reach a producing well via the least resistance path before some of the reservoir has been contacted. The areal fraction which has been contacted at the time of breakthrough of injected fluid is defined as the areal sweep efficiency.

The volumetric sweep efficiency is a measure of the three-dimensional effect of reservoir heterogeneities. It is the product of the pattern areal sweep and vertical sweep. Stated another way, the volumetric sweep efficiency is the pore volume of the reservoir contacted by the injected fluid divided by the total pore volume (Townsend et al, 1977).

It should be noted that the vertical sweep efficiency is a function of reservoir characteristics alone, while areal sweep efficiency is a function of reservoir characteristics and well locations. The geometric pattern for injection and producing wells affects the areal sweep efficiency. Improper placement of wells can lower this efficiency even in the absence of detrimental reservoir heterogeneities.

Polymers can reduce the detrimental effect of permeability variations and fractures and thereby improve both the vertical and areal sweep efficiency.

Mobility Ratios

Even in the absence of reservoir heterogeniety, sweep efficiency may be low because of an unfavorable mobility ratio. The mobility of a fluid in a reservoir is defined as the permeability of the media to that fluid divided by the viscosity of that fluid. The mobility ratio of water to oil is the mobility of water in the reservoir divided by the mobility of oil in the reservoir.

$$M = (k_w/\mu_w)/(k_o/\mu_o) = \lambda_w/\lambda_o \qquad (1)$$

where:
- M = mobility ratio of water to oil
- k_w = relative permeability to water
- μ_w = viscosity of water
- k_o = relative permeability to oil
- μ_o = viscosity of oil
- λ_w = mobility of water
- λ_o = mobility of oil

The permeabilities are not constant. They depend on relative fluid saturations in the reservoir and thus change as oil is displaced in the reservoir. Because the permeability of water increases as the oil saturation decreases, the mobility ratio will increase as oil is produced. A common practice is to define the mobility ratio using the effective water permeability at residual oil saturation and the effective oil permeability at interstitial water saturation.

If the mobility ratio is one or less, the displacement of oil by water will be efficient. In effect, the displacement occurs in a piston-like fashion (Mungan et al, 1966). On the other hand, if the mobility ratio is greater than one, the more mobile water will finger through the oil leaving it in place in the reservoir.

Polymers can improve the mobility ratio and thus increase the displacement efficiency in a reservoir. The mechanisms by which polymer solutions improve mobility ratio as well as reservoir heterogeneity are discussed later.

Polymer Chemistry

There are countless chemicals which can be dissolved in water to decrease water mobility in a reservoir. However, almost all of them must be used in relatively high concentration and/or are costly. Glycols, glycerin, and sugar are examples of these materials (Dauben and Menzie, 1967). When constrained by economics, there are only two basic types of polymers which are presently used: polysaccharides and polyacrylamides.

Polysaccharide

The polysaccharide or "biopolymer" typically used in enhanced oil recovery processes is xanthan gum. The chemical structure for the polymer is illustrated in Figure 1 (Jennings, 1977). This material has a molecular weight of approximately 5 million. The molecular structure of xanthan gum gives a degree of rigidity to the polymer chain which provides excellent resistance to mechanical breakage. However, it is highly susceptible to bacterial action. In fact, microbes are responsible for the formation as well as the destruction of the polymer molecule.

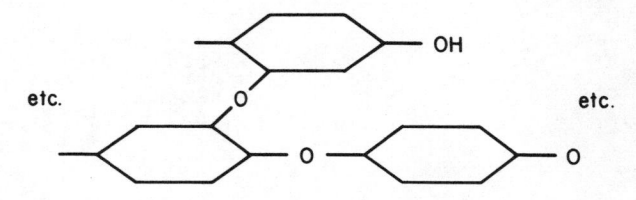

Fig. 1. Polymer chemical formula. (after Jennings, 1977)

Xanthan gum is produced by the microbial action of xanthomonas campestris on a substrata of carbohydrate media, protein supplement and an inorganic source of nitrogen. The biopolymer is an extracellular slime which forms on the surface of the microbe cell. This fermentation broth is pasteurized to kill the xanthomonas campestris. Finally the polymer is precipitated from the broth by a suitable alcohol (Lipton, 1974).

After the biopolymer has been dissolved for use as a mobility control agent, extreme care must be taken to preserve the polymer from bacterial attack. Proper use of biocides and oxygen scavengers is essential. Most of the aerobic bacteria which attack xanthan are of the pseudomonad type. These microbes are particularly troublesome because in addition to degrading the polymer, they produce cells which have a typical diameter of 1 micron and length of up to 4 microns. These cells are significantly larger than the polymer and may aggravate formation plugging of the reservoir at the injection well.

Polyacrylamide

The polyacrylamide molecule is made up of a very long chain of acrylamide monomer molecules. The basic acrylamide unit has the following formula:

$$\begin{array}{c} CH - CH_2 \\ | \\ C = O \\ | \\ NH_2 \end{array}$$

When chemically combined to form the polymer chain the structure is as follows (Jennings, 1977):

$$\begin{array}{c} CH - CH_2 - CH - CH_2 - CH - CH_2 \\ |\qquad\qquad |\qquad\qquad | \\ C = O \quad\; C = O \quad\; C = O \\ |\qquad\qquad |\qquad\qquad | \\ NH_2 \quad\;\; NH_2 \quad\;\; NH_2 \end{array}$$

Because of competing mechanisms in the polymer formation, there is a wide range of chain lengths. The average molecular weight of commercial polyacrylamides ranges from approximately 1-10 million. The typical molecular weight distribution follows the curve shown in Figure 2 (Willhite et al, 1977).

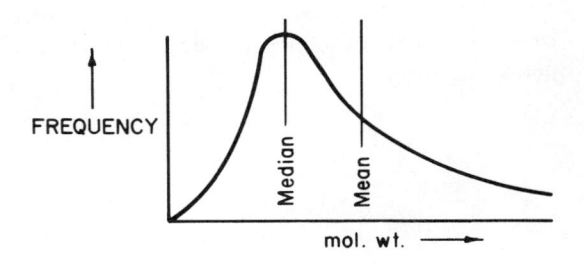

Fig. 2. Typical molecular weight distribution.
(after Willhite et al, 1977)

Unlike biopolymers, the polyacrylamide molecule is very flexible. It is very long with a relatively small diameter which makes the polymer susceptible to mechanical breakage or shear degradation. On the other hand, it is relatively immune to bacterial attack.

The polymer is generally modified chemically by replacing some of the amide groups with a carboxyl group as shown below:

$$
\begin{array}{c}
| \\
C = 0 \\
| \\
NH_2 \\
\text{amide}
\end{array}
\longrightarrow
\begin{array}{c}
| \\
C = 0 \\
| \\
0 - Na^+ \\
\text{carboxyl}
\end{array}
$$

This process, which is called hydrolysis, is accomplished by treating a polyacrylamide solution with a strong base, e.g. sodium hydroxide. The percentage of amide groups which have been changed to carboxyl groups is denoted as the per cent hydrolysis (White et al, 1972). Typically, the hydrolysis ranges from 0 - 30%.

Both hydrolyzed and unhydrolyzed polyacrylamide are highly polar due to the amide and carboxyl groups. This gives the polymer a great affinity for water but not oil.

Displacement Mechanisms

The importance of reducing the water to oil mobility ratio has been recognized for many years. Such a reduction has been the goal of several other secondary recovery processes, including heat, gas resaturation and miscible flooding. Most of these processes rely on reducing the effective viscosity of the oil to achieve the lower mobility ratio. The use of polymer, on the other hand, achieves the lower mobility ratio by increasing the effective viscosity of water, the driving fluid (Pye, 1964). Although the increased effective viscosity is the most obvious way that polymer solutions achieve a lower mobility ratio, it is not the most important way. Several mechanisms will be discussed; some have been clearly proven while others are still somewhat speculative.

Rheology

Polymers are said to increase the viscosity of water, but this is not strictly correct in a technical sense. Polymer solutions are non-Newtonian fluids for all concentration ranges of commercial interest. They range from approximately 50 - 2000 ppm. They are classified as non-Newtonian fluids because their flow behavior is too complex to be characterized by the single parameter, viscosity.

When a Newtonian fluid is subjected to a shearing force, it deforms or flows. There is a resistance to this flow which is defined as the ratio of the shearing force (shear stress) to the rate of flow (shear rate). For a Newtonian fluid, this ratio is constant and is called viscosity. The equation follows:

$$\mu = \frac{\tau}{\gamma} \qquad (2)$$

where:
μ = viscosity
τ = shear stress
γ = shear rate

Non-Newtonian fluids cannot be characterized by a viscosity because the ratio of shear stress to shear rate is not a constant. The flow behavior of these non-Newtonian fluids may follow one of several complex flow models. Polymer solutions are generally classified as pseudoplastic fluids under most conditions. A pseudoplastic material is one which exhibits a smaller resistance to flow as the shearing rate increases. Mathematically, the formula is known as the Power Law model:

$$\tau = K\gamma^n; \; n<1.0 \text{ for pseudoplastic fluids} \qquad (3)$$

where K and n are the two parameters used to define the flow behavior of the fluid. Note that if n = 1.0, the equation reduces to the Newtonian case with K equivalent to μ.

If the ratio of shear stress to shear rate is considered as an "apparent viscosity" for a pseudoplastic fluid, inspection of the formula above shows that the apparent viscosity decreases as the shear rate increases. Stated another way, a pseudoplastic fluid exhibits a larger apparent viscosity when flowing at low velocities and a lower apparent viscosity when flowing at high velocities.

At extremely high shear rates ($>\sim 10^5$ sec^{-1}) the flow curve for a polymer solution approaches linearity and is called viscosity at infinite shear. Similarly, the slope at very low shear rates ($<\sim 10^{-2}$ sec^{-1}) approaches linearity and is called viscosity at zero shear rate (Skelland, 1967). Fortunately, these deviations from the mathematical formula occur at shear rates which are of little practical interest in most cases.

Despite the complex flow behavior of polymer solutions, the apparent viscosities of these fluids are significantly higher than the viscosity of water, even at high shear rates. Figure 3 illustrates a typical case.

In addition to pseudoplastic behavior, there are several other factors which influence the apparent viscosities of polymer solutions.

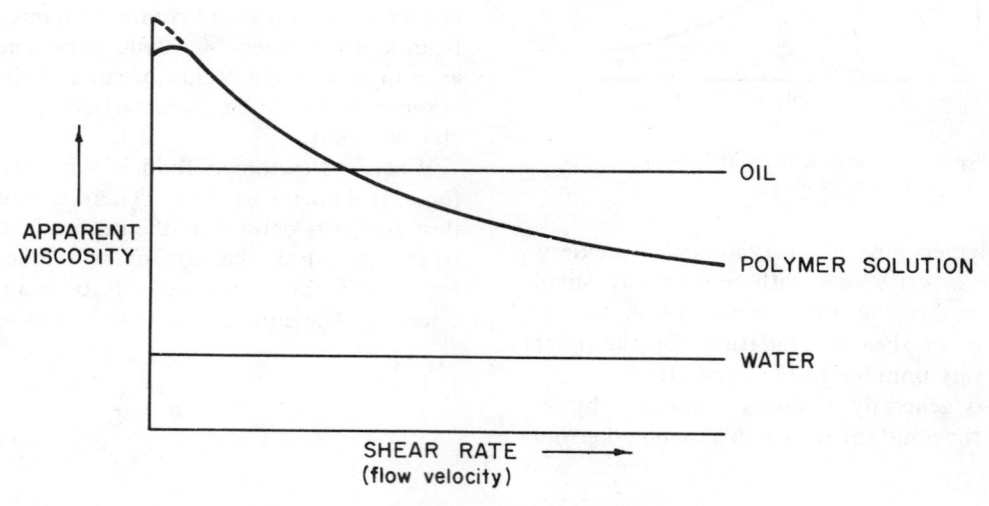

Fig. 3. Apparent viscosity vs. shear rate (flow velocity).
(after Skelland, 1967)

Solvent

The polymer molecule can be visualized as a fibrous aggregate. In a good solvent the polymer molecule extends fully to maximize its contact with the solvent. This gives the polymer a flexible, gel-like appearance. With the polymer molecule extended, polymer-polymer entanglements are maximized. All of this tends to increase the apparent viscosity of the polymer (Mungan et al, 1966).

In a poorer solvent, the polymer minimizes its contact with the solvent (Ford and Kelldorf, 1975). Electron micrographs show reduced entanglements and a more rigid polymer structure (Herr and Routson, 1974).

Distilled water is a good solvent for most polymers. However, as salt is added, the electrolyte neutralizes the charge of the polymer molecule. With the net charge diminished, the force which helped to extend the polymer molecule decreases (Mungan et al, 1966). Thus, as salt concentrations are increased, the polymer molecule contracts, reducing the solution viscosity as shown on Figure 4.

Because of the ionic interaction between solvent and polymer, it is not surprising to find the sensitivity of solution viscosity to salt concentration is greater for ionic polymers, as shown on Figure 5.

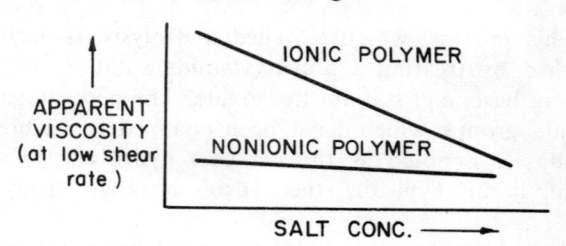

Fig. 5. Apparent viscosity (at low shear rate) vs. salt concentration.
(after Mungan et al, 1966)

The pseudoplastic nature of polymer solutions adds still another variable to solvent effect. At very low shear rates, the polymer has a better opportunity to extend itself in a good solvent, increasing the apparent viscosity. At high shear rates, the opportunity for polymer entanglements is reduced so the detrimental effect of salt is less pronounced. (See Figure 6.)

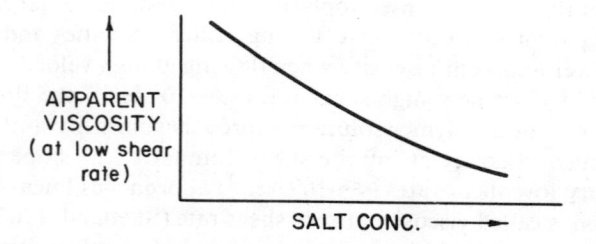

Fig. 4. Apparent viscosity (at low shear rate) vs. salt concentration.
(after Mungan et al, 1966)

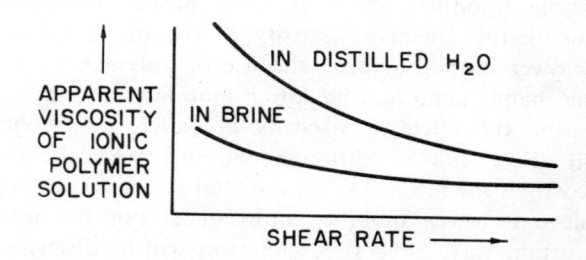

Fig. 6. Apparent viscosity of ionic polymer solution vs. shear rate.
(after Mungan et al, 1966)

Molecular Weight

If polymer molecules were compact, non-interacting spheres, the molecular weight would have little effect on solution viscosity. However, the fibrous nature of the polymer with its ability to extend and entangle makes molecular weight an important factor (Mungan et al, 1966). The high molecular weight polymers exhibit a greater apparent viscosity than their low molecular weight counterparts under similar conditions.

Hydrolysis

The extent of hydrolysis affects polymer rheology and thus performance in the reservoir. Martin and Sherwood (1975) have completed studies with polyacrylamides and acrylamide terpolymers in which the degree of hydrolysis ranged from 0-35%. They discovered that the apparent viscosity of a partially hydrolyzed polyacrylamide was much greater than the apparent viscosity of a corresponding unhydrolyzed polyacrylamide. The apparent viscosity increased with the degree of hydrolysis although this difference was quite small compared to the effect of the initial hydrolysis. (See Figure 7.)

As salt was added, the magnitude of the hydrolysis effect was diminished. The divalent $CaCl_2$ had a much greater effect than monovalent NaCl. In fact, at $CaCl_2$ concentrations greater than 0.1 wt% the hydrolyzed polymer displayed a lower apparent viscosity than the unhydrolyzed polymer. This phenomenon was never observed with NaCl even at concentrations as high as 10%. (See Figures 8 and 9.)

These observations are consistent with the ionic effect of solvents described in a preceding section. In distilled water, the charges on the hydrolyzed molecule maximize the "uncoiling" of the molecular chain, increasing the viscosity. As the salt concentration of the solvent increases, this uncoiling by repulsion is diminished.

Martin and Sherwood (1975) also related their observations regarding polymer retention and mobility reduction in porous media. They found that polymer retention decreased as the degree of hydrolysis increased. In spite of this diminished retention of hydrolyzed polymer, the mobility for the hydrolyzed polymer was less than the mobility for unhydrolyzed polymer.

The explanation offered for this behavior involved relative molecular size. The hydrolyzed polymer has a larger effective diameter which leads to a higher resistance factor or reduced mobility. At the same time, the larger diameter of the hydrolyzed polymer decreases the number of potential adsorption sites and increases the inaccessible pore volume. These mechanisms are discussed in subsequent sections.

Polymer Concentration

An increase in polymer concentration will increase the apparent viscosity of the solution. This is clearly a mass effect as more polymer molecules are dissolved. However, the increase in apparent viscosity may not be proportional to the increase in concentration at low shear rates. As polymer concentration increases, the opportunity for intermolecular entanglement rises sharply. This entanglement translates to increased shear stress and more pronounced pseudoplastic behavior. Conversely, at very low concentrations of polymer (<50 ppm) the opportunity for entanglement is greatly reduced. Not only is the apparent viscosity reduced, but the solution approaches Newtonian flow behavior (Mungan et al, 1966). (See Figure 10.)

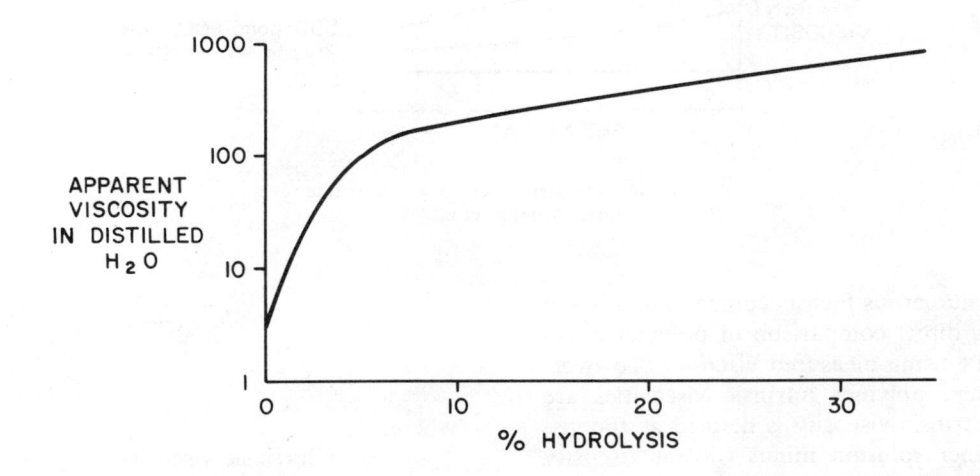

Fig. 7. Apparent viscosity in distilled water vs. hydrolysis.
(after Martin and Sherwood, 1975)

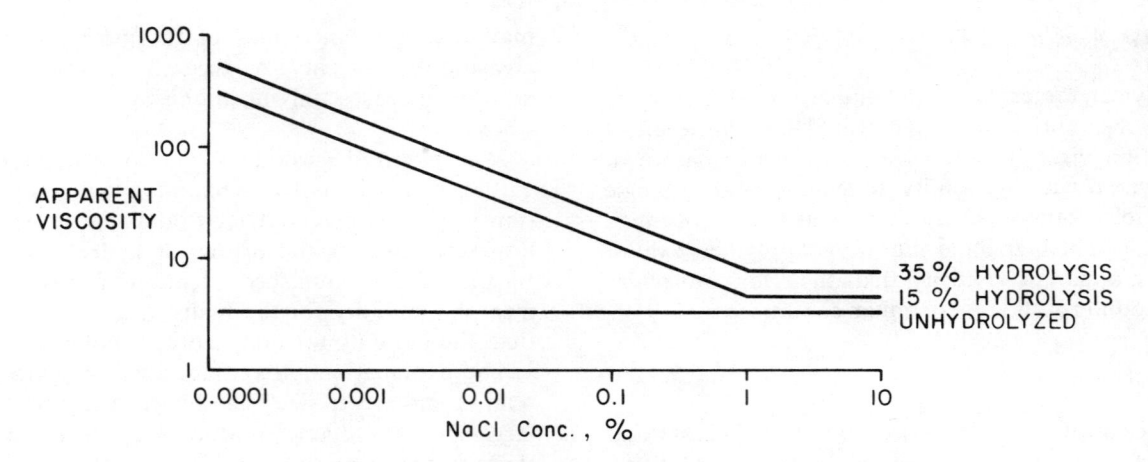

Fig. 8. Apparent viscosity vs. NaCl concentration.
(after Martin and Sherwood, 1975)

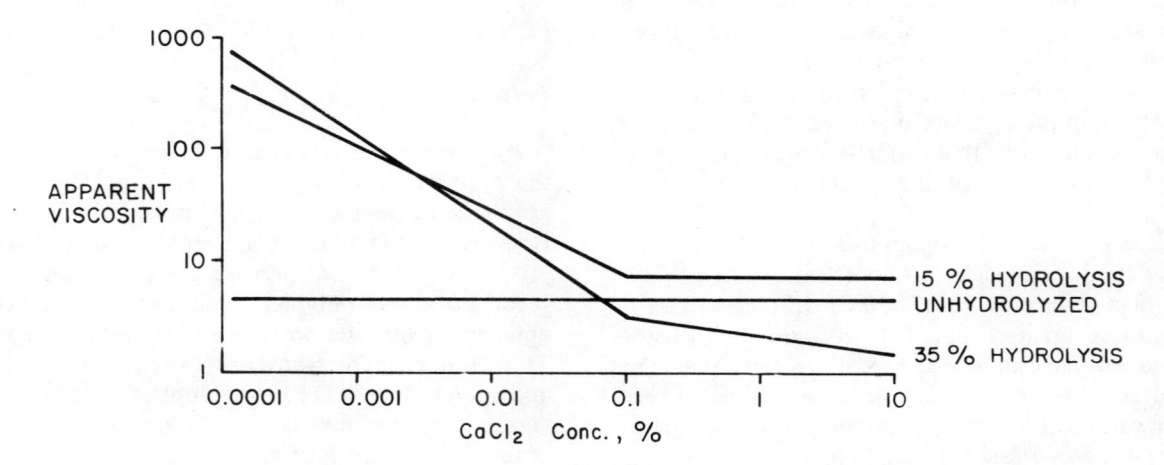

Fig. 9. Apparent viscosity vs. CaCl$_2$ concentration.
(after Martin and Sherwood, 1975)

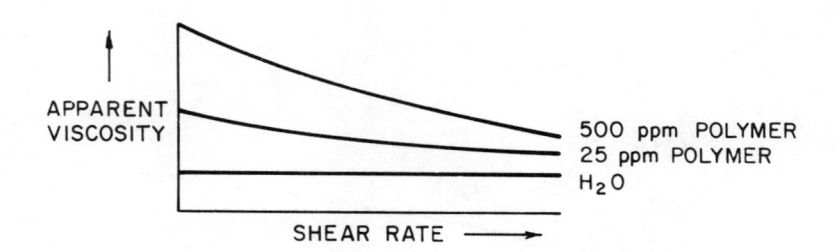

Fig. 10. Apparent viscosity vs. shear rate.
(after Mungan et al, 1966)

Because of the numerous factors controlling polymer solution viscosity, direct comparison of polymer effectiveness is difficult using measured viscosity. To overcome this problem, polymer intrinsic viscosities are compared. The intrinsic viscosity is defined as the viscosity of a polymer solution minus solvent viscosity divided by the product of solvent viscosity and polymer concentration as concentration approaches zero (Ford and Kelldorf, 1975).

$$\eta = \lim_{c \to 0} \frac{\mu - \mu_s}{c \, \mu_s} \tag{4}$$

where:

η = intrinsic viscosity
μ = polymer solution viscosity
μ_s = solvent viscosity
c = polymer concentration

Intrinsic viscosity also has importance in determining the molecular size of a polymer. The relationship is described in a later section.

To this point, the discussion of polymer fluid properties has been general in nature. The effects of shear rate, concentration, molecular weight and solvent which have been noted can be observed in either rotational or capillary viscometers. However, the behavior of polymers in porous media is more complex than the capillary flow considered until now, especially if the pore size in the media approaches the same order of magnitude as the size of the polymer molecule.

Reduced Permeability

The Darcy equation is used to describe the flow of Newtonian fluids in a porous media.

$$q = \frac{kA\phi \Delta P}{\mu L} = \frac{\lambda A \phi \Delta P}{L} \qquad (5)$$

where:

q = volumetric flow rate
k = absolute permeability
A = cross sectional area
ϕ = porosity
μ = viscosity
ΔP = pressure drop
L = length
λ = fluid mobility

For polymer solutions and other non-Newtonian fluids, this equation must be modified because viscosity, μ, is not a constant (Dauben and Menzie, 1967). Rather it is a function of the flow parameter, q. Nevertheless, for a given set of flow conditions an apparent viscosity can be calculated using the Power Law model and applied to the Darcy equation. Surprisingly, the behavior of polymers in porous media would suggest a much higher viscosity than would be predicted by viscosity measurements in capillary flow (Pye, 1964). Stated another way, the mobility of the polymer is less than would have been predicted. (See Figure 11.)

Inspection of the Darcy equation suggests that this unexpected decrease in mobility must be due to a significant decrease in the relative permeability of the porous media to polymer (Mungan et al, 1966). The reasons offered for this permeability reduction are numerous. The most commonly accepted ones will be discussed later. But first, the significance of this phenomenon should be mentioned. The combined mobility reduction resulting from an increased apparent viscosity and a decreased permeability is quite dramatic, even for very low concentrations of polymer. The measure of the mobility reduction is known as the "resistance factor". Mathematically, the relationship is:

$$R = \lambda_w / \lambda_p = \left(\frac{k_w}{\mu_w}\right) / \left(\frac{k_p}{\mu_p}\right) \qquad (6)$$

The large values of R which can be obtained with low concentrations give polymers a potential economic advantage over other methods of obtaining mobility reduction in many reservoirs (Pye, 1964).

Polymer Retention

The most clearly demonstrated reason for reduced permeability in porous media is polymer retention. Numerous experimenters have shown that polymer solutions which are forced through core samples or simulated reservoir media suffer significant reduction in polymer concentration (Mungan et al, 1966; Mungan, 1969; Smith, 1970; Sparlin, 1975; Szabo, 1975; Willhite et al, 1977).

Adsorption

Static equilibrium adsorption tests have been performed with numerous polymers, particularly polyacrylamides. The polymers have been shown to adsorb on the surface of most reservoir minerals. The typical experiment used to determine the static adsorption provides gentle agitation of a polymer solution in contact with the adsorption mineral. The reduced polymer concentration at the conclusion of the agitation is used as a measure of adsorption. Adsorption is found to increase with polymer concentration, at least up to 500 ppm polymer, as shown in Figure 12 (Gogarty, 1966).

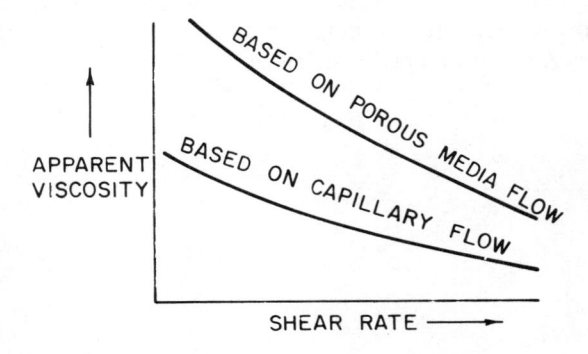

Fig. 11. Apparent viscosity vs. shear rate.
(after Pye, 1964)

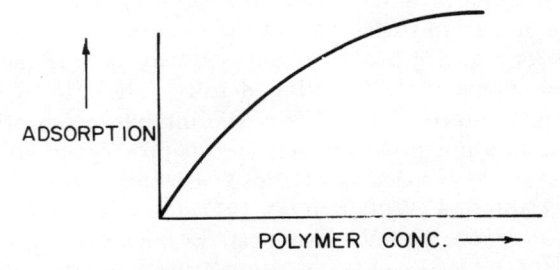

Fig. 12. Adsorption vs. polymer concentration.
(after Gogarty, 1966)

The adsorption varies among minerals used in the tests. Calcium carbonate appears to have a much greater affinity for polymer than does silica (Smith, 1970).

Solvent salinity also plays an important role, although there is some disagreement among experimenters. Mungan (1969) reports that adsorption is reduced with brine solutions of polymers. However, Smith (1970) and Szabo (1975) both indicate that polymer adsorption increases with salt concentration. Even here agreement is not perfect, because Smith indicates a sensitivity of adsorption to salt concentrations as high as 10%. On the other hand, Szabo (1975) shows no increase in adsorption for brine concentrations above 2%.

Care must be exercised in comparing data among the experimenters because test conditions varied from case to case, and other uncontrolled parameters are probably responsible for the apparent experimental discrepancies. Still, the consensus of opinion is that adsorption increases with salt concentration. (See Figure 13.)

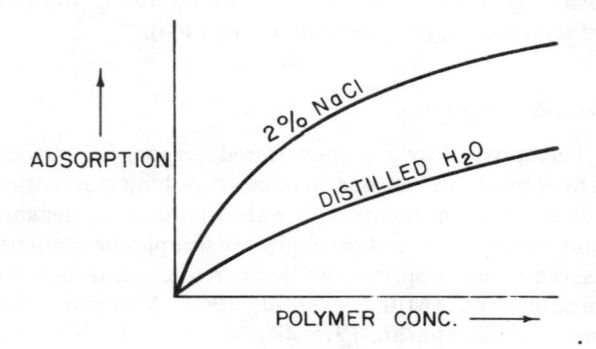

Fig. 13. Adsorption vs. polymer concentration in distilled water and brine.
(after Smith, 1970 and Szabo, 1975)

Dynamic adsorption studies have been completed by observing polymer retention during continuous flow through porous media. Again, the results of several experimenters are difficult to compare because of variations in experimental procedures (Mungan et al, 1966; Mungan, 1969; Smith, 1970; Szabo, 1975). Both Mungan (1969) and Szabo (1975) concluded that polymer retention is reduced significantly in the dynamic studies. However, the differences do not reflect a change in the adsorption mechanism, but result from restriction to polymer flow in the porous media.

The experimental data suggest that the polymer molecules adsorb to the surface of the rock in a monolayer. The ionic charges of the polymer may be one factor which limits the adsorption density. The size of the polymer molecule also limits the ultimate adsorption density although studies indicate the polymer molecule occupies less space when adsorbed than when in a dilute solution (Willhite et al, 1977).

Numerous models have been proposed to predict adsorption characteristics. Among these are the Langmuir isotherm and the capillary tube model.

Still another theory claims that the polymer coats the rock surfaces with a hydrophilic film. As water passes over the polymer the film swells, reducing the effective permeability. In the presence of oil the swelling does not occur. Thus the polymer should reduce mobility to a greater extent in zones with low oil saturation, thereby improving potential oil recovery (Sparlin, 1975).

Inaccessible Pore Volume

A primary cause for reduced adsorption in porous media is the existence of inaccessible pore volume. Whether the media are consolidated or not, there are many pores which have very small openings. These openings allow flow of brine but restrict the passage of the polymer molecule which is considerably larger. These inaccessible volumes do not exist with the static adsorption experimental procedures described earlier.

The existence of inaccessible pore volumes has been clearly demonstrated by Dawson and Lantz (1972) and others. If brine is injected into porous media, the time required for passage can be calculated using the following formula:

$$t_w = \frac{L}{V_w} = \frac{A\phi_w L}{q} \qquad (7)$$

where:
- L = length
- V_w = brine velocity
- ϕ_w = brine effective porosity
- A = area
- q = volumetric flow rate
- t_w = brine passage time

The concentration front arrives at the far end of the porous media at the time required to inject exactly one pore volume of brine.

If this simple test is repeated using polymer solution, adsorption effects are eliminated by preflushing the media with polymer at a concentration at least as high as used for detecting the concentration front. In this case the time required for passage is less than required for brine. Despite equal rates of injection, the concentration front for polymer arrives earlier than the concentration front for brine.

$$t_p = \frac{L}{V_p} = \frac{A\phi_p L}{q} \qquad (8)$$
$$t_p < t_w$$

where:
- ϕ_p = polymer effective porosity
- t_p = polymer passage time

Because the injection rate is unchanged and porous media geometry is unchanged, the velocity of the polymer solution is greater. This reduces to the fact that the effective porosity for polymer solution is less than the effective porosity for brine. Dawson and Lantz (1972) determined ϕ_p to be 22% smaller than ϕ_w for the system they investigated. (See Figure 14.)

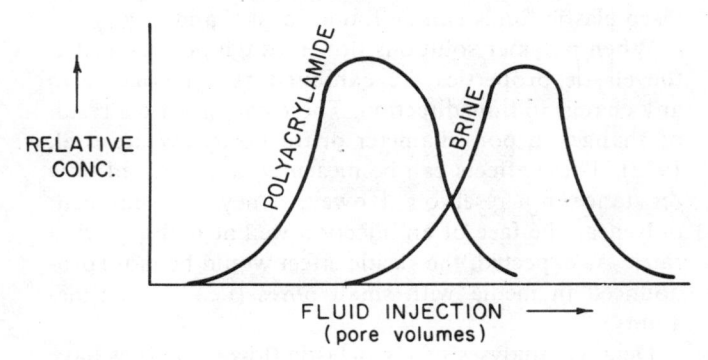

Fig. 14. Relative concentration vs. fluid injection (pore volumes) for polyacrylamide and brine.
(after Dawson and Lantz, 1972)

Entrapment

Still another important cause for permeability reduction with polymer solutions is entrapment. This can be visualized as a variation of inaccessible pore volume. Instead of small pore openings which do not allow flow of polymer, consider a pore space which has a small opening at one end and a relatively large opening at the other end. Whenever the large opening faces the upstream side of the porous media the polymer can enter the pore opening, but cannot leave. It has been trapped. The probability of this mechanism is enhanced by a characteristic of the polyacrylamide molecule. While flowing, the molecule is elongated. It conforms to flow patterns and can easily enter relatively small pore openings. Once in the pore opening where flow is restricted, this molecule coils up as it is known to do when in a low shear rate environment (Sparlin, 1975). By coiling up, the effective diameter is increased, reducing the chances that the molecule can be flushed back out of the pore space.

Mungan et al (1966) have shown the effect of polymer entrapment in alundum and sandstone cores. Szabo (1975) has completed similar studies, and he concludes that adsorption is the dominant mechanism in medium permeability cores, but entrapment is more important in low permeability rock.

Entrapment must be distinguished from plugging, because the entrapped polymer still has sufficient freedom to permit the flow of oil or other non-aqueous fluid, while restricting aqueous flow. Physical plugging is an irreversible blockage of flow passages to all flow (White et al, 1972).

To better understand the roles of adsorption, inaccessible pore volume and entrapment, comprehension of polymer size and pore size is necessary.

Polymer Size

Determination of polymer size has been addressed in two ways: mathematically and experimentally. There is fair agreement between the two approaches. The mathematical relationship employs two parameters: "r," the mean end to end distance, and "s," the mean radius of gyration or distance from the elements of the chain to its center of gravity. Flory (1953) developed the following equations for non-ionic polymers:

$$\sqrt{\bar{r}^2} = 8\,(W\eta)^{1/3} \qquad (9)$$

For a linear polymer:

$$\sqrt{\bar{r}^2} = 6\,\sqrt{\bar{s}^{-2}} \qquad (10)$$

where:
r = polymer molecule end to end distance
W = polymer molecular weight
η = intrinsic viscosity
s = polymer molecule radius of gyration

Lynch and Mac Williams (1969) applied Flory's equation to an ionic polymer and calculated an "r" value of 0.28 microns for a 3 million molecular weight polyacrylamide in a 3% brine.

The experimental approach was utilized by Gogarty (1966) who filtered various polyacrylamide solutions through millipore filters with different size openings. There was very little retention of polymer for filters with mean opening diameters of 1 micron or greater. Gogarty concluded that the effective size of polymer units was between 0.45-0.8 microns. These dimensions are probably larger than the polymer size due to water of hydration effects.

Smith (1970) has performed similar tests and produced similar data. In this case polymer retention became significant with filter openings of 0.7 micron or less. This suggests a slightly smaller polymer size than that determined by Gogarty.

Pore Size

Investigations to determine pore size of reservoir media have shown the smaller pores to be of the same magnitude as the polymer molecule. A mercury porosimeter test by Thomas (1975) on a Berea sandstone showed about 14% of the pore volume to be less than 1 micron in diameter, as shown in Figure 15.

Gray and Rex (1966) performed migratory clay studies in Berea sandstone. The maximum size clay particles found in the effluent were 0.3 microns. Some mica needles displaced were 0.3 microns wide and 1-5 microns long. A similar study was completed by Rhudy (1966). A clay suspension which passed a 1.2 micron millipore filter severely reduced the permeability of a Berea sandstone core. The permeability reduction was greatest near the injection face indicating progressive plugging.

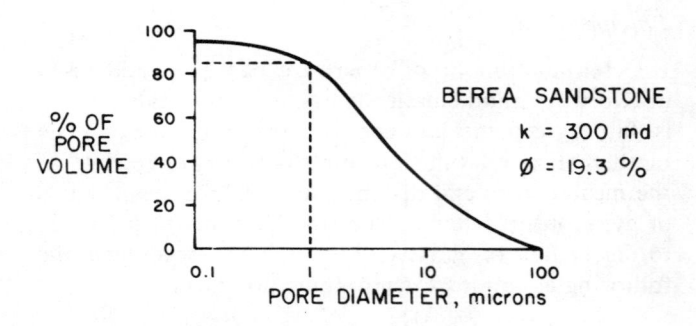

Fig. 15. Pore volume vs. pore diameter. (after Thomas, 1975)

These data support the entrapment and inaccessible pore volume mechanisms described previously. Although the porous media generally have a very broad distribution of pore diameters ($>10^3$), there generally is a significant fraction which is close to the polymer molecular size. If the porous media have very few pores in the 1 micron diameter range, the effectiveness of the polymer to reduce permeability is greatly diminished. A study by Jennings et al (1971) demonstrated this relationship clearly. (See Figure 16.)

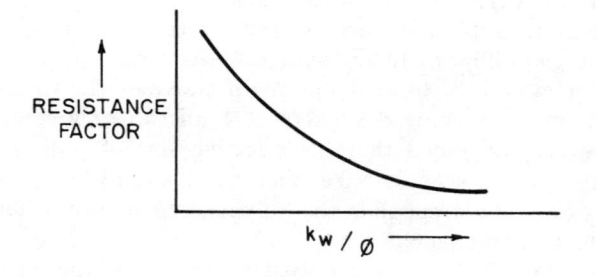

Fig. 16. Resistance factor vs. k_w/ϕ. (after Jennings et al, 1971)

Viscoelastic Effects

Polymer solutions have been characterized as pseudoplastic fluids which display a decreasing apparent viscosity with increasing shear rate. This model fairly represents the flow behavior of polymers at all shear rates except the very high rates which may exist very near the injection well. At these high rates of shear, the polymer solution loses its pseudoplastic nature and displays an increasing apparent viscosity with increasing shear rate. Such a fluid is normally classified as dilatant.

In fact, the polymer solution is not a true dilatant fluid. It only appears to have dilatant flow properties at very high shear rates in a porous media, because of the viscoelastic properties of the polymer. A viscoelastic fluid behaves like a viscous liquid at low shear rates and like an elastic solid at high shear rates (Maerker, 1974).

To provide a clearer comparison of the viscous and elastic properties, a mechanical model may be used. A viscous fluid is analogous to a shock absorber where the force on the shock absorber is proportional to the rate of deformation. An elastic solid is analagous to a spring where the force on the spring is proportioned to the deformation. A more detailed description of visco elastic fluids can be found in Skelland (1967).

When polymer solutions flow through porous media, the elastic properties are exhibited as a resistance to any change in flow direction. These changes are a result of changes in pore diameter or tortuosity (White et al, 1972). These effects can be measured as increased flow resistance in a reservoir. However, they are significant only near the face of an injection well at high injection rates. As expected, the elastic effect would be more pronounced in media with small pores (i.e., low permeability).

Detailed analyses of viscoelastic flow properties have been presented by numerous investigators (Dauben and Menzie, 1967; Hirasaki and Pope, 1974; Jennings et al, 1971; Marshall et al, 1967; Savins, 1969).

Residual Resistance

Because polymer adsorption and entrapment are only partially reversible processes, it follows that much of the polymer within a reservoir will remain long after polymer injection has ceased. Tests have shown that the effects of the polymer are partly retained even after large volumes of brine have been injected following the polymer. This residual effect is shown in Figure 17.

Studies have also shown that very high injection rates of brine following polymer injection will lower the residual permeability reduction. The high pressure gradients and high velocities are responsible for dislodging more polymer than would otherwise occur (Sparlin, 1975).

The economic importance of residual resistance becomes clear when the time scale of a field injection project is considered. The cost of chemicals generally precludes the feasibility of injecting polymer solution for the entire life of the project. With residual resistance, this is not necessary. Most of the benefits of polymer are realized long after the polymer injection has ceased.

Shear Degradation

All the discussion to this point has been directed toward factors which tend to decrease the mobility of polymer solutions in porous media. These mechanisms are favorable to the goal of achieving mobility control through the use of polymers. However, there is one significant negative factor which needs to be considered.

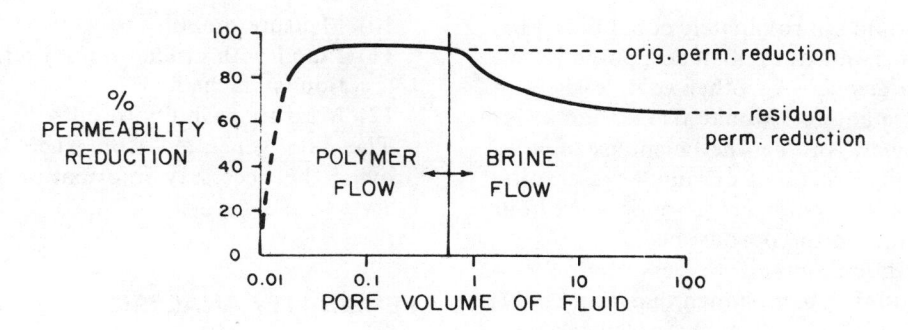

Fig. 17. Permeability reduction vs. pore volumes of fluid.
(after White et al, 1972)

The flexible long-chain polymers, particularly polyacrylamide, are susceptible to shear degradation. This degradation involves breakage of the polymer chain into several shorter molecular chains. This degradation increases as shear rate increases. However, the form of the shear has a great effect on the amount of degradation. Maerker (1974) discovered that if the shear was a viscous deformation, the shear rate had to be more than 100 times greater than a viscoelastic deformation to produce the same degradation. A viscous deformation is a type of shear which occurs in capillary flow and is measured with rotational and tube viscometers. In contrast, viscoelastic deformation is a type of shear which occurs in porous media flow and is measured with a screen factor viscometer. With the discovery that viscous shear degrades polymer less than a corresponding viscoelastic shear, it is not surprising to learn that a screen factor viscometer is more sensitive to a degraded polymer solution than a capillary viscometer. (See Figure 18.)

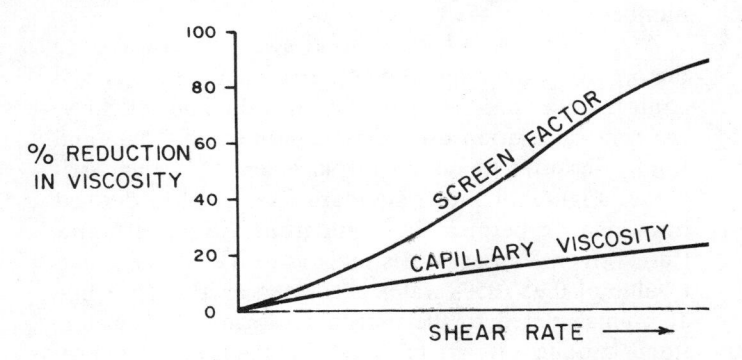

Fig. 18. Reduction in viscosity vs. shear rate.
(after Maerker, 1974)

Maerker (1974) and Smith (1970) and others determined that the loss of mobility reduction in porous media is almost always intermediate to the screen factor reduction and the capillary viscosity reduction. This observation appears reasonable because the flow in porous media is a combination of the viscous and viscoelastic flow mechanisms. The lower the permeability of the porous media, the more important the viscoelastic mechanism becomes. Thus the loss of mobility reduction due to shear is greater in low permeability media.

It should be noted that these conclusions are contradicted by the studies of Sparlin (1975) who found no significant loss of mobility reduction in Berea cores despite polymer degradation as indicated by 50% reduction in viscosity. These data not withstanding, the consensus conclusion is that shear degradation does reduce the effectiveness of polymer injection.

The investigators agree that the greater the salinity of the polymer solution, the more susceptible the solution to shear degradation. It was also found that $CaCl_2$ has a more pronounced effect than NaCl, in excess of the difference in ionic strengths between the two salts. Although plausible explanations for the effect of salt are offered, no reason for the disproportionate effect of $CaCl_2$ is suggested.

Polymer concentration has been shown to have little if any effect on polymer degradation at least in the 300 to 600 ppm range.

RESERVOIR SELECTION

There are several broad guidelines which can be used to eliminate reservoirs as poor candidates for polymer flooding. These guidelines have been developed largely on the basis of past mistakes in field tests (Jennings, 1977).

Polymer flooding offers significant potential over existing waterflooding if the existing mobility ratios are poor (2 to 20) and/or significant permeability distribution variation exists. If the existing mobility ratio is greater than 20, the economics of the process are almost certain to be unfavorable. On the other hand, if the existing ratio is close to 1, very little will be gained by a polymer flood (Jennings, 1977). If the existing waterflood is performing poorly for reasons other than poor mobility ratio or permeability variation, the polymer flood is unlikely to solve the problem. (Dow, 1965).

Stated another way, if the high water-oil ratio is due to water coning, a high permeability zone or high viscosity oil (up to approximately 300 cp) the use of polymer should be economically attractive.

Fluid injection should approximately equal fluid production. If there is a significant imbalance, polymer will be wasted filling up gas caps or other void zones. The existence of extensive aquifers would also lead to a significant loss of polymer. Another inefficient use of polymer would be in highly fractured or vugular reservoirs. These formations allow polymer to bypass without decreasing the mobility in the porous media.

Reservoir temperature should be less than 250 - 300°F. Laboratory studies have shown that many polymer solutions degrade noticeably at these elevated temperatures, losing their effectiveness.

The mobile oil saturation must be reasonably high to afford economic potential for polymer injection. Very low porosities, high residual oil or high recovery from primary or secondary operations would be some of the factors which would limit the mobile oil saturation. (Dow, 1965).

Water drive reservoirs which had little or no water production initially are good candidates for polymer flooding (White et al, 1972).

LABORATORY DESIGN

An effective laboratory design program should be structured to match a polymer injection program to the existing reservoir. To accomplish this goal, core samples must be evaluated to ensure that the reservoir is a likely candidate for polymer flooding. This means that sufficient data must be available to identify the likely cause for present production difficulties. Polymer injection is not a cure-all for declining performance.

Once it has been determined that the production performance can be improved through the use of polymers, then the polymer system must be chosen to maximize the effect in the reservoir. This selection process demands the determination of several variables. Among these are the following: polymer type; polymer concentration variations with time; injection rate variations with time; and total time of injection.

The expected performance of a polymer solution can be predicted based on laboratory flooding studies. A typical test program to determine reservoir parameters and polymer effectiveness is given below:

1) Measure core porosity and permeability to nitrogen.
2) Saturate core with connate water.
3) Flood with reservoir crude oil until residual water saturation is reached.
4) Measure mobility to oil.
5) Flood with connate water until residual oil saturation is reached.
6) Measure mobility to water.
7) Flood with polymer solution being tested.
8) Flood with crude oil until residual water saturation is reached.
9) Flood with connate water until residual oil saturation is reached.
10) Measure mobility to water.
11) Flood with crude oil until residual water saturation is reached.
12) Measure mobility to oil.

The data generated from such a laboratory study provide the necessary information to prepare a feasibility analysis.

FEASIBILITY ANALYSIS

The feasibility analysis is basically an economic test weighing the value of the incremental oil which is expected against the cost of the treatment program. The greatest uncertainties in this test are associated with the prediction of incremental oil.

The technique normally used is to predict the expected oil recovery through continued waterflooding by means of one of the common calculation procedures (Stiles, Dykstra-Parson, Johnson, Buckley-Leverett, etc.). This calculation is then repeated for the polymer flood using the modified flow properties expected from polymer usage. The difference in recovered oil represents the incremental oil due to the polymer project.

Two sample methods are presented:

Example 1

Use simple, established methods to calculate a waterflood performance, then calculate polymer flood performance assuming it will be the same as that of a waterflood of a lower viscosity oil. The comparison should be made only at a point near the economic limit — 95% water cut is traditional. The results will not be exact; the idea is to get some comparative numbers.

At a water cut of 95%, areal sweep is not a large factor, so only displacement efficiency and vertical conformance need to be considered. The Buckley-Leverett calculation uses relative permeability data and fluid viscosities to obtain displacement efficiency estimates. Figure 19 shows standard f vs. S_w plots derived from relative permeability data from a Squirrel Sand Reservoir in Kansas. Lines tangent to the curves at the f value of 0.95 (95% water cut) intersect the f = 1 line at points related to the percentage recovery of waterflood mobile oil as indicated by the supplementary scale. The results are shown in Table I for several viscosities of oil. If the assumption is made that the mobility of water can be economically decreased by a factor of 10, then the resulting change in displacement efficiency can be determined by comparing the recoveries at all viscosities differing by a factor of 10. The data in Table I show that the displacement efficiency of a flood of 120 cp oil could be improved markedly. A waterflood of 6 cp oil already has a displacement performance which is 93%, and polymer could not improve this significantly.

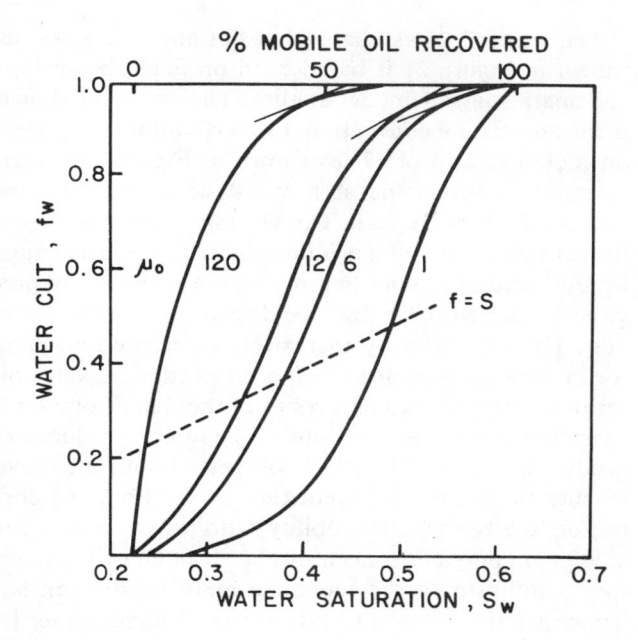

Fig. 19. Buckley-Leverett plot - Squirrel Sand example.

TABLE I
ESTIMATES OF OIL RECOVERY AT 95% WATERCUT
BY SIMPLE CALCULATION TECHNIQUES

Oil Viscosity Centipose	Displacement Efficiency (Buckley-Leverett)	Vertical Conformance (Stiles-Core Log)	Product
1	1.000	0.950	0.950
6	0.925	0.806	0.745
12	0.850	0.785	0.667
120	0.550	0.514	0.283

Recovery As Per Cent of Waterflood Mobile Oil

The effect of polymer on vertical conformance, which is due to reservoir heterogeneity, can be evaluated by a simple Stiles calculation if only relative values are required. Stiles calculations of recoveries using core log data from the same Squirrel Sand Reservoir are shown in Table I for a water cut of 95%. These results show that a large improvement in vertical conformance might be obtained by the use of a mobility control polymer in the 120 cp oil case. A modest improvement would be expected from the use of polymer if the oil in the reservoir were 6 cp viscosity. Since the Stiles calculation ignores the effects of displacement efficiency, it also overestimates recovery unless corrected. It is not unusual to see the displacement efficiency factor and the vertical conformance factor multiplied together to give an overall efficiency factor (also shown in Table I). This is probably a little too pessimistic.

The situation in the field from which the data were taken approximated that for the 120 cp oil examples. Combined primary and waterflood production amounted to about 40% of the 515 Bbls of mobile oil-in-place, and polymer flooding, with a mobility reduction of about 8, resulted in recovery of about 80% of the mobile oil. The calculations are therefore not exact, but the indications with respect to the desirability of polymer flooding correspond to the results in the field test. Had the field contained 12 cp oil, polymer flooding calculations would still have looked favorable. For a 6 cp oil, the results would have been mildly encouraging. Polymer flooding calculations for 1 cp oil in this reservoir would have appeared unfavorable (Smith and Rinehart, 1966).

Example 2

The feasibility study is not intended to be an exhaustive reservoir study, nor is it usually an adequate basis for recommending a field application. Such a study is considered to be a second-stage screening in which are examined such parameters as: 1) the amount of remaining mobile oil, particularly as it compares with current voidage; 2) the degree of permeability variation and its effect upon floodout performance; 3) the existing water-oil mobility ratio which would control conventional waterflood behavior; 4) the extent to which improved mobility ratio would enhance both areal sweep and vertical conformance; and 5) the overall unit economics of a range of polymer slug sizes. These may show a prospect to be either marginal or unattractive and further study would not be recommended for either of these conditions.

Since this type of study is basically a screening device, fairly simple methods of analysis are used. Usually, at this stage, no laboratory work is performed on cores to determine the resistance factor. From experience, it is usually possible to estimate a realistic resistance factor and chemical concentration, both of which are subsequently verified by laboratory measurements if further study is undertaken.

Using this estimated resistance factor together with the reported rock and fluid properties, theoretical performance calculations are then made for: 1) a conventional waterflood operating under the existing water-oil mobility ratio and 2) a continuous polymer flood in which the controlling mobility ratio has been reduced (improved) by the numerical value of the estimated resistance factor. The calculation procedures used by Dow are those commonly in use throughout the oil industry (such as Stiles, Dykstra-Parsons, Johnson, Prats et al, Deppe, etc.). Typical results of such calculations are shown in Figure 20 which depicts the familiar relationship between producing water-oil ratio (WOR) and the cumulative oil recovery in Bbls/acre-ft. The curve on the left labeled "12" is for a conventional waterflood with a water-oil mobility ratio of 12. It had been estimated that for this reservoir a resistance factor of 6 could be achieved with a polymer concentration of 300 ppm. This would improve the mobility ratio to a value of 2 (i.e. 12 ÷ 6). Thus the curve labeled "2" represents the calculated behavior of a continous polymer flood operating with a polymer solution-to-oil mobility ratio of 2.

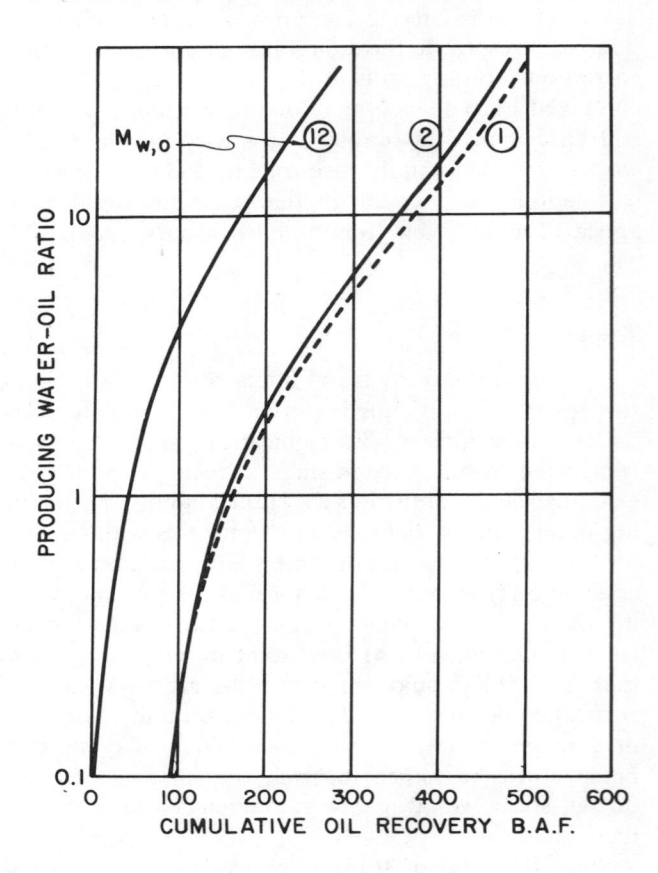

Fig. 20. Producing water oil ratio vs. cumulative oil recovery. (after Dow, 1965)

Two important observations must be made at this point. First, a resistance factor greater than 6 could be achieved in this case by using a more concentrated polymer solution which would further shift the mobility ratio closer to the "ideal" value of 1 (dotted line in Figure 20). However, the increased cost of chemical to do this would be greater than the value of the corresponding increase in oil recovery. Hence, the shift from a mobility ratio of 12 to 2, as shown in Figure 20, is close to an optimum situation. Second, it would not be economical to use polymer continuously over the entire life of the flood, because of the higher cost per Bbl of injection fluid. At some time during the early life of the process, injection of polymer solution must be terminated at some "slug cut-off" point. This polymer slug is then moved through the reservoir by the continued injection of normal brine. Incremental oil recovery over and above conventional waterflood recovery will always be somewhat less for a polymer slug than for continuous polymer injection. However, the chemical cost per incremental Bbl of oil recovered will be less for the slug, thus leading to a better profitability.

The method of estimating an optimum slug size as shown in Figure 21 is based both on field observation and mathematical model studies. The situation shown is for an existing conventional waterflood with a water-oil mobility ratio of 12 (as shown in Figure 20) which currently is producing at a WOR of 0.7 and has recovered 35 Bbls/acres-ft. Conversion to polymer injection at this point will not immediately result in a change in production behavior because a certain amount of previously injected brine must be displaced from the reservoir. However, shortly thereafter, a definite flattening (or in some cases, a decrease) in the producing water-oil ratio will be observed as a result of the change of mobility ratio within those portions of the reservoir which are producing water. This effect will persist until the curve reaches the theoretical prediction curve (Point A) corresponding to the new mobility ratio (in this case 2 for a 300 ppm polymer concentration). If polymer injection were continued to the economic limit, production behavior would follow the extreme right-hand curve. If, however, polymer injection is terminated at Point A, the water-oil ratio trend will gradually diverge from the "100% slug" curve and at the economic limit will have produced less incremental oil. Slug cut-off at Point B will yield a somewhat higher recovery but will also require a greater amount of chemical. Several slug sizes such as points A and B are examined in this manner in order to determine the one which gives the greatest incremental oil recovery per pound of chemical used.

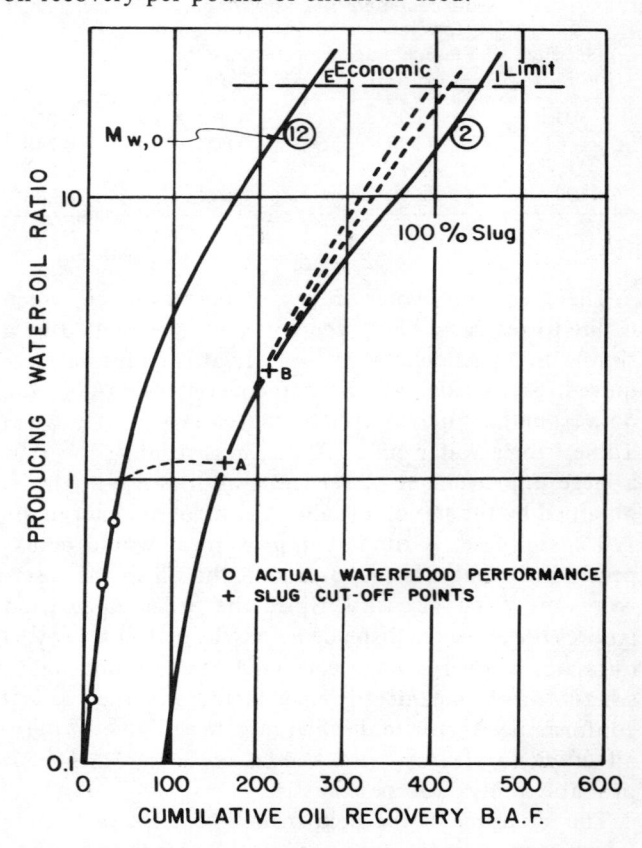

Fig. 21. Producing water oil ratio vs. cumulative oil recovery. (after Dow, 1965)

In those cases where waterflooding has not yet begun, the point at which the WOR curve will begin to deviate from the brine-flood prediciton is less certain. Usually this point is taken at the WOR where the prediction method indicates water breakthrough in the highest permeability layer. (Dow, 1965)

In assessing the feasibility of a prospect by the foregoing methods, the value of the incremental oil recovery must be compared to the chemical cost to determine the economic viability.

FIELD HANDLING SYSTEMS

Polymer Blending System

Polymer mixing is generally accomplished in a facility of the type shown in Figure 22. The heart of this system is the dry polymer mixer which meters the powdered or granulated polymer into a controlled water stream to give a uniform dispersion. The typical arrangement allows the polymer to contact a tangentially swirling stream of water in a funnel-shaped device. The GACO and Dow mixers are of this type. The polymer feed rate to the mixer is controlled by a variable speed feed auger. The water rate is adjusted as necessary to provide the needed mixing in the funnel. The remainder of the water needed to meet the target polymer concentration is introduced as a by-pass stream which mixes with the polymer dispersion immediately down stream of the mixer.

Maintenance of a sufficient dry polymer inventory in the feed hopper is generally accomplished in one of two ways. In small scale operations, fifty pound bags of polymer are emptied into the feed hopper or into a storage bin which is connected to the feed hopper. In large scale operations, the polymer is stored in bulk and transported to the feed hopper pneumatically. Care must be exercised to control the polymer dust in these areas and personnel are required to wear masks and possibly additional protective gear.

Because the rate of dilution of high concentration polymer is quite slow, relatively large mixing tanks with gentle agitation are needed downstream of the polymer mixers. These tanks are generally blanketed with nitrogen to exclude atmospheric oxygen. This is also a common place to inject an oxygen scavenger or biocide as necessary.

The thoroughly mixed polymer from the tank is generally injected directly with a piston-type positive displacement pump. In some cases, when face plugging of injection wells is feared, a filter may be used to ensure the injected polymer contains no gelled agglomeration of high concentration polymer.

Preparation of polymer solution from emulsion polymer supply is less complicated. Only the metered dilution water and chemical additions are required. Polymer dilution can frequently be accomplished with static or in-line mixers eliminating the large mixing tank. High concentration liquid polymer is stored in a tank with a metering pump used to control polymer rate to the mixer.

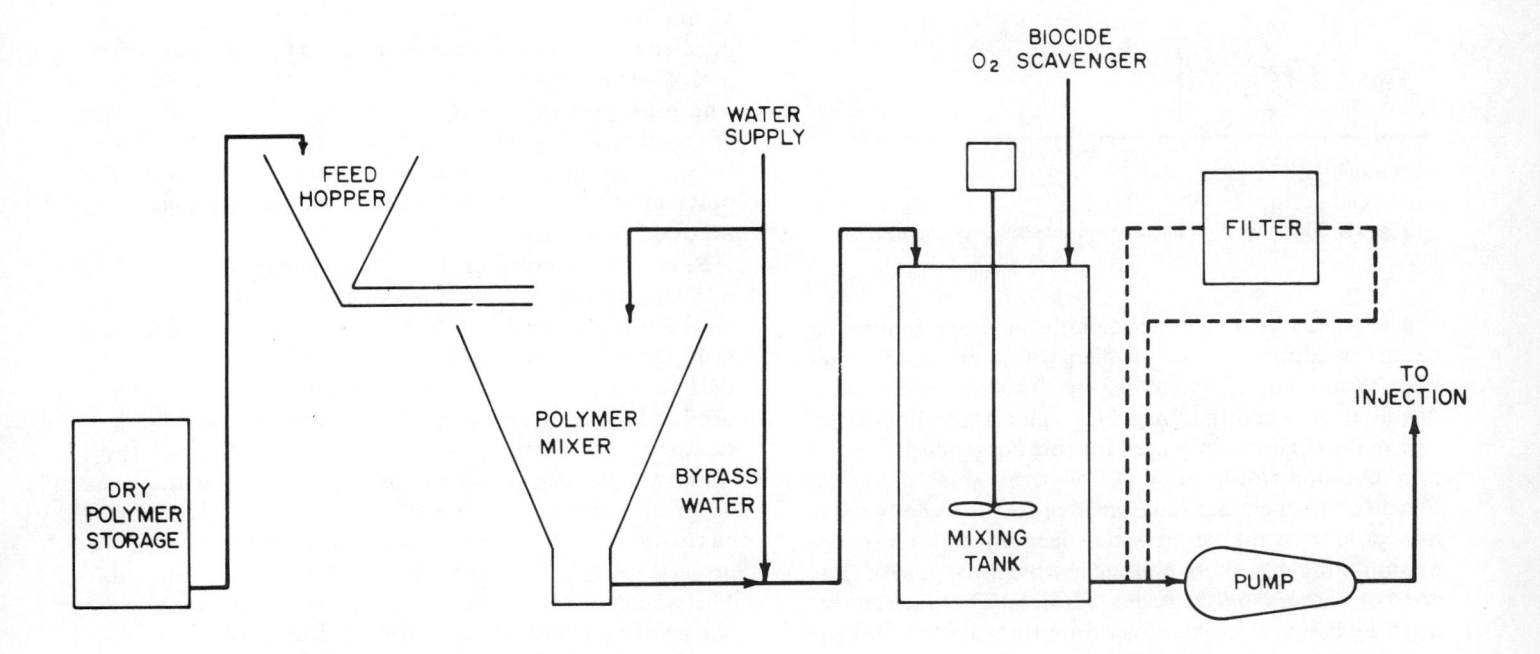

Fig. 22. Dry polymer blending system diagram.

Fluid Injection System

Fluid injection into a reservoir through numerous wells is generally accomplished by means of a manifold system. Figure 23 illustrates a simplified case. Because variable speed positive displacement pumps are generally used to inject fluids into a reservoir, the total volumetric flow rate can be controlled to meet the overall injection program. However, without flow control devices at each well, the relative flow is determined by the flow resistance in each injection well. To compensate for this uncontrolled injection, some type of flow controller is needed at each well.

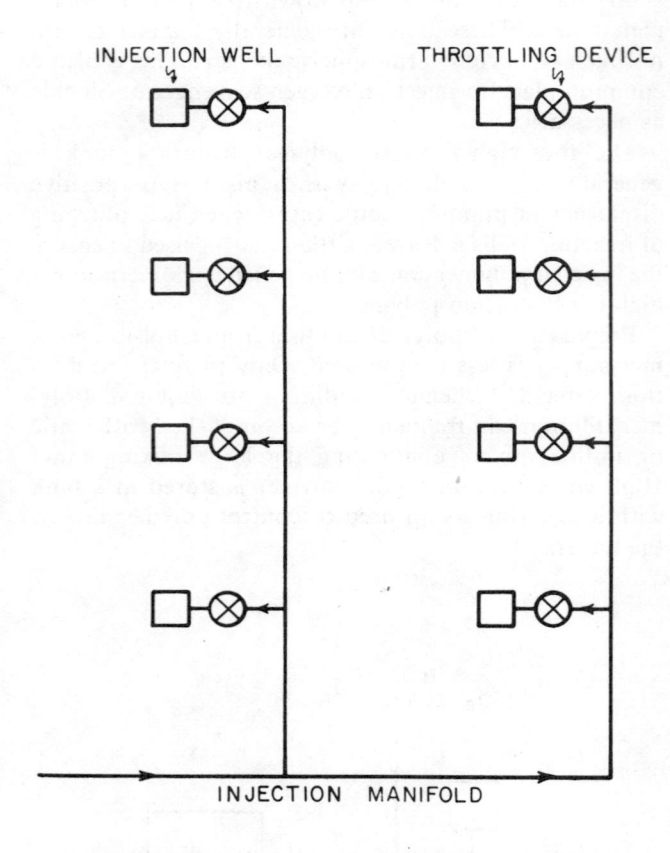

Fig. 23. Fluid injection distribution manifold system diagram.

The injection of polyacrylamide polymer requires a special solution to the problem of controlling injection rates. These polymers are susceptible to shear degradation when they pass through a throttling valve. The method commonly used for rate control is the insertion of a long coil of tubing of relatively small diameter. Because polymers are less sensitive to the viscous shear in a pipe than the viscoelastic shear through an orifice or similar device, these coils accomplish the goal of flow control without polymer degradation. The diameter of the tube coil is calculated based on the shear rate for the desired flow rate, while the length of the coil is calculated based on the pressure which must be dissipated prior to entering the wellhead.

CASE HISTORIES

East Coalinga Field, California (USA)

The Coalinga Field has been producing oil for 75 years. Poor reservoir management early in the life of this field resulted in considerable loss of solution gas. Thus the oil viscosity increased (21° API) giving the reservoir an unfavorable mobility ratio of approximately 15:1. The reservoir is an unconsolidated sand with 24% porosity and permeability ranging from 50-480 md.

The enhanced oil recovery program was begun in June 1976 when a closely monitored water injection phase was initiated to predict long term waterflood performance.

A biopolymer solution was injected into a test well for about 2½ months beginning in April 1978. The mobility ratio was reduced to 1.5 with the 400 ppm polymer solution.

Although there was no sign of increased oil production or decreased water-oil ratio as of July 1978, the projected recovery is 2.6 MMBO, of which 0.6 MMBO would be classified as polymer oil.

North Stanley Field, Oklahoma (USA)

This field has been waterflooded since 1959.

Water-oil ratios had been steadily rising reaching 75:1 at the time of polymer treatment. This high water production was due to a high permeability section along an edge of the field. The polymer injection was preceded by a fresh water preflush buffer for 4 months. Nearly 12 million barrels of Dow Pusher-700 polymer were injected at a mean concentration of 285 ppm over a 1 year period. This represented a 17.3% PV slug of polymer. The peak polymer concentration was 625 ppm shortly after initiation of the treatment. This was reduced in several increments to 100 ppm. After termination of polymer injection a 4.8% PV of fresh water was injected, followed by connate water.

Shortly after polymer injection was begun, the water-oil ratio declined to 50:1. A concurrent increase in oil production was realized. Six months later the peak oil production of 660 BOPD was achieved, of which 130 BOPD was judged to be due to polymer treatment. Oil production has declined since then, but the rate is considerably less than the pre-polymer decline. Thus the enhanced oil production has been increasing. Although oil production has never matched the projected values based on a 2-D reservoir model, results were close to projections until the salt water injection was resumed. The water-oil ratio has climbed sharply since then and has nearly reached the pre-treatment level.

An economic analysis of North Stanley performance indicates the project to be attractive at "new" oil prices, but not at lower tier prices.

BIBLIOGRAPHY

Agnew, H. J.: "Here's How 56 Polymer Oil-Recovery Projects Shape Up," *O. & G. Jour.* (May 1, 1972).

Anderson, E. V.: "Enhanced Oil Recovery: Big Chemical Potential," *Chem. and Eng. News,* Vol. 55 (1948).

Aronofsky, J. S.: "Mobility Ratio - Its Influence on Flood Patterns during Water encroachment," *Trans.,* AIME (1952) 195.

Aronofsky, J. S. and Ramey, H. J., Jr.: "Mobility Ratio - Its Influence on Injection or Production Histories in Five-Spot Waterflood," *Trans.,* AIME (1956) 207.

Barnes, A. L.: "The Use of Viscous Slug to Improve Waterflood Efficiency in a Reservoir Partially Invaded by Bottom Water," *Jour. Pet. Tech.* (Oct., 1962).

Bilhartz, H. L., Jr. and Charlson, G. S.: "Field Polymer Stability Studies," SPE 5551 prepared for the SPE-AIME 50th Annual Fall Meeting, Dallas, TX. (Oct., 1975).

Burcik, E. J.: "The Mechanism of Microgel Formation in Partially Hydrolyzed Polyacrylamide," *Jour. Pet. Tech.* (April, 1969).

Burcik, E. J.: "Pseudo Dilitant Flow of Polyacrylamide Solutions on Porous Media," *Producers Monthly.* (March, 1967).

Burcik, E. J.: "The Use of Polymers in the Recovery of Petroleum," *Earth Mineral Sci.* (April, 1968).

Burcik, E. J. and Thakur, G. C.: "Reaction of Polyacrylamide with Commonly Used Additives," *Jour. Pet. Tech.* (Sept., 1972).

Burcik, E. J. and Thackur, G. C.: "Some Reactions of Microgel in Polyacrylamide Solutions," *Jour. Pet. Tech.* (May, 1974).

Burcik, E. J. and Walrond, K. W.: "Microgel in Polyacrylamide Solutions and Its Role in Mobility Control," *Producers Monthly.* (Sept., 1968).

Burnett, D. B.: "Laboratory Studies of Bipolymer Injectivity Behavior Effectiveness of Enzyme Clarification," SPE 5372 prepared for the SPE-AIME 45th Annual California Regional Meeting, Ventura, CA. (April, 1975).

Caudle, B. H. and Witte, M. D.: "Production Potential Changes During Sweep-Out in a Five-Spot System," *Trans.,* AIME (1959) 216.

Chauveteau, G. and Kohler, N.: "Polymer Flooding — The Essential Elements for Laboratory Evaluation," The 3rd SPE Symposium on Improved Oil Recovery, Tulsa, OK. (April, 1974).

Clampitt, R. L. and Reid, T. B.: "An Economic Polymer Flood in the North Burbank Unit, Osage County, Ok.," SPE 5552 prepared for the SPE-AIME 50th Annual Fall Meeting, Dallas, TX. (Oct., 1975).

Clark, R. K., Scheuerman, R. F., Rath, H. and van Lear, H.: "Polyacrylamide — Potassium Chloride Mud for Drilling Water and Sensitive Shales," SPE 5514 prepared for the SPE-AIME 50th Annual Fall Meeting, Dallas, TX. (Oct., 1975).

Craig, F. F., Jr.: "The Reservoir Engineering Aspects of Waterflooding," *SPE Monograph No. 3,* Dallas, TX. (1971).

Dabbous, M. K. and Elkins, L. E.: "Preinjection of Polymers to Increase Reservoir Flooding Efficiency," SPE 5836 presented at the SPE-AIME 4th symposium on Improved Oil Recovery, Tulsa, OK. (March, 1976).

Dauben, D. L. and Menzie, D. E.: "Flow of Polymer Solutions Through Porous Media," *Jour. Pet. Tech.* (Aug., 1967).

Davis, B. W.: "Estimation of Surface Free Energies of Polymeric Materials," *Journal of Colloid and Interface Science.* (May, 1977).

Dawson, R. and Lantz, R. B.: "Inaccessible Pore Volume in Polymer Flooding," *Soc. Pet. Eng. Jour.* (Oct., 1972).

Dominguez, J. G. and Willhite, G. P.: "Polymer Retention and Flow Characteristics of Polymer Solutions in Porous Media," SPE 5835 prepared for the SPE-AIME 4th Symposium on Improved Oil Recovery, Tulsa, OK. (March, 1976).

Dow Chemical Company: *The Dow Flooding Process for Secondary Oil Recovery Pushes Chemicals for Mobility Control.* Houston, TX. (1965).

DuBois, B. M., Johnson, J. P. and Cunningham, J. W.: "Polymer Enhanced Waterflooding: A Status Report of the North Stanley Project, Osage County, Ok., *ERDA Enhanced Oil, Gas Recovery and Improved Drilling Methods, Vol. 1 - Oil,* Tulsa, OK. (Sept., 1977).

Dyes, A. B., Caudle, B. H. and Erickson, R. A.: "Oil Production After Breakthrough as Influenced by Mobility Ratio," *Trans.,* AIME (1954) 81.

Dykstra, H. and Parsons, R. L.: "The Prediction of Oil Recovery by Water Flood," *Secondary Recovery of Oil in the United States,"* 2d ed. API, New York (1950).

Elata, C., Burger, J., Michlin, J. and Takserman, U.: "Dilute Polymer Solutions in Elongational Flow," *The Physics of Fluids,* Vol. 20. (Oct., 1977).

Ershaghi, I. and Handy, L. L.: "Mobility of Polymer Solutions in Porous Media," SPE 3683 prepared for the SPE-AIME 42nd Annual California Regional Meeting, Los Angeles, CA. (Nov., 1971).

Ferrer, J.: "Some Mechanistic Features of Flow of Polymers through Porous Media," SPE 4029 prepared for the SPE-AIME 47th Annual Fall Meeting, San Antonio, TX. (Oct., 1972).

Ferrer, G. J. and Larreal, B. J.: "Microscopic Observations of Polyacrylamide Structures," *Jour. Pet. Tech.* (Jan., 1973).

Ford, W. O., Jr. and Kelldorf, W. F. N.: "Field Results of a Short Setting Time Polymer Placement Technique," SPE 5609 prepared for the SPE-AIME 50th Annual Fall Meeting, Dallas, TX. (Oct., 1975).

Fox, V. G., Gogarty, W. B. and Levy, G. L.: "Viscoelastic Effects in Polymer Flow through Porous Media," SPE 4025 prepared for the SPE-AIME 47th Annual Fall Meeting, San Antonio, TX. (Oct., 1972).

Friedman, R. H.: "Reversible Cross Linking in Oil Recovery," SPE 4357 prepared for the SPE-AIME Oilfield Chemistry Symposium (1973).

Garland, T. M.: "Selective Plugging of Water Injection Wells," *Jour. Pet. Tech.* (Dec., 1966).

Goddard, J. E. and Vanlandingham, J. V.: "New Material and Treatment Design Improve Water Control in Oil Wells and Water Injection Wells," SPE 5365 prepared for the SPE-AIME 45th Annual California Regional Meeting, Ventura, CA. (April, 1975).

Gogarty, W. B.: "Mobility Control with Polymer Solutions," *Soc. Pet. Eng. Jour.* (June, 1967).

Gogarty, W. B.: "Rheological Properties of Pseudoplastic Fluids in Porous Media," *Soc. Pet. Eng. Jour.* (June, 1967).

Gogarty, W. B. and Poettmann, F. H.: "The Use of Non-Newtonian Fluids in Oil Recovery," 8th World Petroleum Congress. (1971).

Graue, D. J.: "Prediction Method for Reservoir Flooding with Fluids of Reduced Mobility," SPE 2257 prepared for the SPE-AIME 43rd Annual Fall Meeting, Houston, TX. (Sept., 1968).

Harvey, A. H. and Menzie, D. E.: "Polymer Solution Flow in Porous Media," *Soc. Pet. Eng. Jour.* (June, 1970).

Herbeck, E. F., Heintz, R. C. and Hastings, J. R.: "Fundamentals of Tertiary Recovery: Polymers," *Pet. Eng. Int'l.* (July, 1976).

Herr, J. W. and Routson, W. G.: "Polymer Structure and Its Relationship to the Dilute Solution Properties of High Molecular Weight Polyacrylamides," SPE 5098 prepared for the SPE-AIME 49th Annual Fall Meeting, Houston, TX. (Oct., 1974).

Hill, H. J. et al: "The Behavior of Polymers in Porous Media," SPE 4748 prepared for the 3rd Symposium on Improved Oil Recovery, Tulsa, OK. (April, 1974).

Hill, H. J. and Lake, L. W.: "Cation Exchange — Chemical Flooding Experiments," SPE 6770 prepared for the SPE-AIME 52nd Annual Fall Meeting, Denver, CO. (Oct., 1977).

Hirasaki, G. J. and Pope, G. A.: "Analysis of Factors Influencing Mobility and Adsorption in the Flow of Polymer Solution Through Porous Media," *Soc. Pet. Eng. Jour.* (Aug., 1974).

Jeanes, A., Pittsley, J. E. and Senti, F. R.: "Polysaccharide B-1459: A New Hydroculloid Polyelectrolyte Produced from Glucose by Bacterial Fermentation," *Jour. App. Pol. Sci.* 5, No. 17. (1961).

Jennings, R. R., Rogers, J. H. and West, J. J.: "Factors Influencing Mobility Control by Polymer Solutions," *Jour. Pet. Tech.* (March, 1971).

Jewett, R. L. and Schurz, G. F.: "Polymer Flooding — A Current Appraisal," *Jour. Pet. Tech.* (June, 1970).

Kelley, D. L. and Caudle, B. H.: "The Effect of Connate Water on the Efficiency of High Viscosity Waterfloods," *Jour. Pet. Tech.* (Nov., 1966).

Knight, B. L.: "Reservoir Stability of Polymer Solutions," *Jour. Pet. Tech.* (May, 1973).

Knight, B. L. and Rhudy, J. S.: "Recovery of High Viscosity Crudes by Polymer Flooding," CIM 75-39 prepared for the 46th Annual Meeting of Pet. Soc. of CIM Tech. (1975).

Lee, K. S. and Claridege, E. L.: "Areal Sweep Efficiency of Pseudoplastic Fluids in a Five-Spot Hele-Shaw Model," *Soc. Pet. Eng. Jour.* (March, 1968).

Lipton, D.: "Improved Injectability of Biopolymer Solution," SPE 5099 prepared for the SPE-AIME 49th Annual Fall Meeting, Houston, TX. (Oct., 1974).

Lynch, E. J. and MacWilliams, D. C.: "Mobility Control with Partially Hydrolyzed Polyacrylamide — A Reply to Emil Burcik," *Jour. Pet. Tech.* (Oct., 1969).

MacWilliams, D. C., Rogers, J. H. and West, T. J.: "Water-Soluble Polymers in Petroleum Recovery," 164th Meeting of the Nat. Amer. Chem. Soc. (1972).

Maerker, J. M.: "Dependence of Polymer Retention on Flow Rate," *Jour. Pet. Tech.* (Nov., 1973).

Maerker, J. M.: "Mechanical Degradation of Partially Hydrolyzed Polyacrylamide Solutions in Unconsolidated Porous Media," *Soc. Pet. Eng. Jour.* (Aug., 1976).

Maerker, J. M.: "Shear Degradation of PAA Solutions," SPE 5101 prepared for the SPE-AIME 49th Annual Fall Meeting, Houston, TX. (Oct., 1974).

Martin, F. D.: "Laboratory Investigation in the Use of Polymers in Low Permeability Reservoirs," SPE 5100 prepared for the SPE-AIME 49th Annual Fall Meeting, Houston, TX. (Oct., 1974).

Martin, F. D. and Sherwood, N. S.: "The Effect of Hydrolysis of Polyacrylamide on Solution Viscosity, Polymer Retention and Flow Resistance Properties," SPE 5339 prepared for the SPE-AIME Rocky Mountain Regional Meeting, Denver, CO. (April, 1975).

Mungan, N.: "Improved Waterflooding through Mobility Control," *Cdn. Jour. Chem. Eng.* (Feb., 1971).

Mungan, N.: "Programmed Mobility Control in Polymer Floods," 3rd Symposium of the Assn. de Recherche sur les Techniques de Forage et de Production." Pau, France. (June, 1968).

Mungan, N.: "Rheology and Adsorption of Aqueous Polymer Solutions," *Jour. Cdn. Pet. Tech.* (April-June, 1969).

Mungan, N., Smith, F. W. and Thompson, J. L.: "Some Aspects of Polymer Floods," *Jour. Pet. Tech.* (Sept., 1966).

Muskat, M.: *Physical Principles of Oil Production,* McGraw-Hill, New York (1949).

Naudascher, E. and Killen, J. M.: "Onset and Saturation Limit of Polymer Effects in Porous Media Flows," *The Physics of Fluids,* Vol. 20. (Oct., 1977).

Neale, M. J., Penton, J. R. and Routson, W. G.: "A New Blocking Agent for Waterflood Channeling," SPE 3992 prepared for the SPE-AIME 47th Annual Fall Meeting, San Antonio, TX. (Oct., 1972).

Needham, R. B., Threkeld, C. B. and Gall, J. W.: "Control of Water Mobility Using Polymers and Multivalent Cations, "SPE 4747 prepared for the 3d Symposium on Improved Oil Recovery, Tulsa, OK. (April, 1974).

Noran, D.: "Sophisticated Logging Program Monitors California Polymer Flood," *O. & G. Jour.* (April 4, 1977).

Norton, C. J. and Falk, D. O.: "Synergism in Thickened Water Systems For Improved Oil Recovery," SPE 5550 prepared for the SPE-AIME 50th Annual Fall Meeting, Dallas, TX. (Oct., 1975).

Nouri, N. H. and Root, P. J.: "A Study of Polymer Solution Rheology, Flow Behavior, and Oil Displacement Processes," SPE 3523 prepared for the SPE-AIME 46th Annual Fall Meeting, New Orleans, LA. (Oct., 1971).

Ogandzhanyants, V. G. and Polishchuk, A. M.: "Lag of the Front of the Polymer Behind the Carrier Liquid with the Filtration of a Polymer Through a Porous Media," Mekh Zhidkosti Gaza. (July-Aug., 1976).

Patton, J. T.: "Chemical Treatment Enhances Xanflood Polymer," SPE 4670 prepared for the SPE-AIME 48th Annual Fall Meeting, Las Vegas, NV. (Sept., 1973).

Pope, G. A., Lake, L. W. and Helfferich, F.: "Cation Exchange in Chemical Flooding — Basic Theory Without Dispersion," SPE 6771 prepared for the SPE-AIME 52nd Annual Fall Meeting, Denver, CO. (Oct., 1977).

Pye, D. J.: "Improved Secondary Recovery by Control of Water Mobility," *Jour. Pet. Tech.* (Aug., 1964).

Rodriquez, J. R. and Marsden, S. S.: "Flow of Aqueous Polymer Solutions through Porous Media at Reservoir Temperature," CIM 75-40 prepared for the 26th Annual Pet. Soc. of CIM Tech. (1975).

Sadowski, T. J.: "Non-Newtonian Flow through Porous Media: I. Theoretical," *Trans., Soc. of Rheo.,* 9, No. 2. (1965).

Sadowski, T. J.: "Non-Newtonian Flow through Porous Media: II. Experimental," *Trans., Soc. of Rheo.,* 9, No. 2. (1965).

Sandiford, B. B.: "Laboratory and Field Studies of Water Floods Using Polymer Solutions to Increase Oil Recoveries," *Jour Pet. Tech.* (Aug., 1964).

Sarem, A. M.: "On the Theory of Polymer Solution Flooding Process," SPE 3002 prepared for the SPE-AIME 45th Annual Fall meeting, Houston, TX. (Oct., 1970).

Savins, J. G.: "Non-Newtonian Flow through Porous Media," *Ind. and Eng. Chem.*, 61, No. 10. (1969).

Schurz, G.: "Field Preparation of Polymer Solutions Used to Improve Oil Recovery," SPE 4254 prepared for the SPE-AIME Handling of Oilfield Waters Symposium, Los Angeles, CA. (Dec., 1972).

Seyer, F. A. and Vossoughi, S.: "Pressure Drop for Flow of Polymer Solution in a Model Porous Medium," *Cdn. Jour. Chem. Eng.* (Oct., 1974).

Sherborne, J. E., Sarem, A. M. and Sandiford, B. B.: "Flooding Oil-Containing Formations with Solutions of Polymer in Water," 7th World Petroleum Congress. (1967).

Sloat, B.: "Choosing the Right Floods for Polymer Treatment," *Pet. Eng.* (May, 1971).

Sloat, B.: "How Six Polymer Floods are Faring," *O. & G. Jour.* (Dec. 11, 1974).

Sloat, B.: "Polymer Treatment Boosts Production on Four Floods," *World Oil* (March, 1969).

Sloat, B.: "Polymer Treatment Should Be Started Early," *Pet. Eng.* (July, 1970).

Sloat, B.: "When Should Polymer Treatment Be Started on Waterfloods," No. 851-44-E prepared for the Mid-Continent District Spring Meeting of API Prod. Div. (1970).

Sloat, B., Fitch, J. P. and Taylor, J. T.: "How to Produce More Oil and More Profit with Polymer Treatments," SPE 4185 prepared for the SPE-AIME California Regional Meeting, Bakersfield, CA Nov., 1972).

Smith, F. W.: "The Behavior of Partially Hydrolyzed Polyacrylamide Solutions in Porous Media," *Jour. Pet. Tech.* (Feb., 1970).

Smith, C. R. and Rinehart, R. D., eds.: "Thermal Recovery Techniques Short Course," Petroleum Engineering Associates, Laramie, WY. (1966).

Snell, G. W. and Schurz, G. F.: "Polymer Chemical Aid in Unique Recovery Process," *Pet. Eng.* (Feb., 1966).

Sparlin, D.: "An Evaluation of Polyacrylamides for Reducing Water Production," SPE 5610 prepared for the SPE-AIME 50th Annual Fall Meeting, Dallas, TX. (Oct., 1975).

Stahl, C. D.: "Displacement of Oil by Polymer Flooding, Part 1," *Producers Monthly.* (May, 1966).

Stahl, C. D.: "Displacement of Oil by Polymer Flooding, Part 2," *Producers Monthly.* (June, 1966).

Stahl, C. D.: "Displacement of Oil by Polymer Flooding, Part 3," *Producers Monthly.* (July, 1966).

Stiles, W. E.: "Use of Permeability Distribution in Waterflood Calculations," *Trans.*, AIME (1949) 186.

Szabo, M. T.: "Factors Influencing Oil Recovery and Polymer Retention During Polymer Floods," SPE 4668 prepared for the SPE-AIME 48th Annual Fall Meeting, Las Vegas, NV. (Sept., 1973).

Szabo, M. T.: "Laboratory Investigation of Factors Influencing Polymer Flood Performance," *Soc. Pet. Eng. Jour.* (Aug., 1975).

Szabo, M. T.: "Molecular and Microscopic Interpretation of the Flow of Hydrolyzed Polyacrylamide Solution through Porous Media," SPE 4028 prepared for the SPE-AIME 47th Annual Fall Meeting, San Antonio, TX. (Oct., 1972).

Szabo, M. T.: "Some Aspects of Polymer Retention in Porous Media Using C^{14}-Tagged Hydrolyzed Polyacrylamide," *Soc. Pet. Eng. Jour.* (Aug., 1975).

Thakur, G. C.: "Prediction of Resistance Effect in Porous Media," SPE 4956 prepared for the SPE-AIME Permian Basin Oil Recovery Conference, Midland, TX. (March, 1974).

Thomas, C. P.: "The Mechanism of Reduction of Water Mobility by Polymers in Glass Capillary Arrays," SPE 5556 prepared for the SPE-AIME 50th Annual Fall Meeting, Dallas, TX. (Sept., 1975).

Tinker, G. E. et al: "Coalinga Demonstration Project Oil Recovery by Polymer Flooding," *ERDA Enhanced Oil, Gas Recovery and Improved Drilling Methods*, Vol. 1 - Oil, Tulsa, OK. (Sept., 1977).

Townsend, W. R., Becker, S. A. and Smith, C. W.: "Polymer Use in Calcareous Formations," SPE 6382 prepared for the SPE-AIME 1977 Permian Basin Oil and Gas Recovery Conference, Midland, TX. (March, 1977).

Ustik, R. E. and Hillhouse, J. D.: "Comparison of Polymer Flooding and Waterflooding at Huntington Beach, CA." SPE 1734 prepared for the SPE-AIME 96th Annual Meeting, Los Angeles, CA. (1967).

Wade, J. H. T. and Kumar, P.: "Electron Microscope Studies of Polymer Degradation," *Jour. of Hydronautics.* (Jan., 1972).

Wang, G. C. and Caudle, B. H.: "Effects of Polymer Concentrations, Slug Size and Permeability Stratification in Viscous Waterfloods," SPE 2927 prepared for the SPE-AIME 45th Annual Fall Meeting, Houston, TX. (Oct., 1970).

Watson, W. D.: "Method of Recovering Oil Using a Chemical Blending System," U.S. Patent No. 3,902,558. (Dec., 1973).

White, J. L., Phillips, H. M., Goddard, J. E. and Baker, B. D.: "Use of Polymers to Control Water Production in Oil Wells," SPE 3783 presented at the SPE-AIME 2nd Symposium on Improved Oil Recovery, Tulsa, OK. (April, 1972).

Yost, M. E. and Stokke, O. M.: "Filtration of Polymer Solutions," *Jour. Pet. Tech.* (Oct., 1978).

Chapter 6
CAUSTIC FLOODING

INTRODUCTION

The first patent on the use of caustic for enhanced oil recovery was issued to H. Atkinson in 1927. However, even before the issuance of the Atkinson patent, it was noted in 1917 by F. Squires that the displacement of oil may be made more complete by introducing alkali into the water. Other early workers described the use of salts such as sodium carbonate, sodium silicate and dilute solutions of sodium and potassium hydroxides. In 1942, P. Subkow patented the injection of aqueous emulsifying agents for recovering heavy oil or bitumen. In spite of early recognition that caustic would enhance oil recovery, there is no record of successful field applications.

The earliest field trial of a caustic flooding process was in 1925. It was a test of sodium carbonate injection in the Bradford Field. Results were disappointing and were never reported in detail.

TECHNICAL DISCUSSION
Rock Properties

Rock properties can be a major factor in caustic flooding, although the rock effects on the recovery process are unclear. The reaction with the reservoir rock may be responsible for favorable wettability changes, but the reaction also consumes caustic.

When present in sufficient quantity, divalent ions in the connate water can deplete a caustic slug by precipitation of insoluble hydroxides.

Gypsum or anhydrite, when present in more than trace quantities, will cause precipitation of $Ca(OH)_2$ and render the NaOH slug ineffective.

Clays with high ion exchange capacity can deplete a NaOH slug by exchanging hydrogen for sodium.

Limestone and dolomite are unreactive and the reaction with the silica component in sandstone is too slow and incomplete to present a problem.

The caustic reactivity with reservoir rock can be determined in the laboratory following the procedure outlined in the Laboratory Design section.

In screening a large number of target reservoirs some general criteria can eliminate the obviously poor candidates. Some screening criteria that should be considered are as follows:

1. Reservoir should be susceptible to waterflooding. The addition of caustic will increase the oil recovered above that which would be expected from water injection alone.

2. There should be no extensive fractures or thief zones. The caustic is not a blocking and diverting type process and should not be applied for such purposes. If emulsification and entrapment is the mechanism, improvement in volumetric sweep efficiency can result from better conformance due to flow of emulsion in porous media. Major gross heterogenieties cannot be corrected by caustic flooding.

3. Reservoir should not contain a gas cap. The concern in this regard is that a substantial amount of the mobilized oil may move to resaturate the gas cap. The effect of the presence of a gas cap can be determined by a thorough reservoir evaluation.

4. Injectivity should be adequate. The lower mobility of any emulsions formed and plugging due to formation damage or scale formation will generally mean some decrease in injectivity. Consequently, if injectivity is a problem in conventional waterflooding, it may become more of a problem in alkaline flooding.

5. Reservoir temperature should be less than 200°F. If temperature is greater, excessive caustic consumption and emulsion stability become problems.

6. Sandstone reservoirs are preferred. The only reported application of alkaline floodings in a carbonate reservoir has been in Hungary, involving the injection of ammonium hydroxide. (Ban et al, 1977).

7. Oil viscosity should be less than 200 cp. For higher viscosity oils, thermal processes may be more applicable.

CAUSTIC FLOODING

8. The acid number of the crude oil should be larger than 0.2 mg KOH/gm of crude.
9. Interfacial tension between the crude oil and caustic solution should be less than 0.01 dyne/cm.

Displacement Mechanisms

Although caustic flooding appears to be a simple process and relatively inexpensive to apply, the mechanisms involved in displacement of oil are complicated. An explanation and description of six functioning mechanisms follow:

1. The ratio of viscous to capillary forces, commonly referred to as the capillary number ($v\mu/\sigma\emptyset$), plays an important role in the immiscible displacement of oil by water. In most waterfloods, the capillary number is about 10^{-6}. If this number can be increased to the range of $10^{-4} - 10^{-2}$ by lowering the interfacial tension by two or three orders of magnitude, significant incremental recovery could result. Alkaline solutions can, with some crude oils, give extreme lowering of interfacial tension.

   The lowering of interfacial tension with caustic solution is one mechanism that can influence recovery. This mechanism correlates with acid number, gravity and viscosity. Acid number is the number of milligrams of potassium hydroxide required to neutralize one gram of crude oil to pH = 7.0. For a good candidate, the acid number should be 0.5 mg KOH/gm crude or greater. However, further testing should be done with acid numbers of 0.2 mg KOH/gm crude or greater.

   Most crude oils with gravities of 20°API or lower produced a caustic solution - crude oil interfacial tension of less than 0.01 dyne/cm. Most interfacially active crudes will reach a maximum measureable surface activity at a caustic concentration of about 0.1% by weight. Sodium chloride in solution reduces the caustic concentration required to give maximum surface activity. Conversely, the calcium ion suppresses surface activity.

2. A second possible mechanism that will improve oil recovery by addition of caustic is the reversal of the rock wettability from oil-wet to water-wet. It was recognized that the improvement in oil recovery resulted from favorable changes in relative oil and water permeability that accompany reversal from oil-wet to water-wet in the region where oil is still flowing. Some laboratory data found this mechanism to be temperature dependent, working well at 160°F, but not at 70°F. It was also demonstrated that wettability reversal from oil-wet to water-wet improved the water-oil mobility ratio even though water saturation reached higher values.

3. A third mechanism that can benefit oil recovery is the wettability change from water-wet to oil-wet. A discontinuous non-wetting residual oil is converted to a continuous wetting phase, providing a flow path for what would otherwise be trapped oil. At the same time, low interfacial tension induces formation of an oil-external emulsion of water droplets in the continuous wetting oil phase. These emulsion droplets tend to block flow and induce a high pressure gradient in the region where they form. The high pressure gradient overcomes the capillary forces already decreased by low interfacial tension, thus reducing oil saturation further. Salinity of the caustic is important in this process because it helps to make the sand oil-wet and it promotes formation of water-in-oil emulsions.

4. A fourth mechanism present in caustic flooding is emulsification and entrapment. Laboratory experiments have shown that if interfacial tension were low enough, residual oil in a water-wet core could be emulsified and move downstream and be entrapped again by pore throats too small for the oil emulsion droplets to penetrate. This mechanism results in a reduced water mobility that improves both vertical and areal sweep efficiency. This mobility ratio improvement is particularly important in waterflooding viscous oils where sweep efficiency is poor.

5. In situ emulsification of the crude oil and entrainment into a continuous flowing caustic water phase is a fifth mechanism that has been suggested for caustic flooding. The caustic has an ability to prevent adherence of oil to sand surfaces. The condition necessary for continuous entrainment appears to be maintaining the interfacial tension at a low level while moving the mixture through the reservoir.

6. A sixth mechanism that has been suggested for the caustic flooding process is the solubilization of rigid films that may form at the oil-water interface. If such a rigid film is formed, the viscous forces during water injection may not be sufficient to displace the residual oil. Upon injection of a caustic solution the films are broken up or solubilized leading to the mobilization of the residual oil.

Chemicals

The most commonly used chemical has been sodium hydroxide. Sodium orthosilicate, ammonium hydroxide, potassium hydroxide, trisodium phosphate, sodium carbonate, sodium silicate and polyethylenimine, an organic, have also been suggested. Cost is an important consideration, and NaOH and sodium orthosilicate are the least expensive and the most effective in increasing the incremental oil recovery.

Laboratory Design

From the discussion presented on caustic flooding mechanisms it appears necessary that a laboratory testing program be designed which will result in grading of prospective fields. The economic results of any field application are related directly to a properly designed laboratory program.

A laboratory program should be designed in which some simple bench tests could be obtained on the crude oil and caustic solutions that would define the reservoirs that were potential candidates. Further work could be done by flooding sand packs or Berea sandstone. On the most promising fields, core flooding could be performed on actual reservoir rock.

A laboratory program should consider the following:
1. Acid number - This is the number of milligrams of KOH required to neutralize 1 gram of crude oil to pH = 7.0. For a good candidate, the acid number should be 0.5 mg KOH/gm crude or greater. However, further testing should be done with acid numbers of 0.2 mg KOH/gm crude or greater. It is important that the crude oil sample be free of emulsion breakers, inhibitors or other oilfield chemicals.
2. Interfacial tension lowering - Interfacial tension between the crude oil and the caustic solution should be less than 0.01 dyne/cm. Measurements can be made with spinning drop (at ambient pressure-temperature conditions with dead oil) or pendant drop apparatus (at reservoir pressure-temperature with live oil). Fluids used in the measurements should be representative of the reservoir fluids and proposed injection water.
3. Wettability alterations - If the reservoir is oil wet, the NaOH should make it water wet. Imbibition tests or contact angle measurements can be performed to study wettability. The former is more meaningful and can conveniently be performed on the bench in glassware.
4. Emulsion formation - To study the formation of emulsions with NaOH solutions, the simplest experiments consist of shaking measured volumes of NaOH solutions of known concentration with the crude oil in glassware. The type of emulsions formed are determined and their viscosity measured. The stability of the emulsions will be determined by allowing them to age.
5. Rigid films - Some crude oils form rigid films upon contact with brines. These can be studied in the same apparatus as contact angles or interfacial tension. If rigid interfacial films are formed between the crude oil and the water, the viscous force during water injection may not be sufficient to displace the residual oil. Upon injection of a caustic solution, the films are broken up or solubilized, leading to the mobilization of the residual oil.
6. Reaction with reservoir rock and minerals - Jennings and Johnson, (1974) have recommended the following procedures for determining the caustic reactivity of reservoir rock:
   a. Clean about 600 gm of lightly pulverized reservoir rock by toluene extraction and dry.
   b. Pack the clean, dry sand of weight (W) into an 18-in long 1¼-in. I.D. Lucite column equipped with suitable end fittings to prevent loss of sand and permit flooding of the column. Saturate the pack with distilled water and determine its pore volume (PV).
   c. After saturation with distilled water, begin injection of caustic (NaOH) solution of concentration (C).
   d. Continue injection of caustic solution until the pH of the effluent from the column reaches a pH approaching that of the injected solution. Measure the total volume (V) of effluent solution collected to this point.
   e. The caustic rock reactivity (R) is then calculated using the following expression:
   $$R = 100 (V - PV) C/W$$
   where R is in units of meq NaOH consumed per 100 gm of rock, V and PV are milliliters, C is in meq NaOH/ml, and W is in gm of rock. Jennings and Johnson (1974) have made 69 measurements of caustic rock reactivity on samples from 20 different reservoirs. While most of the values ranged between 0.4 and 5.0 meq/100gm, several were higher, up to 22.6 meq/100 gm. One of the interesting findings was that samples taken from the same reservoir at different depths showed considerable variation, such as from 2.5 to 20.6 meq/100 gm for one reservoir.

In addition to the above procedure, the alkalinity of the effluent (to determine the caustic consumption) should be obtained during oil displacement experiments. Also, slim tube displacement tests can be used and are ideally suited for determining the reactivity of rock with the caustic solution. They permit long residence time between the NaOH solution and the reservoir rock and can easily be placed in a thermostated bath to conduct the experiment at reservoir temperature.

Laboratory Core Flooding

All of the preliminary laboratory bench testing should result in information which would allow testing of the caustic process in reservoir core material. If the quantity of preserved reservoir rock is at a minimum, the coreflood experiments could be conducted in linear rock, butted end to end to obtain the desired length. Unaltered crude oil, actual connate and injection water should be used. The floods should be run at reservoir temperature and flooding rates should be close to those expected in the field. Recovery tests should compare the caustic flood with a regular waterflood.

During the above displacement tests, the following data should be obtained:
1. Permeability
2. pH and NaOH concentration in the produced water
3. Emulsions formed, their rheological properties and stability
4. Oil recovery as a function of pore volume injected.

Field Operations

The field application of caustic flooding can be simple and relatively inexpensive to apply. If a reservoir is currently being waterflooded, it can easily be converted to a caustic flood by mixing caustic materials, at the laboratory determined optimum concentration, into the injection water. If a precipitate forms during the mixing of the caustic solution it can be removed from the system by settling or filtration.

A major operational problem reported has been a severe acceleration of plugging of production wells with gypsum. It appears that gypsum in the reservoir was dissolved by the injected caustic solution and subsequently deposited at or near the wellbore.

CASE HISTORIES

Singleton Field, Nebraska (USA)

This field has been under a peripheral waterflood since 1962. The residual oil saturation was estimated to be about 40% in the watered out portion of the reservoir. The pilot area encompassed 50 acres and had a pore volume of 756 MBbls. The average permeability is about 330 md. The operator injected a slug volume of 60650 barrels of water containing 2% by weight NaOH. Although three producing wells were affected by the caustic solution, the largest production response was observed in Well No. 8. The oil production went from no oil to a maximum of 28 BOPD and produced about 17600 Bbls of tertiary oil. This represents a recovery of about 6% of the oil-in-place.

North Ward-Estes Field, Texas (USA)

This field had been under a successful waterflood for a number of years when, in 1973, a pilot test of NaOH flooding was initiated. The pilot area was a regular five-spot pattern, 5 acres in areal extent. Extensive laboratory tests with native-state cores had been carried out to define the parameters which would influence the field pilot application. The NaOH concentration used was 4.85% by weight, making this the highest caustic concentration ever employed in a field test.

The pilot test has been well described in the literature (Raimondi et al, 1976). The application appears to have recovered additional oil although it is difficult to estimate how much of this oil would not have been recovered by waterflooding. Some plugging of the production wells with gypsum occurred.

Midway - Sunset Field, California (USA)

The alkaline flooding pilot test was carried out primarily to improve volumetric sweep efficiency by the injected water. The test involved the injection of oil-in-water emulsion prepared at the surface using lease crude oil and fresh water containing 1% by weight NaOH. The pilot consisted of 100 acres and contained eleven production and three emulsion injection wells.

Aside from utilizing an emulsion slug as the principal displacement fluid, the Midway-Sunset pilot is interesting from many aspects, such as:
1. Radioactive tracers were used and showed that the injection of emulsion decreased water channelings.
2. Fresh water was injected both before and after the injection of the emulsion slug to buffer it from the saline injection water. The emulsion slug was 3% PV of the pilot area.

The Midway-Sunset test is regarded as successful. Estimated additional oil recovery is 12.5% of the oil-in-place (McAuliffe, 1973).

BIBLIOGRAPHY

Alvarado, D. A. and Marsden, S. S.: "Flow of Oil-in-Water Emulsions Through Tubes and Porous Media," SPE 5859 presented at the SPE-AIME 46th Annual California Regional Meeting, Long Beach, CA. (April, 1976).

Atkinson, H.: "Recovery of Petroleum From Oil Bearing Sands," U.S. Patent No. 1,651,311. (Nov., 1927).

Ban, A., Balint, V. and Pach, F.: "Ammonia Enhances Production From Hungarian Field," *O. & G. Jour.* (July 11, 1977).

Becher, P.: *Emulsions: Theory and Practice,* Reinhold Publishing Corporation, New York (1966).

Beckstrom, R. C. and Van Tuyl, F. M.: "The Effect of Flooding Oil Sands With Alkaline Solutions," *Bull.,* AAPG (1927).

Bernard, G. G.: "Method for Waterflooding Heterogeneous Petroleum Reservoirs," U.S. Patent 3,530,937. (Sept., 1970).

Bernard, G. G. and Holbrook, O. C.: "Recovery of Oil From Subterranean Formations," U.S. Patent No. 3,185,214. (May, 1965).

Bérubé, Y. G. and De Bruyn, P. L.: "Adsorption at the Rutile-Solution Interface I. Thermodynamic and Experimental Study," *Journal of Colloid and Interface Science.* (June, 1968).

Bérubé, Y. G. and De Bruyn, P. L.: "Adsorption at the Rutile-Solution Interface II. Model of the Electrochemical Double Layer," *Journal of Colloid and Interface Science.* (Sept., 1968).

Biles, J. W.: "Selective Plugging Method," U.S. Patent No. 3,658,131. (April, 1972).

Binder, G. G., Jr., Clark, N. A. and Russell, C. D.: "Method of Secondary Recovery," U.S. Patent No. 3,208,517. (Sept., 1965).

Bobek, J. E., Mattax, C. C. and Denekas, M. O.: "Reservoir Rock Wettability — Its Significance and Evaluation," *Trans.,* AIME (1958) 213.

Buckley, S. E. and Leverett, M. C.: "Mechanism of Fluid Displacement in Sands," *Trans.,* AIME (1942) 146.

Burdyn, R. F., Chang, H. L. and Cook, E. L.: "Oil Recovery by Alkaline-Surfactant Waterflooding," U.S. Patent No. 4,004,638. (Jan., 1977).

Burdyn, R. F., Chang, H. L. and Foster, W. R.: "Alkaline Waterflooding Process," U.S. Patent No. 3,927,716. (Dec., 1975).

Carmichael, J. D., Moyer, E. H., Alpay, A. O. and Boyle, P. R.: "Caustic Waterflooding Demonstration Project, Ranger Zone, Long Beach Unit, Wilmington Field, California," *DOE Symposium on Enhanced Oil, Gas Recovery and Improved Drilling Methods,* Vol. 1 - Oil, Tulsa, OK. (Aug., 1978).

Cartmill, J. C. and Dickey, P. A.: "Flow of a Disperse Emulsion of Crude Oil in Water Through Porous Media," *Am. Assoc. Pet. Geol. Bull.* (1970).

Caudle, B. H. and Witte, M. D.: "Production Potential Changes During Sweep-Out in a Five-Spot System," *Trans.,* AIME (1959) 216.

Chang, H. L.: "Oil Recovery by Alkaline-Sulfonate Waterflooding," U.S. Patent No. 3,977,470. (Aug., 1976).

Chan, P., Flock, D. L. and Gardiner, H. D.: "The Use of Slug Injection of High pH Water in the Recovery of Viscous Oils," 28th Annual Petroleum Society of CIM Technical Meeting, Preprint 8. (1977).

The City of Long Beach, Dept. of Oil Properties, Technical Proposal: "Improved Secondary Oil Recovery by Controlled Water Flooding Pilot Demonstration," *Contracts and Grants for Cooperative Research on Enhancement of Recovery of Oil and Gas.* (April, 1977).

Cooke, C. E., Jr.: "Method of Oil Recovery," U.S. Patent No. 3,251,412. (May, 1966).

Cooke, C. E., Williams, R. E. and Kolodzie, P. A.: "Oil Recovery by Alkaline Waterflooding," *Jour. Pet. Tech.* (Dec., 1974).

Cooper, R. J.: "The Effect of Temperature on Caustic Displacement of Crude Oil," SPE 3685 presented at the SPE-AIME 41st Annual California Regional Meeting, Los Angeles, CA. (Nov., 1971).

Craig, F. F., Jr.: *The Reservoir Engineering Aspects of Waterflooding,* Monograph Series 3. SPE-AIME, Dallas, TX. (1971).

Craig, F. F., Jr.: "Waterflooding," *Secondary and Tertiary Oil Recovery Processes,* Interstate Oil Compact Commission, Oklahoma City, OK. (Sept., 1974).

Cuiec, L. E.: "Restoration of the Natural State of Core Samples," SPE 5634 presented at the SPE-AIME 50th Annual Fall Meeting, Dallas, TX. (Sept., 1975).

Culberson, C. H., Pytkowicz, R. M. and Atlas, E. L.: "Hydrogen Ion Exchange on Amorphous Silica in Sea Water," *Mar. Chem.* (1975).

Devereux, O. F.: "Emulsion Flow in Porous Solids I and II," *Chem. Eng. Jour.* (1974).

Donaldson, E. C. and Thomas, R. D.: "Microscopic Observations of Oil Displacement in Water-Wet and Oil-Wet Systems," SPE 3555 presented at the SPE-AIME 46th Annual Fall Meeting, New Orleans, LA. (Oct., 1971).

Doscher, T. M. and Reisberg, J.: "Oil Recovery From Tar Sands," Canadian Patent No. 639,050. (March, 1962).

Dranchuk, P. M., Scott, J. D. and Flock, D. L.: "Effect of Addition of Certain Chemicals on Oil Recovery During Waterflooding," *Jour. Cdn. Pet. Tech.* (July-Sept., 1974).

Dykstra, H. and Parsons, R. L.: *The Prediction of Oil Recovery by Waterflood: Secondary Recovery of Oil in the U.S.,* 2nd ed., API (1950).

Ehrlich, R.: "Wettability Alteration During Displacement of Oil by Water From Petroleum Reservoir Rock," paper presented at 48th National Colloid Symposium, Austin, TX. (June, 1974).

Ehrlich, R., Hasiba, H. H. and Raimondi, P.: "Alkaline Waterflooding for Wettability Alteration — Evaluating a Potential Field Application," *Jour. Pet. Tech.* (Dec., 1974).

Ehrlich, R. and Wygal, R. J.: "Interrelation of Crude Oil and Rock Properties with the Recovery of Oil by Caustic Waterflooding," SPE 5830 presented at the SPE-AIME Improved Oil Recovery Symposium, Tulsa, OK. (March, 1976).

Emery, L. W., Mungan, N. and Nicholson, R. W.: "Caustic Slug Injection in the Singleton Field," *Jour. Pet. Tech.* (Dec., 1970).

England, D. C. and Berg, J. C.: "Transfer of Surface Active Agents Across a Liquid-Liquid Interface," *AIChE Journal.* (March, 1971).

Friedman, R. H.: "Recovery of Acidic Crude Oils With PEI (Polyethyleneimine)," SPE 5374 presented at the SPE-AIME 45th Annual California Regional Meeting, Ventura, CA. (April, 1975).

Froning, H. R.: "Plugging Highly Permeable Zones of Underground Formations," U.S. Patent 3,386,509. (June, 1968).

Froning, H. R. and Leach, R. O.: "Determination of Chemical Requirements and Applicability of Wettability Alteration Flooding," *Jour. Pet. Tech.* (June, 1967). *Trans.,* AIME (1967) 240.

Fuerstenau, D. W.: "Interfacial Processes in Mineral/Water Systems," *Pure and Applied Chem.* (1970).

Gardescu, I. I.: "Variation of Pressure Gradient with Distance of Rectilinear Flow of Gas-Saturated Oil and Unsaturated Oil Through Unconsolidated Sands," *Trans.,* AIME (1930) 86.

Graue, D. J. and Johnson, C. E., Jr.: "A Field Trial of the Caustic Flooding Process," *Jour. Pet. Tech.* (Dec., 1974).

Grim, R. E.: "Properties of Clays," *Recent Marine Sediments,* AAPG. (1939).

Hardy, W. C., Shepard, J. C. and Reddick, K. L.: "Secondary Recovery of Petroleum With a Preformed Emulsion Slug Drive," U.S. Patent No. 3,294,164. (Dec., 1966).

Harkins, W. D. and Zollman, H.: "Interfacial Tension and Emulsification," *Jour. Am. Chem. Soc.* (1926).

Herzig, J. P., Leclerc, D. M. and Le Goff, P.: "Flow of Suspensions Through Porous Media-Application to Deep Filtration," *I. & E. C.* (1970).

Holbrook, O. C.: "Waterflooding Process," U.S. Patent No. 3,036,631. (May, 1962).

Hudson, H. E., Jr.: "A Theory of the Function of Filters," *Jour. Am. Water Works Assoc.* (1948).

Jamin, J.: "On the Equilibrium Motion of Liquids in Porous Bodies," *Phil. Mag. 4th Series,* Taylor and Francis, Lmtd., London, England (1860).

Jennings, H. Y., Jr.: "Apparatus for Measuring Interfacial Tension by the Pendent Drop Method," *Rev. Scientific Instruments.* (1968).

Jennings, H. Y., Jr.: "A Study of Caustic Solution-Crude Oil Interfacial Tensions," *Soc. Pet. Eng. Jour.* (June, 1975). *Trans.,* AIME (1975) 259.

Jennings, H. Y., Jr.: "Waterflood Behavior of High Viscosity Crudes in Preserved Soft and Unconsolidated Cores," *Jour. Pet. Tech.* (Jan., 1966). *Trans.,* AIME (1966) 116.

Jennings, H. Y., Jr., Johnson, C. E., Jr. and McAuliffe, C. D.: "A Caustic Waterflooding Process for Heavy Oils," *Jour. Pet. Tech.* (Dec., 1974).

Johnson, C. E., Jr.: "Status of Caustic and Emulsion Methods," *Jour. Pet. Tech.* (Jan., 1976).

Kremesic, V. J. and Treiber, L. E.: "Effect of System Wettability on Oil Displacement by Micellar Flooding," SPE 6001 presented at the SPE-AIME 51st Annual Fall Meeting, New Orleans, LA., (Oct., 1976).

Leach, R. O.: "Use of Partially Oxidized Oil in a Waterflooding Process," U.S. Patent No. 3,195,629. (July, 1965).

Leach, R. O., Wagner, O. R., Wood, H. W. and Harpke, C. F.: "A Laboratory and Field Study of Wettability Adjustment in Waterflooding," *Jour. Pet. Tech.* (Feb., 1962). *Trans.,* AIME (1962) 225.

Lochte, H. L. and Littman, E. R.: *The Petroleum Acids and Bases,* Chemical Publishing Co., New York (1955).

Mansfield, W. W.: "The Spontaneous Emulsification of Mixtures of Oleic Acid and Paraffin Oil in Alkaline Solutions," *Australian Journal of Scientific Research Series A, Physical Sciences.* (1952).

McAuley, R. G.: "Sodium Hydroxide Enhanced Waterflood in the Wainwright Field," 28th Annual Petroleum Society of CIM Technical Meeting, Preprint 10. (1977).

McAuliffe, C. D.: "Crude-Oil-in-Water Emulsions to Improve Fluid Flow in an Oil Reservoir," *Jour. Pet. Tech.* (June, 1973).

McAuliffe, C. D.: "Oil-in-Water Emulsions and Their Flow Properties in Porous Media," *Jour. Pet. Tech.* (June, 1973).

McCaffery, F. G.: "Interfacial Tensions and Aging Behaviour of Some Crude Oils Against Caustic Solutions," *Jour. Cdn. Pet. Tech.* (July-Sept., 1976).

McCaffery, F. G. and Bennion, D. W., "The Effect of Wettability on Two-Phase Relative Permeabilies," *Jour. Cdn. Pet. Tech.* (Oct.-Dec., 1974).

McCardell, W. M.: "Method of Secondary Oil Recovery Using Surfactants Formed In Situ," U.S. Patent No. 3,298,436. (Jan., 1967).

Meadors, V. G.: "Method of Recovering Oil From Underground Reservoirs," U.S. Patent No. 3,208,515. (Sept., 1965).

Melrose, J. C. and Brandner, C. F.: "Role of Capillary Forces in Determining Microscopic Displacement Efficiency for Oil Recovery for Waterflooding," *Jour. Cdn. Pet. Tech.* (Oct.-Dec., 1974).

Michaels, A. S. and Porter, M. C.: "Water-Oil Displacements from Porous Media Utilizing Transient Adhesion-Tension Alterations," *AIChE Jour.* (July, 1965).

Michaels, A. S., Stancell, A. and Porter, M. C.: "Effect of Chromatographic Transport in Hexylamine on Displacement of Oil by Water in Porous Media," *Soc. Pet. Eng. Jour.* (Sept., 1964).

Michaels, A. S. and Timmins, R. S.: "Chromatographic Transport of Reverse-Wetting Agents and Its Effect on Oil Displacement in Porous Media," *Trans.*, AIME (1960) 219.

Mungan, N.: "Certain Wettability Effects in Laboratory Water Floods," *Jour. Pet. Tech.* (Feb., 1966). *Trans.*, AIME (1966) 237.

Mungan, N.: "Interfacial Effects in Immiscible Liquid-Liquid Displacement in Porous Media," *Soc. Pet. Eng. Jour.* (Sept., 1966). *Trans.*, AIME (1966) 237.

Mungan, N.: "Permeability Reduction Through Changes in pH and Salinity," *Jour. Pet. Tech.* (Dec., 1965).

Nutting, P. G.: "Chemical Problems in the Water Driving of Petroleum From Oil Sands," *I. & E. C.* (1925).

Nutting, P. G.: "Petroleum Recovery by Soda Process," *O. & G. Jour.* (1928).

Nutting, P. G.: "Principles Underlying Soda Process," *O. & G. Jour.* (1927).

Nutting, P. G.: "Soda Process for Petroleum Recovery," *O. & G. Jour.* (1927).

Oh, S. G. and Slattery, J. C.: "Interfacial Tension Required for Significant Displacement of Residual Oil," SPE 5992 (March, 1977).

Owens, W. W. and Archer, D. L.: "The Effect of Rock Wettability on Oil-Water Relative Permeability Relationships," *Jour. Pet. Tech.* (July, 1971). *Trans.*, AIME (1971) 251.

Radke, C. J. and Somerton, W. H.: "Enhanced Recovery with Mobility and Reactive Tension Agents," *ERDA Enhanced Oil, Gas Recovery and Improved Drilling Methods,* Vol. 1 - Oil, Tulsa, OK. (Sept., 1977).

Raimondi, P. et al: "Alkaline Waterflooding: Design and Implementation of a Field Pilot," SPE 5831 presented at the SPE-AIME Improved Oil Recovery Symposium, Tulsa, OK. (March, 1976).

Rappoport, L. A. and Leas, W. J.: "Properties of Linear Waterfloods," *Trans.*, AIME (1953) 198.

Reisberg, J.: "Secondary Recovery Method," U.S. Patent No. 3,174,542. (March, 1965).

Reisberg, J. and Doscher, T. M.: "Interfacial Phenomena in Crude-Oil-Water Systems," *Producers Monthly.* (Sept., 1956).

Sarem, A. M., "Method for Improving the Injectivity of Water Injection Wells," U.S. Patent No. 3,920,074. (Nov., 1975).

Sarem, A. M., "Mobility Controlled Caustic Flood," U.S. Patent No. 3,805,893. (April, 1974).

Sarem, A. M., "Mobility Controlled Caustic Flooding Process for Highly Heterogeneous Reservoirs," U.S. Patent No. 3,871,453. (March, 1975).

Sarem, A. M., "Mobility Controlled Caustic Flooding Process for Reservoirs Containing Dissolved Divalent Metal Cations," U.S. Patent No. 3,871,452. (March, 1975).

Sarem, A. M., "Secondary and Tertiary Recovery of Oil by MCCF (Mobility-Controlled Caustic Flooding) Process," SPE 4901 prepared for the SPE-AIME California Regional Meeting, San Francisco, CA., (April, 1974).

Sarem, A. M., "Waterflooding Process," U.S. Patent No. 3,876,002. (April, 1975).

Schechter, R. S. et al: "Spontaneous Emulsification - A Possible Mechanism for Enhanced Oil Recovery," SPE 5562 presented at the SPE-AIME 50th Annual Fall Meeting, Dallas, TX. (Sept., 1975).

Scott, G. R., Collins, H. N. and Flock, D. L.: "Improving Waterflood Recovery of Viscous Crude Oils by Chemical Control," *Jour. Cdn. Pet. Tech.* (Oct.-Dec., 1965).

Seifert, W. K.: "Carboxylic Acids in Petroleum Sediments," *Progress in the Chemistry of Natural Products*, Springer-Verlag New York, Inc., New York (1975).

Seifert, W. K., Gallegos, E. J. and Teeter, R. M.: "Proof of Structure of Steroid Carboxylic Acids in a California Petroleum by Deuterium Labeling, Synthesis and Mass Spectrometry," *Journal of American Chemical Society.* (Aug., 1972).

Seifert, W. K. and Howells. W. G.: "Interfacially Active Acids in a California Crude Oil: Isolation of Carboxylic Acids and Phenols," *Analytical Chemistry.* (April, 1969)

Shah, D. O., Tamjeedi, A., Falco, J. W. and Walker, R. D.: "Interfacial Instability and Spontaneous Formation of Microemulsions," *AIChE Journal.* (Nov., 1972).

Shapiro, I. and Kolthoff, I. M.: "Aging of Precipitates and Coprecipitation," *Jour. Amer. Chem. Soc.* (1950).

Spielman, L. A. and Goren, S. L.: "Progress in Induced Coalescence and a New Theoretical Framework for Coalescence by Porous Media," *I. & E.C.* (1970).

Squires, F.: "Method of Recovering Oil and Gas," U.S. Patent No. 1,238,355. (Jan., 1917).

Su, Y. P. and Spielman, L. A.: "Coalescence of Oil-in-Water Suspension by Flow Through Porous Media," *I. & E.C.* (May, 1977).

Subkow, P., "Process for the Removal of Bitumen from Bituminous Deposits," U.S. Patent No. 2,288,857. (July, 1942).

Taber, J. J.: "Dynamic and Static Forces Required to Remove a Discontinuous Oil Phase From Porous Media Containing Both Water and Oil," *Soc. Pet. Eng. Jour.* (March, 1969).

Treiber, L. E., Archer, D. L. and Owens, W. W.: "A Laboratory Evaluation of the Wettability of Fifty Oil-Producing Reservoirs," *Soc. Pet. Eng. Jour.* (Dec., 1972). *Trans.*, AIME (1972) 253.

Wagner, O. R. and Leach, R. O., "Improving Oil Displacement by Wettability Adjustment," *Trans.*, AIME (1959) 216.

Wnek, W. J., Gidaspow, D. and Wasan, D. T.: "The Role of Colloid Chemistry in Modeling Deep Bed Liquid Filtration," *Chem. Eng. Sci.* (1975).

Wotring, D. H.: "Orcutt Hill Field," *Improved Oil-Recovery Field Reports*, SPE-AIME, Dallas, TX. (June, 1975).

Part III

MISCIBLE DISPLACEMENT PROCESSES

Chapter 7
MISCIBLE HYDROCARBON DISPLACEMENT

INTRODUCTION

Miscible fluids are completely soluble in one another. Unlike immiscible fluids, the interfacial tension between miscible fluids is zero, and there is no distinct fluid-fluid interface. Liquid hydrocarbon such as naphtha, kerosene, gasoline, alcohol and liquified petroleum gas products (LPG), such as ethane, propane and butane, are miscible with reservoir oils immediately on contact. The LPG products are miscible with oil only as long as they remain in the liquid state. At temperatures above their critical temperatures (Table I), LPG products change into gas regardless of pressure, and thus become immiscible with oil.

TABLE I
TEMPERATURE/PRESSURE RELATIONSHIP
TO MAINTAIN LIQUID STATE

Ethane		Propane		N-Butane	
Temp (°F)	Press (psia)	Temp (°F)	Press (psia)	Temp (°F)	Press (psia)
50	460	50	92	50	22
90*	709	100	190	100	52
		150	360	150	110
		200	590	200	198
		206*	617	250	340
				300	530
				305*	550

*Critical Temperature

(after Herbeck et al, 1976)

When ethane, propane or butane are in the liquid state, they are not readily miscible with natural gas except at a high pressure (Figure 1). For example, at a temperature of 160°F, butane is miscible with lean gas (dry natural gas essentially composed of methane) only at pressures greater than 1600 psi, and it is miscible with nitrogen (air) at pressures greater than 3600 psi. Thus, if a slug of butane is injected in a reservoir at a temperature of 160°F and displaced with lean gas, it would be necessary to maintain a pressure greater than 1600 psi

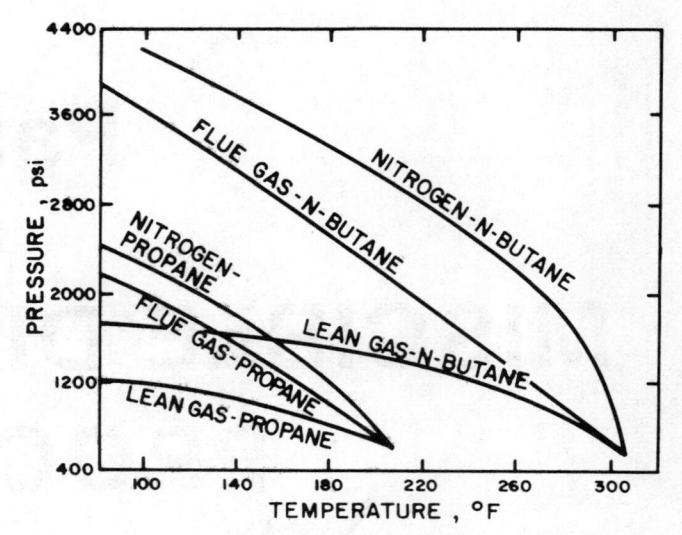

Fig. 1. Minimum pressure for miscibility, miscible slug-driving gas combinations.
(after Herbeck et al, 1976)

in order to assure miscibility of the butane with the displacing lean gas, even though the miscibility of butane with oil is attained at a pressure of about 125 psi.

There are three different miscible hydrocarbon displacement processes for oil recovery. The first process, known as the Miscible Slug Process, consists of injecting a slug of liquid hydrocarbon equivalent to about 5% PV followed by natural gas, or gas and water, in order to drive the slug through the reservoir. The second process, known as the Enriched Gas Process, consists of injecting a slug of enriched natural gas followed by lean gas or lean gas and water. The size of the slug usually ranges between 10-20% PV, and it is enriched with ethane through hexane ($C_2 - C_6$). The third process, known as the High Pressure Lean Gas Process, consists of injecting lean gas at a high pressure in order to cause retrograde evaporation of the crude oil and formation of a miscible phase, consisting of $C_2 - C_6$, between the gas and oil. Thus, the main difference between Enriched Gas and High Pressure Lean Gas Processes is that in the former, C_2-C_6 components are transferred from the gas to the oil, whereas in the latter,

C_2 - C_6 components are transferred from the oil to the gas.

The Miscible Slug Process is schematically represented as shown in Figure 2. The depth at which the process can be applied must exceed 2000-2500 ft. This is because of the high pressure needed to maintain miscibility between the slug and the displacing natural gas. If air or flue gas is used to displace the slug, the pressure requirement would even be higher (Figure 1), and the process must then be applied at greater depths. Under any set of conditions, the reservoir temperature must be less than the critical temperature of the miscible slug.

The Enriched Slug Process also requires high pressures. Figure 3 is a schematic representation of the process, and Figure 4 shows the relationship between pressure and concentration of enriching agent at a temperature of 100°F. From Figure 4, it is evident that if the reservoir temperature is 100°F, if the displacement pressure is 1500 psi and if the enriching agent is propane, the enriched slug must contain at least 42 mol % of propane in order to maintain miscibility. At higher pressures, Figure 4 shows that miscibility can be maintained at lower propane concentrations.

The High Pressure Lean Gas Process is schematically represented in Figure 5. This process can only be applied

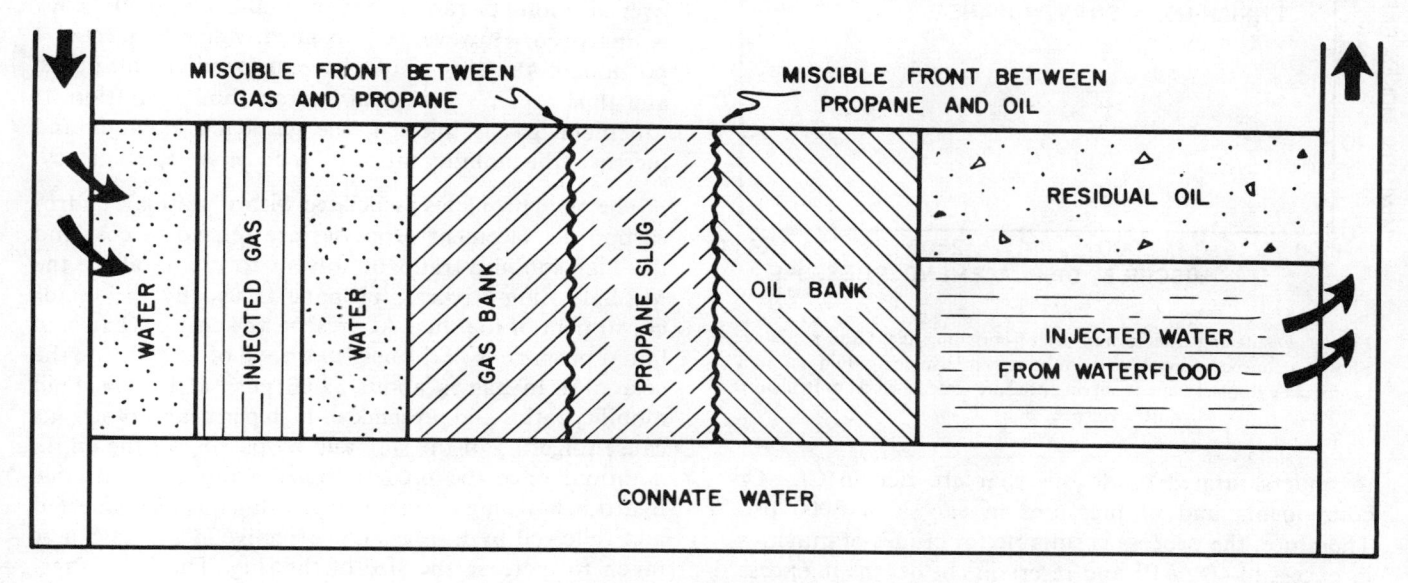

Fig. 2. Schematic representation of propane slug process.
(after Herbeck et al, 1976)

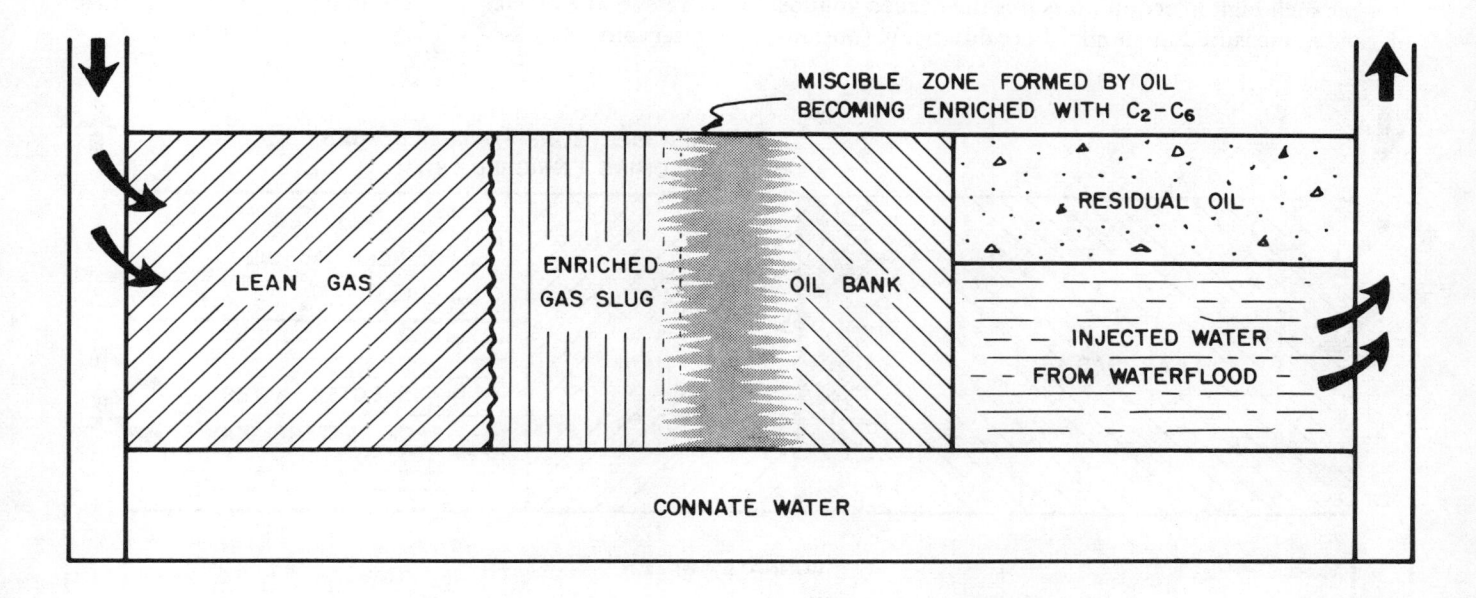

Fig. 3. Schematic representation of enriched gas process used in tertiary recovery.
(after Herbeck et al, 1976)

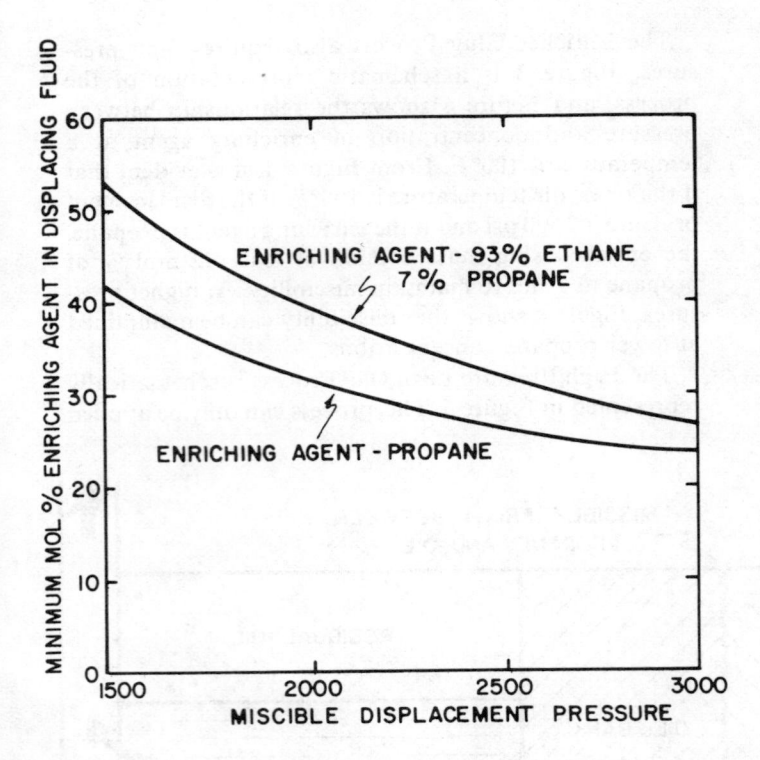

Fig. 4. Displacing fluid is mixture of enriching agent and methane. Shown is concentration of enriching agent in displacing fluid required for miscible displacement at varying pressure (calculated from Benham). (after Herbeck et al, 1976)

to undersaturated crude oils that are rich in C_2 - C_6 components and at pressures in excess of 3000 psi. Therefore, the process is suitable for crudes of gravities in excess of 40° API and reservoirs at depths in excess of 5000 ft. As Figure 6 suggests, oil recovery by this process is a function of the displacement pressure. However, at such high injection pressures the needed volume of gas, as measured at standard conditions, is substantial. In addition, since the injected gas must replace both crude oil and movable water, several MCF of gas are required for the recovery of one barrel of oil. For these reasons, it is necessary to investigate the possibility of injecting gas alternately with water.

There is still another important reason for injecting water alternately with gas, especially in the Miscible Slug and Enriched Slug Processes. The mobility of the injected slug is higher than the mobility of the oil. This high mobility combines with the high mobility of the displacing gas to give low sweep and displacement efficiencies. Caudle and Dyes (1958) found that when injection of a slug is followed by gas and water the over-all mobility ratio is reduced, and sweep efficiency is improved. However, if too much water is injected a portion of the slug could be entrapped in the pores, and thus become a residual hydrocarbon saturation. If too much gas is injected, the gas bank expands and increases the mobility ratio and reduces sweep efficiency.

The main problems associated with all miscible hydrocarbon displacement processes are related to cost and the high mobility ratio of solvent to crude oil. In the Miscible Slug Process, propane is usually the major constituent of the slug. At present, the cost per barrel of liquid propane is in the neighborhood of $12.00, and the cost of lean gas is about $2.00 per MCF. The high mobility ratios and presence of high permeability streaks cause fingering of the slug and by-passing of the oil. In addition, once the predetermined volume of miscible hydrocarbon slug or enriched gas slug has been injected and followed by lean gas no corrective measures can be taken to increase the size of the slug. The High Pressure Lean Gas Process is not subject to this latter drawback, since there is no slug involved. However, the High Pressure Lean Gas Process is not applicable to most reservoirs.

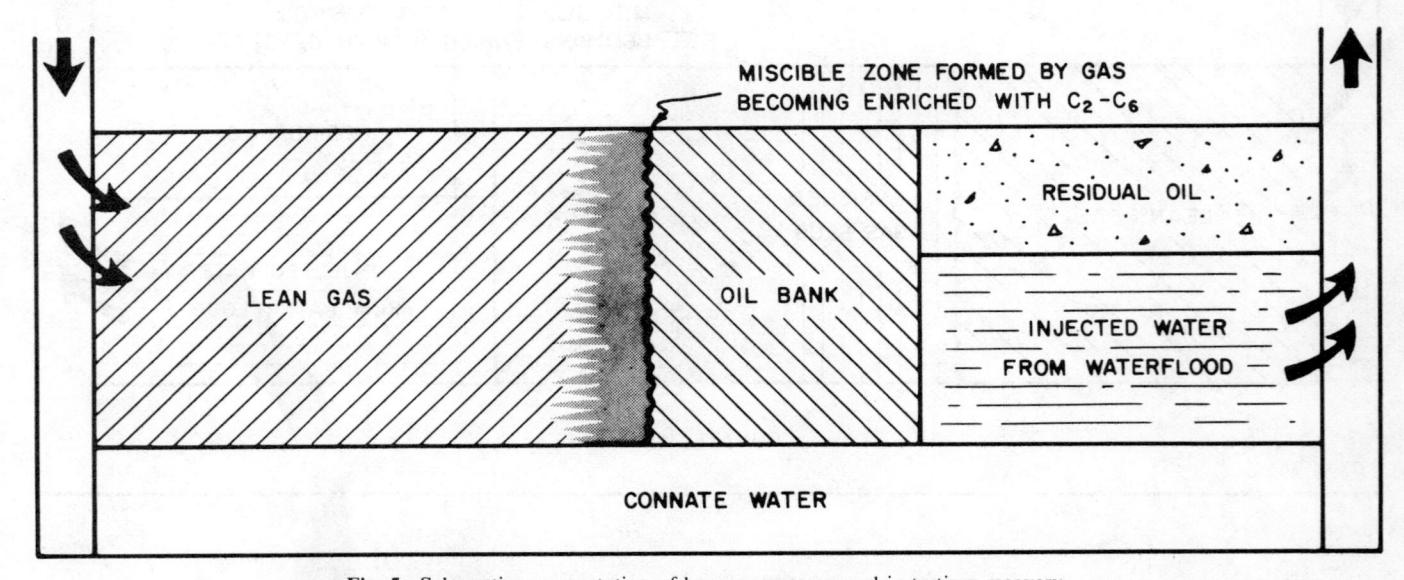

Fig. 5. Schematic representation of lean gas process used in tertiary recovery. (after Herbeck et al, 1976)

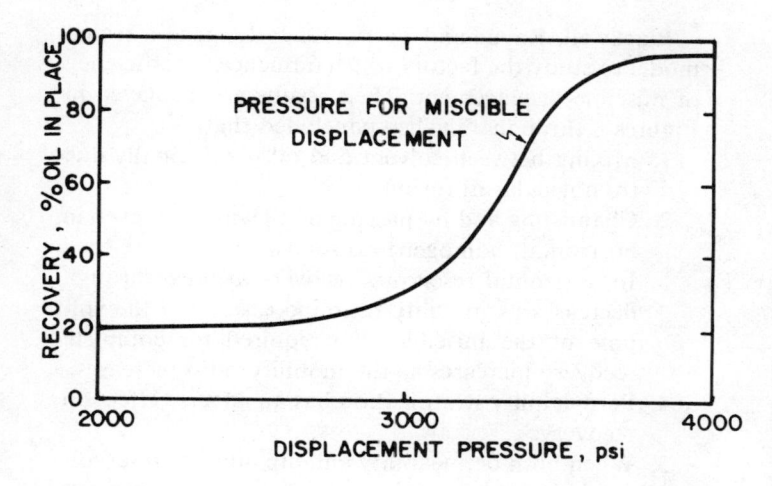

Fig. 6. Graph showing laboratory recovery for different displacement pressures. The pressure needed for miscibility can be estimated from this type graph.
(after Herbeck et al, 1976)

considerably. Prior to 1966 most projects utilized the Miscible Slug method and since that time the emphasis has been on the Enriched Gas technique.

Design of a miscible hydrocarbon displacement process requires laboratory model studies followed by field pilot tests. The reservoir parameters involved in the design are: depth and thickness of reservoir rock, viscosity of oil, reservoir temperature and pressure, permeabilities and dip angle. The studies should yield: optimum slug size, rate of flooding and recovery efficiency. In general, miscible slug flooding is best applied to reservoirs with the following characteristics: oil viscosity less than 5 cp at reservoir conditions, steeply dipping reservoir rock of thickness less than 25 ft, homogeneous permeability less than 100 md, reservoir temperature below the critical temperature of the slug and sufficient reservoir pressure for miscibility of the slug with both oil and driving gas. The slug volume usually ranges between 2-10% PV, and could be as much as 20% PV for the Enriched Gas Process.

The number of active miscible hydrocarbon displacement projects in the Western Hemisphere has fluctuated

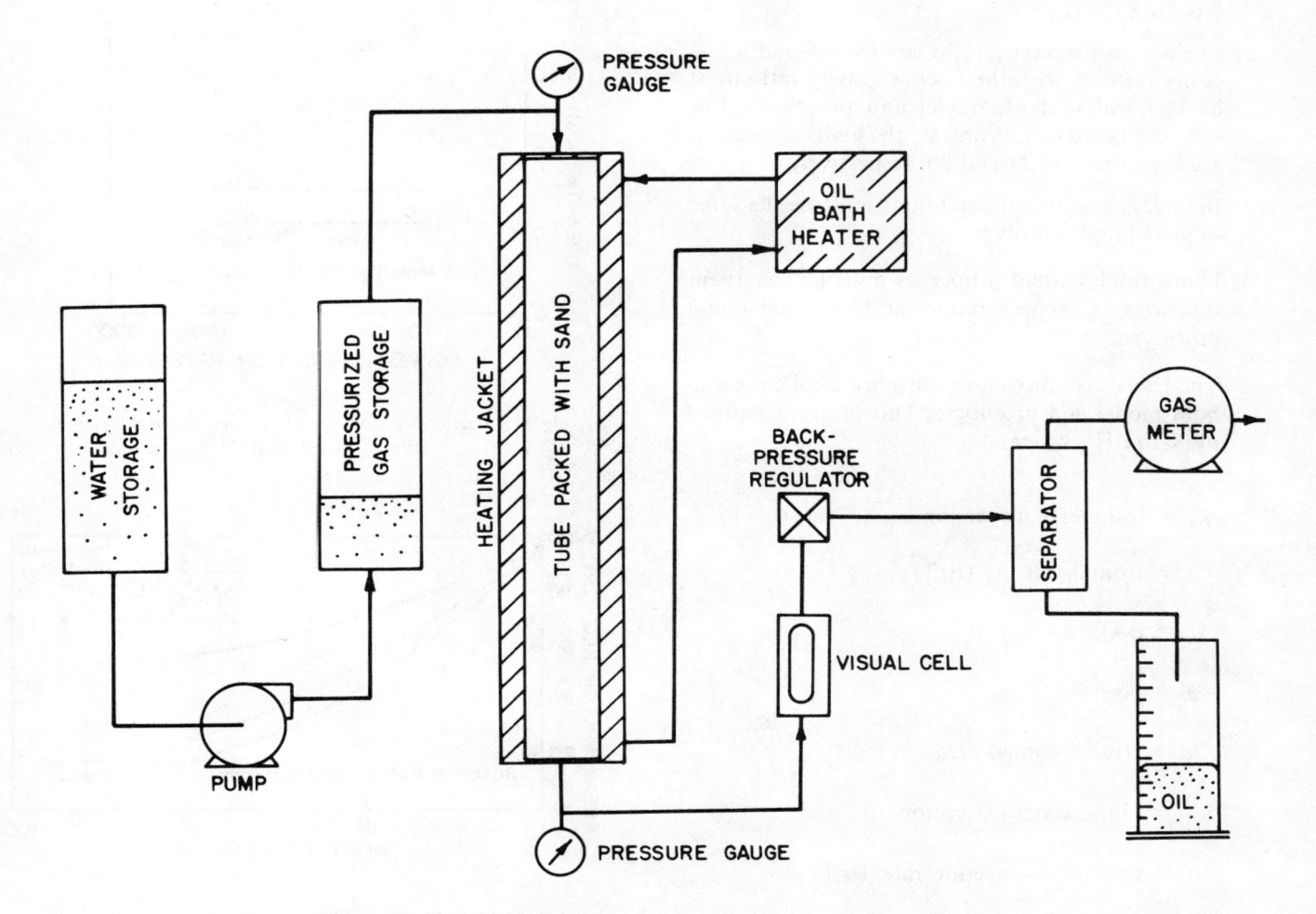

Fig. 7. Schematic diagram of laboratory apparatus used in displacement tests for determining pressure required to miscibly displace oil from sand-packed tube.
(after Herbeck et al, 1976)

TECHNICAL DISCUSSION
Experimental Studies

Laboratory tests are extensively used for evaluation of miscible displacement oil recovery processes. These tests require an apparatus similar to that shown in Figure 7. However, in order to obtain meaningful results, it is necessary to construct a properly scaled model. The literature is replete with experimental studies, but only a few of these studies have used scaled models. The main problem in designing scaled models is that precise scaling of transverse dispersion and strict adherence to the requirement of geometrical similarity result in large laboratory models and long experiment times.

Pozzi and Blackwell (1963) evaluated the relative importance of the various mechanisms which affect the scaling of miscible displacement processes. They developed the following set of criteria which permits design of reasonably sized models, providing that longitudinal mixing is unimportant.

1. The ratio of the model length L and height H need not be scaled when viscous-gravity ratio is less than 0.4 $(L/H)^2$.
2. For a viscous-gravity ratio less than 4 and a viscosity ratio of 16.3, the viscous-gravity ratio must be identical in both model and prototype. For viscosity ratios of 1.85 and 69, the limits of viscous-gravity ratios are 27 and 20, respectively.
3. Boundary and initial conditions must be the same in model and prototype.
4. Dimensionless fluid properties must be congruent functions of concentration in both model and prototype.
5. The transverse mixing group must be identical in both model and prototype. This group is defined by $K_t L/UH^2$, where:

K_t = transverse dispersion coefficient, (L^2/T)

U = frontal velocity (L/T)

U = $\phi AL (1 - S_{wi})/qt$

ϕ = porosity

A = cross-sectional area

S_{wi} = initial water saturation

q = volumetric injection rate, (L^3/T)

t = time, (T)

Blackwell, Rayne and Terry (1959) designed a scaled model to study the factors which influence the efficiency of miscible displacement. Their results are displayed in Figures 8 through 13. They concluded that:

1. Mixing between solvent and oil is principally due to molecular diffusion.
2. Channeling and by-passing of oil will occur even in horizontal, homogeneous sands.
3. In horizontal reservoirs recovery at breakthrough decreases as mobility ratio increases, and the volume of the miscible slug required for complete recovery increases as the mobility ratio increases.
4. Permeability stratification has an adverse effect on recovery.
5. When both permeability and dip angle of reservoir are adequate, gravity segregation can prevent channeling.

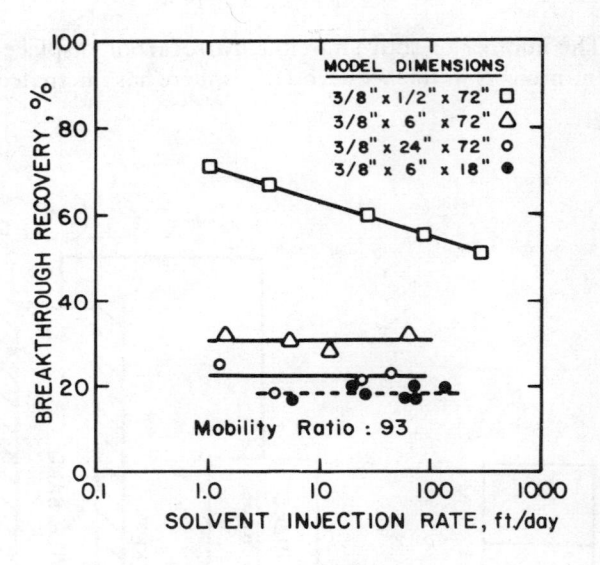

Fig. 8. Effect of injection rate on recovery at breakthrough. (after Blackwell et al, 1959)

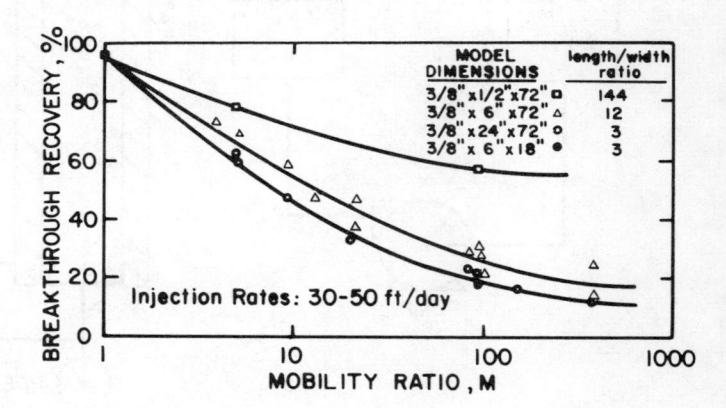

Fig. 9. Effect of mobility ratio on recovery at breakthrough. (after Blackwell et al, 1959)

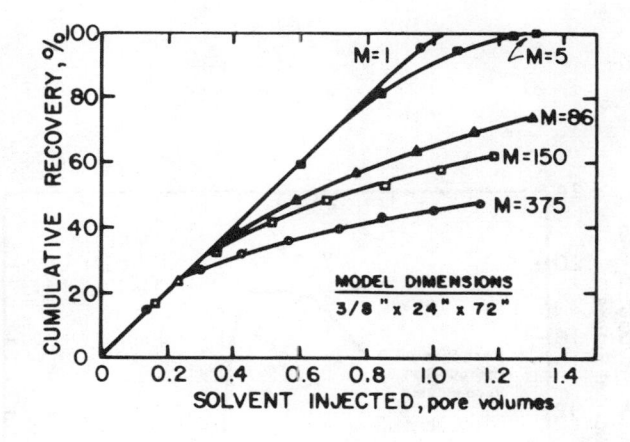

Fig. 10. Effect of mobility ratio on cumulative recovery.
(after Blackwell et al, 1959)

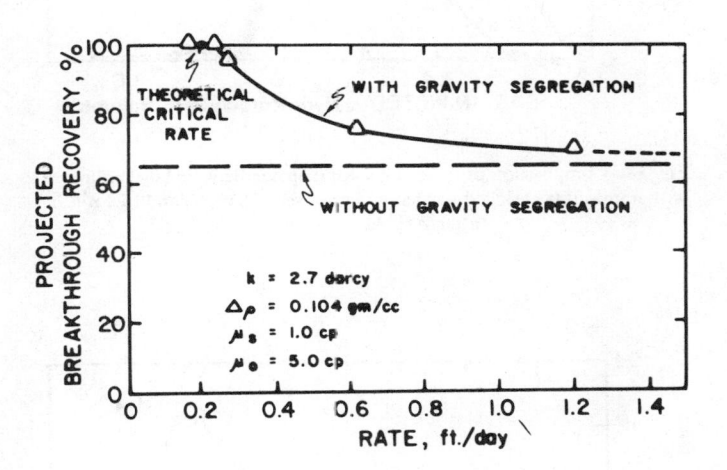

Fig. 11. Effect of rate on recovery at breakthrough with gravity segregation.
(after Blackwell et al, 1959)

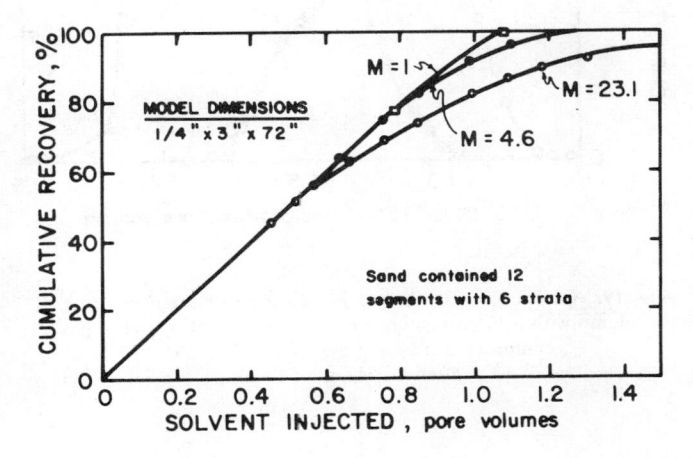

Fig. 12. Effect of mobility ratio on cumulative recovery from segmented-stratified model.
(after Blackwell et al, 1959)

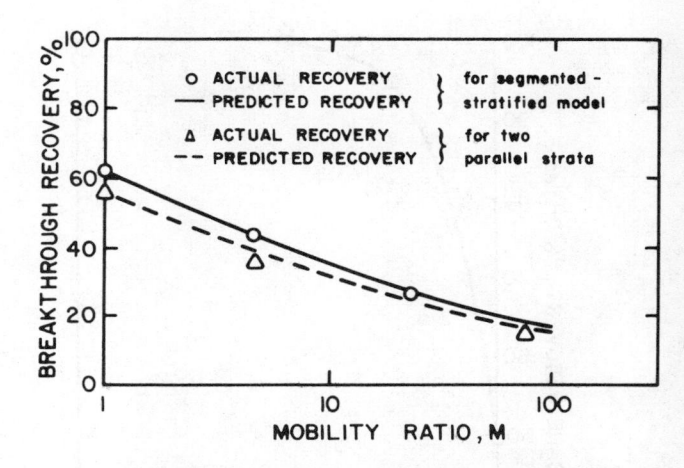

Fig. 13. Comparison of predicted with actual recoveries at breakthrough for stratified models.
(after Blackwell et al, 1959)

Arnold, Stone and Luffel (1960) studied displacement of oil by the Enriched Gas Process. They used a sand-packed tube 20-ft long and ½-in. in diameter. The oil in the reservoir was simulated by decane saturated with methane. The enriched gas was a mixture of methane and butane. The scavenging gas was methane. The model temperature and pressure were kept at 160°F and 2500 psia, respectively. The results are shown in Figures 14 through 20. The authors concluded that:

1. When major by-passing of oil by enriched gas does not occur, a small bank of enriched gas driven by methane is sufficient to obtain an oil recovery comparable to that obtained by continuous injection of enriched gas.
2. The dispersion of an oil-miscible enriched gas bank is controlled by molecular diffusion and conventional mixing.
3. Displacement with an oil-miscible bank results in increased oil recovery. In addition, the total amount of injected gas would be less than that required in immiscible displacement.

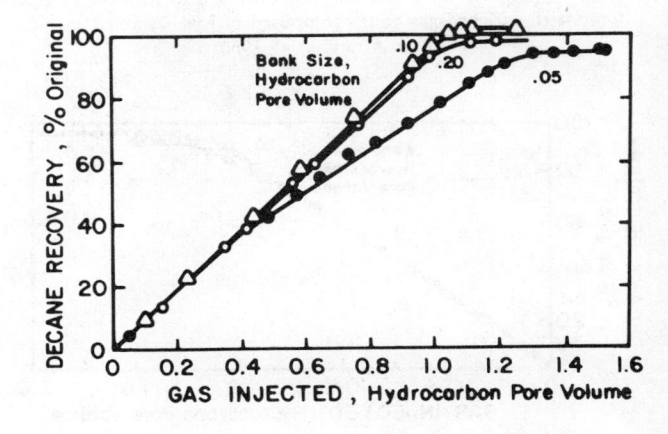

Fig. 14. Decane recovery during displacement in 20-ft column with enriched-gas banks composed of 75% C_1 and 25% n-C_4.
(after Arnold et al, 1960)

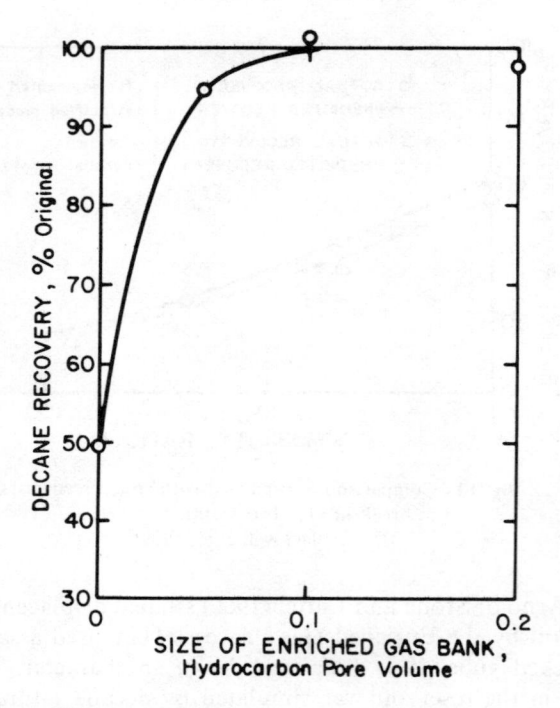

Fig. 15. Ultimate decane recovery as a function of bank size for displacements in 20-ft column with enriched-gas banks composed of 75% C_1 and 25% $n\text{-}C_4$.
(after Arnold et al, 1960).

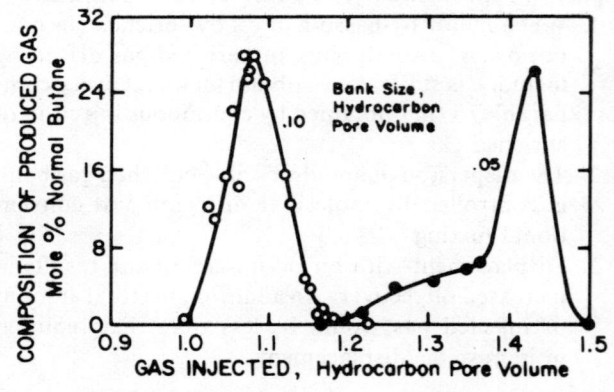

Fig. 16. Composition of flashed gas during displacement in 20-ft column with enriched-gas banks composed of 75% C_1 and 25% $n\text{-}C_4$.
(after Arnold et al, 1960)

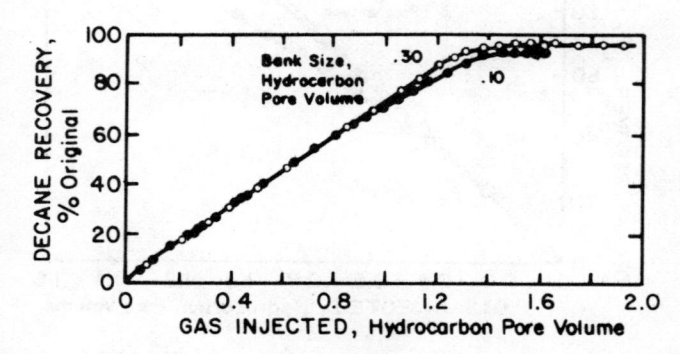

Fig. 17. Decane recovery during displacement in 20-ft column with immiscible enriched-gas banks composed of 85% C_1 and 15% $n\text{-}C_4$.
(after Arnold et al, 1960)

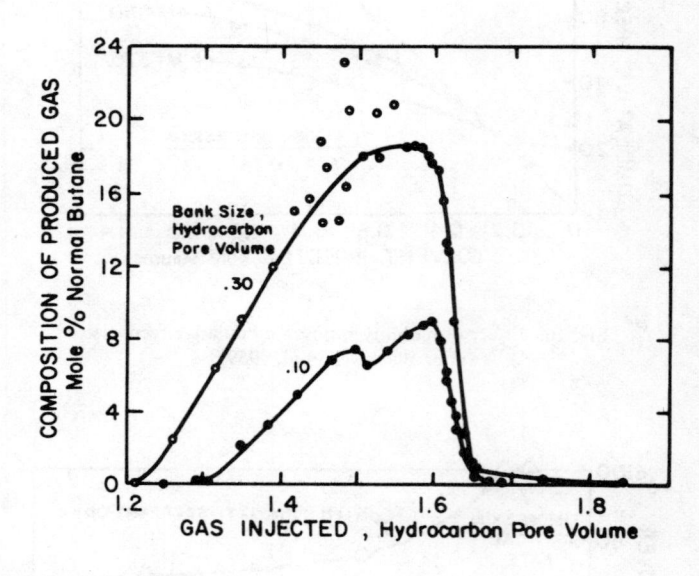

Fig. 18. Composition of flashed gas for displacements in 20-ft column with immiscible enriched-gas banks composed of 85% C_1 and 15% $n\text{-}C_4$.
(after Arnold et al, 1960)

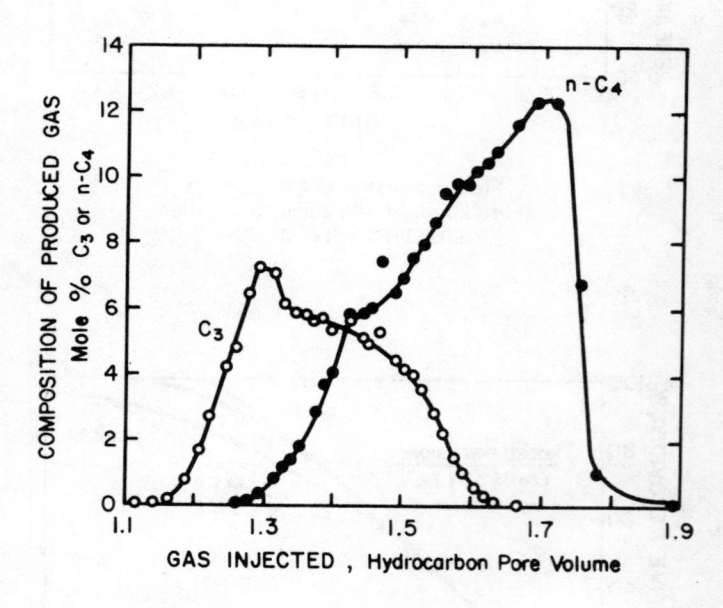

Fig. 19. Composition of flashed gas during displacements in 20-ft column with 0.30-pore volume bank of immiscible enriched gas composed of 85% C_1, 5% C_3 and 10% $n\text{-}C_4$.
(after Arnold et al, 1960)

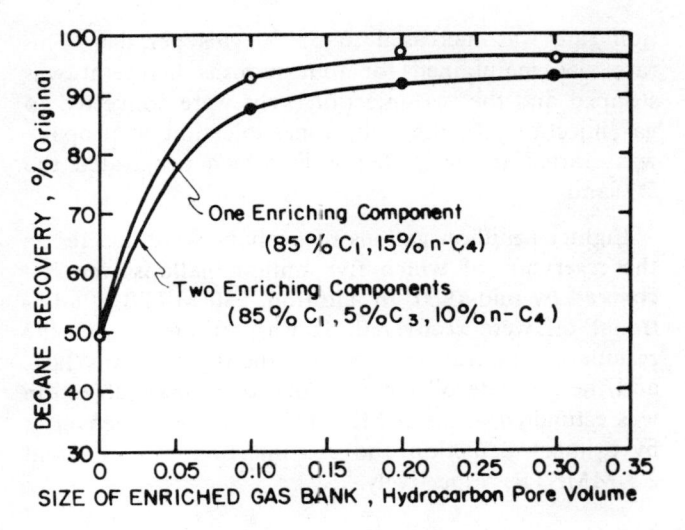

Fig. 20. Ultimate recovery as a function of bank size for immiscible displacements in 20-ft column.
(after Arnold et al, 1960)

Model Studies

The differential equations which describe multi-dimensional miscible displacement are (Peaceman and Rachford, 1962):

$$\nabla \cdot (k/\mu)(\nabla p + \rho g \nabla h) = 0 \quad (1)$$

and,

$$KX(\partial^2 C/\partial x^2) + KY(\partial^2 C/\partial y^2) + \nabla \cdot (Ck/\mu)(\nabla p + \rho g \nabla h) = \phi \partial C/\partial t \quad (2)$$

where p is pressure, ρ is density, g is gravitational acceleration, h is height above a datum, k is local permeability, μ is viscosity, C is volume concentration of solvent, x and y are distances and KX and KY are the dispersivities in the x and y directions, respectively. It is assumed that the first two terms in Equation 2 represent the dispersion.

Equations 1 and 2 were solved for both one and two-dimensional systems by the method of characteristics (Garder, Peaceman and Pozzi, 1964). The results agreed quite well with the experimental results obtained earlier by Pozzi and Blackwell (1963). These experimental results have been presented in this chapter.

Settari, Price and Dupont (1977) developed and applied variational methods for simulation of miscible displacement in porous media. They showed that variational methods give accurate results when the dispersion levels are low and the slugs are small. Variational methods are particularly useful where finite difference approximations would require 10^4 to 10^5 grid nodes.

CONCLUSIONS

Miscible hydrocarbon displacement techniques have been shown to be quite effective in removing oil from the pores. However, vertical and horizontal sweep efficiencies of miscible floods using gas or LPG have generally been poor (Holm, 1976). The injected slugs are subject to excessive dispersion and deterioration. Gravity segregation, high mobility ratios, irregular permeability and trapping of solvent by dead-end pores combine to cause deterioration and loss of miscibility. Therefore, uneconomically large slugs are needed. Process variations have been developed to improve conformance of miscible displacements. These techniques are: alternate and prewater injection, vertical displacement (effective only in steeply dipping reservoirs), selective injection and displacement by a soluble oil and/or oil-external micellar solution.

CASE HISTORIES

Numerous articles are available on field applications of miscible hydrocarbon displacement techniques. The more recent literature, however, shows that the industry is moving towards application of enriched gas and lean or flue gas techniques. The need for excessively large LPG slugs and the rising costs of liquid propane make the LPG slug method economically unattractive. The lean gas or enriched gas methods are particularly applicable in areas where there are no readily available markets for the gas.

All the case histories presented here have been reported to be technical successes. However, technical success does not necessarily mean economic success. Thus, careful design and economic analysis are necessary to ensure both technical and economic success.

Midway-Sunset Field, California (USA)

The reservoir, which covers 530 acres originally contained 7.5 million STB of 31° API oil (Block and Donovan, 1961). The reservoir is a Miocene stratigraphic trap approximately 5300 ft deep and composed of three major productive intervals. Measured and estimated properties of reservoir rock are summarized in Table II. Saturation data are given in Table III, and properties of reservoir and injected fluids are summarized in Table IV. Before propane injection began, reservoir pressure and temperature were 1500 psi and 180°F, respectively.

TABLE II
PROPERTIES OF RESERVOIR ROCK

Sand Member

Air Permeability	D_1	D_2	D_3	Pool
Arith. Avg., md	39	232	204	181
Geom. Avg., md				96
Range, md	10-160	10-1500	10-1220	10-1500
Variation				0.7
Porosity, %				24.1

(after Block and Donovan, 1961)

TABLE III
AVERAGE FLUID SATURATION DATA
(ALL NUMBERS IN PER CENT OF PORE VOLUME)

Initial Conditions
- S_{oi} 75
- S_{wi} 25
- S_{gi} 0

Conditions at Start of Propane Injection
Secondary Gas Cap
- S_o 44
- S_w 25
- S_g 31

Oil Band
- S_o 68
- S_w 25
- S_g 7

Watered-Out Area
- S_o 47
- S_w 53
- S_g 0

(after Block and Donovan, 1961)

TABLE IV
PROPERTIES OF RESERVOIR AND INJECTION FLUIDS

	Initial	At Start of Propane Injection	April, 1960
Reservoir Pressure, psia	2260	1500	1300
Oil			
p_i, psia	2260	—	—
B_o, rb/STB	1.318	1.239	1.218
R_s, Scf/STB	529	366	323
μ_o, cp	0.663	0.844	0.917
Gravity, °API	31.0	30.5	32.0
Water			
Salinity, grains/gal	1100	—	—
Gas			
B_g, rb/MSCF	1.25	2.00	2.30
Produced Gas Gravity	0.75	0.77	0.99
Propane			
Reid Vapor Pressure, psig	—	200	220

(after Block and Donovan, 1961)

Twenty-one wells were drilled in the reservoir on a 10 acre spacing. The wells were completed with 7-in pipe cemented on bottom and then gun or jet perforated.

Originally there was no gas cap, but a large secondary cap developed. Gas was injected in the upper portion of the pool. Within approximately two years a total of 780 MMCF of gas were injected.

Liquid propane injection began in the D_1 and D_3 sand at a rate of 10000 gals per day through one well, while gas injection was continued in another well in order to maintain the reservoir pressure above that required for miscibility. After six months the gas injection well was converted to propane injection, and the propane injection rate was increased to 52000 gals per day. This rate was maintained for four months before it was stopped and the two injection wells were converted to gas injection. At the same time, injection of propane was started in two other wells which penetrated the D_2 sand.

Eighteen million gallons of propane were injected in the reservoir, of which five million gallons were recovered by mid-1960. In addition, 230 MSTB of additional oil were recovered. It was estimated that the residual oil saturation after propane flooding was 22%, and the ultimate oil recovery due to propane injection was estimated at 3.8 MMSTB. The ultimate recoveries by primary depletion and gas injection were 1.9 and 2.3 MMSTB, respectively.

Pembina Field, Canada

A 10-acre inverted five-spot pilot study was made in an isolated part of the reservoir (Justen et al, 1960). One injection and four production wells were drilled specifically for this pilot test. The production wells were located 467 ft north, east, south and west of the injector. After coring each well, 5½-in casing was run and cemented through the pay section, then perforated opposite the Upper Cardium sand and fraced. The fracturing treatments averaged 1000 lbs of 20 - 40 mesh sand in 700 gals of a low fluid loss sand-carrying agent. Injection rates and wellhead fracturing pressures averaged 5.5 BPM and 2535 psi, respectively. After fracturing, the productivity index of each pilot well was measured, averaging 0.41 BOPD/psi drawdown. Examination of cores showed that the reservoir was stratified with respect to permeability (Table V) and indicated the presence of permeability streaks. For this reason, a slug of 2000 SCF of helium was included in the scavenging gas. The results are shown in Figure 21.

TABLE V
SELECTED PERMEABILITY ZONE FOR WEST QUADRANT

	Injection Well			West Well			Average Value for Quadrant			
Zone	K (md)	h (ft)	φ (%)	K (md)	h (ft)	φ (%)	K (md)	h (ft)	φ (%)	S_w (%)
W-1	113	0.8	23.0	142	0.7	21.6	127	0.75	22.3	5.0
W-2	89	0.6	25.2	203	0.4	23.3	146	0.5	24.2	5.0
W-3	13	0.7	19.3	16	0.5	14.8	15	0.6	17.1	9.8
W-4	98	0.8	22.6	55	0.9	16.3	76	0.85	19.5	6.4
W-5	183	0.6	20.7	167	0.8	17.3	175	0.7	19.0	7.0
W-6	62	0.6	21.4	99	0.4	20.5	81	0.5	21.0	5.0
W-7	155	0.7	27.5	147	0.8	20.3	151	0.75	23.9	5.0
W-8	20	1.8	18.4	28	1.9	18.6	24	1.85	18.5	7.6

(after Justen et al, 1960)

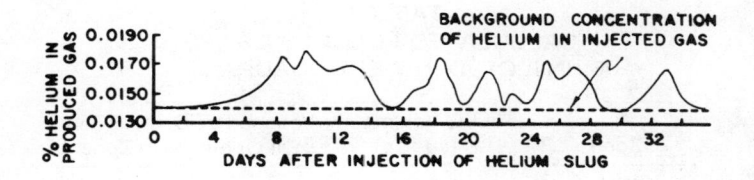

Fig. 21. Helium tracer test, west quadrant Pembina pilot.
(after Justen et al, 1960)

The reservoir fluid characteristics in the pilot area were: saturation pressure 1560 psi, reservoir temperature 118°F, oil gravity 37.2° API, oil viscosity 1.38 cp at 118°F, formation volume factor at saturation pressure 1.22 and solution gas-oil ratio 416 SCF/Bbl. The initial reservoir pressure at the sand face was approximately 69 psi above the saturation pressure. Other reservoir characteristics are given in Table VI.

TABLE VI
RESERVOIR CHARACTERISTICS

Quadrant	Area (acres)	Effective Pay (ft)	Avg. Porosity (%)	Avg. Water Sat. (%)	Effective Oil in Place (STB)
North	2.449	7.4	19.0	5.5	21841
South	2.548	8.5	18.4	7.6	23397
East	2.472	8.1	18.8	7.0	22270
West	2.531	8.0	18.9	6.9	22722
Total	10.000				90230
Weighted Average		8.0	18.8	6.7	

(after Justen et al, 1960)

Injection was carried out in the following order: 7692 reservoir Bbls of LPG, 250 MMSCF of gas, 7500 Bbls of water and 52 MMSCF of gas. The LPG slug consisted of 50:50 mixture by volume of butane and propane. The injection performance is summarized in Figure 22.

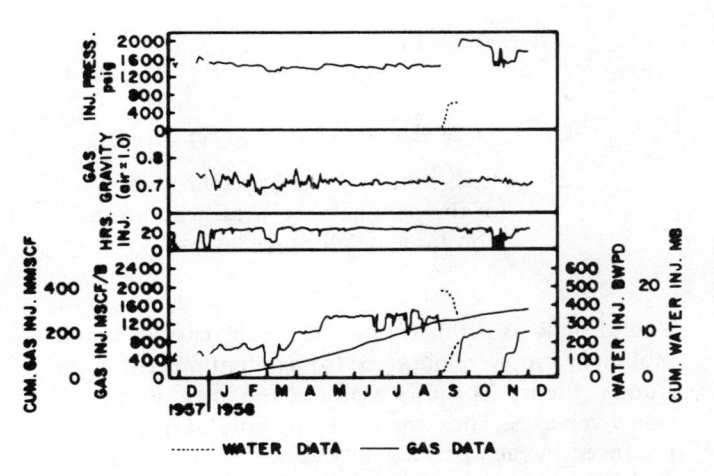

Fig. 22. Injection performance.
(after Justen et al, 1960)

The pilot was conducted at constant reservoir pressure in order to prevent movement of fluids from the pilot area. Therefore, zero net withdrawals were maintained throughout the pilot's life. The production rates were controlled in proportion to the pore volume of each quadrant. The production performance is shown in Figure 23 and a recovery summary is given in Table VII.

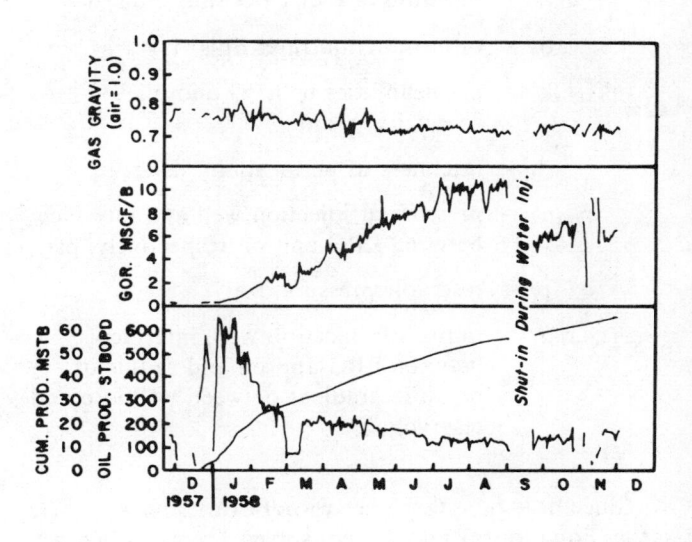

Fig. 23. Composite production performance.
(after Justen et al, 1960)

TABLE VII
RECOVERY SUMMARY

Well	Cum. Recovery to Start of Water Inj. STB	%	Add. Recovery Obtained after Water Inj. STB	%	Cum. Recovery to End of Pilot STB	%
North	12012	55.1	2668	12.1	14680	67.2
South	14596	62.4	1687	7.2	16283	69.6
East	14431	64.2	2361	11.2	16792	75.4
West	15424	68.0	1778	7.7	17202	75.7
Total	56463	62.6	8494	9.4	64957	72.0

(after Justen et al, 1960)

Flow in the strata was calculated assuming single phase with no cross flow between permeable layers. It was also assumed that connate water did not lower the permeability, and that because of miscibility, no relative permeability effects existed behind the flood front.

To calculate the volume of LPG injected into each layer, radial flow and a sharp interface between LPG and oil were assumed.

Thus:

$$q_{LPG} = 7.07\, k_{LPG}\, h(p_{iw} - p_{if})/(\mu_{LPG}\, \ln r_{if}/r_{iw}) \quad (3)$$

$$q_o = 7.07\, k_o h\, (p_{if} - p_r)/(\mu_o \ln r_r/r_{if}) \quad (4)$$

where:

q_{LPG} = injection rate of LPG slug, Bbls/day

q_o = oil production rate, Bbls/day

k_{LPG}, k_o = permeabilities to LPG and oil, respectively, md

h = thickness of permeability layer, ft

p_{iw}, p_{if} = pressure at injection well and interface between LPG and oil respectively, psi

p_r = reservoir pressure, psi

r_{iw}, r_{if}, r_r = radius of: injection well, interface between LPG and oil and radius of pressure gradient between wellbore and reservoir, ft

Since $q_o = q_{LPG}$ (zero net withdrawal), and $k_o = k_{LPG} = k$, Equations 3 and 4 were solved for p_{if}, which was substituted in Equation 3. From the expression of $q_{LPG}dt = dV_{LPG}$, an expression for dV_{LPG} was obtained. But,

$$dV_{LPG} = 2\pi r_{if} h \phi (1 - S_w) dr_{ij}/5.615 \quad (5)$$

also. Thus equating Equation 5 with the expression for dV_{LPG} that was derived earlier and carrying out the integration and solving for t:

$$t = [0.079\, \phi\, (1 - S_w)/k\, (p_{iw} - p_r)][\mu_o r_{if}^2 (\ln r_r - \ln r_{if} + 1/2) +$$
$$\mu_{LPG}\, r_{if}^2\, (\ln r_{if} - \ln r_{iw} - 1/2)] \quad (6)$$

By substituting the different reservoir parameters in Equation 6 and the LPG injection time of 9.5 days, and taking $r_r = 1000$ ft, the following result was obtained:

$$k/\phi(1 - S_w) = (1.622 \times 10^{-2}/\Delta p)\,[r_{if}^2\,(5.149 - 0.64\,\ln r_{if})] \quad (7)$$

where:

$\Delta p = p_{iw} - p_r$, psi

Thus, by substituting the permeability of each layer in Equation 7, the results shown in Table VIII were obtained.

TABLE VIII
CALCULATED VOLUMES OF LPG INJECTED IN WEST QUADRANT

Zone	LPG Volume (cu ft)	Volume Per Cent LPG of Oil Originally in Zone
W-1	2100	12.0
W-2	1625	12.9
W-3	151	1.48
W-4	1399	8.15
W-5	3010	22.0
W-6	860	7.8
W-7	2545	13.6
W-8	810	2.3

(after Justen et al, 1960)

For an isolated, inverted five-spot pattern, LPG breakthrough at the production wells can be calculated by the following equation:

$$q_1 = (28.28\, kh\Delta p)/\mu(5\,\ln(WS/r_w) - \ln 4) \quad (8)$$

where:

WS = distance between injection and production wells, ft

Equation 8 is applicable only in the case of unit mobility ratio. For the pilot, during the gas injection period, it was assumed that: there was no residual oil saturation behind the front; $k_g = k_o = k$; and the mobility ratio μ_o/μ_g approached infinity. Thus, Equation 8 was modified for infinite mobility ratio by introducing $R = q_i/q_\infty$. By letting $q_\infty = dV/dt$ and taking $dV = Ah\phi(1 - S_w)dS$ (where S = percent of pattern area swept by driving fluid), the modified equation was integrated to give:

$$\int_0^S R\, dS = C_1\, k\, \Delta t/\phi(1 - S_w) \quad (9)$$

where:

$C_1 = 28.28\, \Delta p/A\mu\, (5\,\ln(WS/r_w) - \ln 4)$

Δt = least time to LPG breakthrough in one of the permeability stratifications, a constant for all other layers

Values of R as a function of S, for the case of infinite mobility ratio, were obtained from potentiometric model studies. Figure 24 shows a plot of the integral in Equation 9 versus S. Thus, the total pore volume injected and produced in each permeability zone until breakthrough of LPG or until breakthrough of injected gas can be calculated.

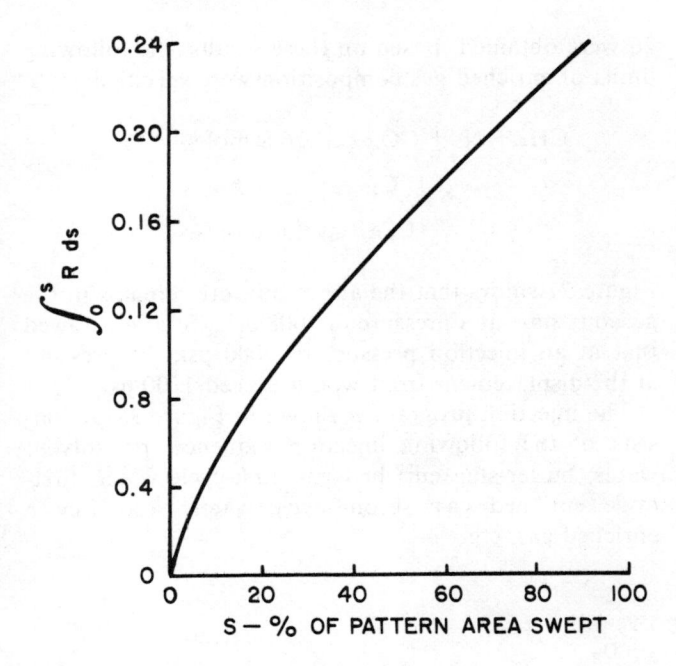

Fig. 24. Relationship between $\int_0^S R\,ds$ and S used for calculations, mobility ratio = ∞.
(after Justen et al, 1960)

Moss et al (1959) determined that for an infinite mobility ratio, in an inverted five-spot pattern, S at breakthrough is 92% of the square pattern area. However, at breakthrough of LPG, the reservoir still contains: LPG-oil mixture, LPG slug, LPG enriched gas and scavenging gas. (The less permeable layers would still, in addition, contain oil.) Therefore, following LPG breakthrough, there would be an interval of increasing LPG production followed by breakthrough of injected gas.

Following breakthrough of gas in each zone, it was assumed that gas flow through each zone could be approximated by Equation 10 (given below):

$$Q_g = 0.702\ kh\ (p_{iw}^2 - p_w^2)/\mu_g\ T(2\ \ln(WS/r_w) - 1.238) \quad (10)$$

which is the flow equation for a developed five-spot (Muskat, 1949). Figure 25 shows the calculated gas-oil ratio versus oil production. No attempts were made to quantitatively interpret recovery and GOR performance after water injection.

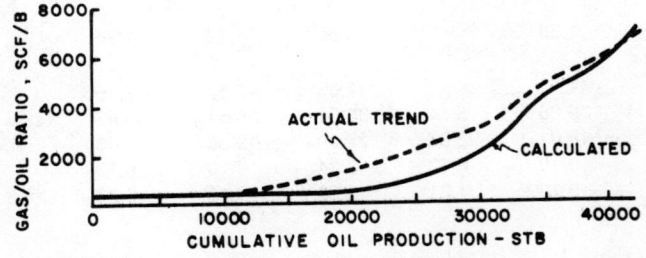

Fig. 25. Gas-oil ratio vs cumulative oil production for total pattern, Pembina pilot.
(after Justen et al, 1960)

In conclusion, the project yielded 72% of the effective oil-in-place in the pilot area, even though early breakthrough of LPG was experienced. Calculated performance compared well with actual pilot behavior. This work revealed that stratification would be a dominant factor in the recovery process and these effects were offset by injecting a slug of water into the partially swept reservoir.

Levelland Field, Texas (USA)

This is a limestone reservoir producing 30° API sour crude oil from a depth of 4800 ft (Ballard and Smith, 1972). The pay consists of discrete sections which appear to be vertically isolated from one another. The net pay thickness of approximately 100 ft is contained within a 250 ft thick section that dips slightly to the southeast. The oil viscosity at reservoir conditions is 2 cp. Other reservoir data are given in Tables IX and X.

TABLE IX
COMPOSITION AND PROPERTIES
OF LEVELLAND FIELD RESERVOIR CRUDE

Component	Mol %
N_2	2.52
C_1	20.92
C_2	6.19
CO_2	0.05
C_3	4.32
iC_4	0.87
nC_4	3.25
iC_5	1.80
nC_5	2.48
C_6's	4.20
C_7+	53.40
Total	100.0
Tank oil gravity, °API	30.0
Reservoir volume factor (1000 psi and 105°F)	1.157
Viscosity (1000 psi and 105°F), cp	1.93
Mol weight C_7+	244
Gravity C_7+, °API	28.0

(after Ballard and Smith, 1972)

TABLE X
RESERVOIR DATA, LEVELLAND FIELD

Type of pay	limestome-dolomite
Type of trap	stratigraphic-structural
Average depth, ft	4800
Gross thickness, ft	250
Net pay thickness, ft	100
Average permeability, md	2.1
Average gas saturation (current), %	9.8
Current pressure, psi	925
Reservoir temperature, °F	105
Primary producing mechanism	solution gas drive

(after Ballard and Smith, 1972)

The primary producing mechanism was solution gas drive, although an inactive aquifer and a gas cap existed at discovery. The project area underwent pressure maintenance by gas injection during the 15 year period which preceded application of miscible hydrocarbon.

An enriched gas which is miscible with the oil and scavenging gas at reservoir temperature must remain a single phase in the reservoir in order to maintain its miscibility. For LPG enriched gas, the necessary pressure is approximately 1400 psi at the present reservoir temperature of 105°F. By using a mixture of deethanizer overhead gas (DEO) and propane, the operating pressure could be reduced to 1100 psi.

A series of miscibility flow tests were performed on the reservoir oil, using several mixtures of DEO and propane. The tests were made in a 22-ft long, small diameter sand pack. The results shown in Table XI and Figure 26 were obtained. Based on these results, the following limits of enriched gas composition were specified:

$$CH_4 + N_2 + CO_2 \leq 36.8 \text{ mol }\%$$
$$C_3 \geq 25.7 \text{ mol }\%$$
$$C_4\text{'s} \leq 1.0 \text{ mol }\%$$

Figure 27 shows that the above mixture remains in the gaseous state at a pressure of 1008 psia. Studies showed that at an injection pressure of 3000 psi, the pressure at the displacement front would exceed 1100 psi.

The injection program is shown in Figure 28. It consists of the following injection sequence: pre-solvent water; buffer-slug enriched gas; first-cycle water; first-cycle enriched gas; second-cycle water; second-cycle enriched gas, etc.

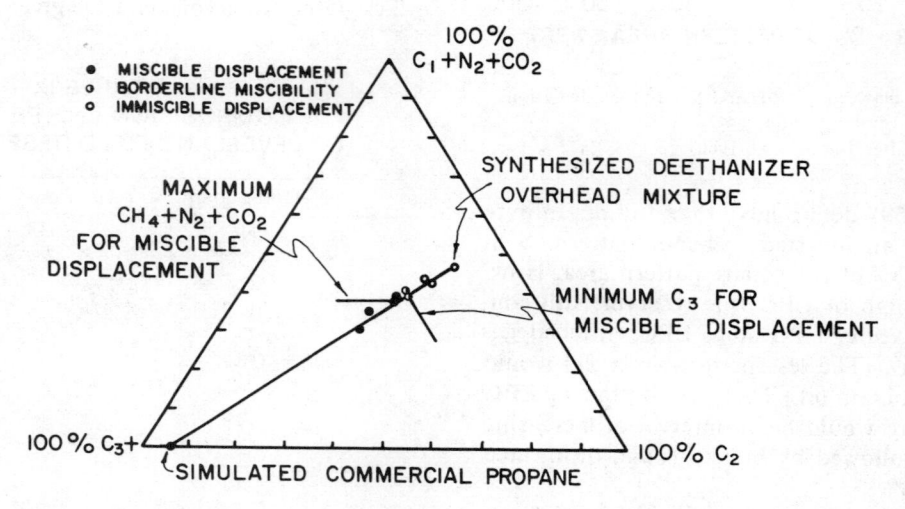

Fig. 26. Displacement tests of Field A crude oil at 105°F and 1100 psig.
(after Ballard and Smith, 1972)

TABLE XI
SUMMARY OF DISPLACEMENT TESTS AT 105°F

Run Number:	1	2	3	4	5	6	7	8	9	10
Exit pressure, psig	1105	1090	1090	1115	1095	1100	1100	1110	1105	1105
Oil produced, % oil-in-place	98.7	98.3	71.0	98.7	98.9	71.5	99.3	97.3	92.1	95.8
Injection gas, mol %										
N_2	1.71	34.95	1.49	37.47	0.64	1.58		0.95	1.32	1.34
C_1	38.28		45.13		29.84	41.12		37.53	42.52	37.15
C_2	34.83	30.88	41.30	33.26	30.43	39.06	5.74	33.61	36.51	36.47
C_3	25.02	34.17	11.97	28.88	38.39	18.01	92.62	27.49	19.36	24.68
iC_4	0.13		0.06	0.14	0.23	0.10	0.33	0.14	0.10	0.13
nC_4	0.03		0.05	0.25	0.47	0.13	1.31	0.28	0.19	0.23
Displacement type*	B	M	I	M	M	I	M	M	I	B

*B = Borderline, M = Miscible, I = Immiscible

(after Ballard and Smith, 1972)

MISCIBLE HYDROCARBON DISPLACEMENT

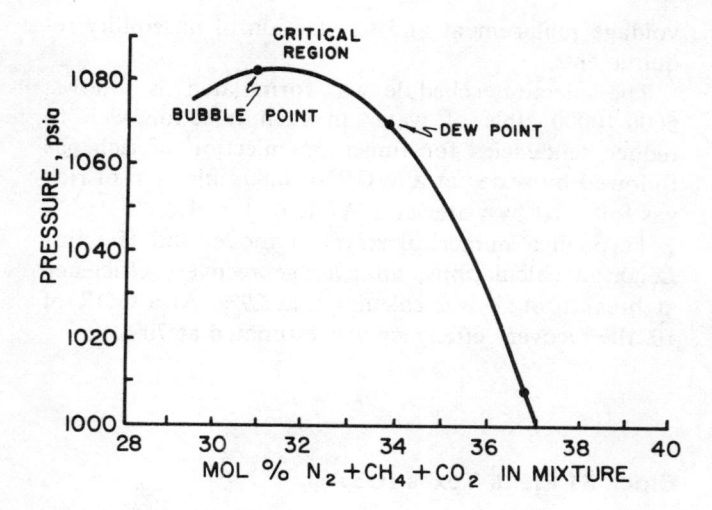

Fig. 27. 7.5:1 DEO-propane phase diagram at 105°F.
(after Ballard and Smith, 1972)

in injectivity was believed to result from a decrease in relative permeability caused by trapped gas saturation resulting from a small amount of residual oil.

Ante Creek Field, Canada

This reservoir produces from the Swan Hills Member of the Beaverhill Lake formation, at a depth of 11000 ft. The oil occupies the updip termination of a porous reefal development (Griffith et al, 1970). Calculations showed that initial oil-in-place was 37.5 MMSTB, and the primary recovery was estimated at 16.2%. Other reservoir properties are shown in Table XII.

Performance predictions were made on the basis of an enriched gas slug of 13% of the hydrocarbon pore volume HCPV. Figure 29 is a plot of predicted performance for a single five-spot for waterflood and for miscible flood. The figure shows that significantly more additional oil will be recovered by miscible flooding than by waterflooding.

Brannan and Whitington (1977) reported on the progress of this project. About 13 BCF of enriched gas have been injected. This represents 11% HCPV. However, after injection of an enriched gas buffer slug, a marked decrease in water injectivity occurred. This decrease

TABLE XII
SUMMARY OF BASIC FACTORS, ANTE CREEK BEAVERHILL LAKE

Discovery date	Nov., 1962
Producing formation	Beaverhill Lake
Depth, ft	11000
Number of wells	19
Well spacing, acres	320
Average net pay, ft	29.5
Average porosity, %	6.2
Average permeability, md	9.3
Connate water saturation, %	20.0
Initial reservoir pressure, psia	5173
Formation volume factor at bubble point	2.16
Bubble-point pressure, psia	4140
Reservoir temperature, °F	233
Solution GOR, Scf/STB	1930
Oil viscosity, cp	0.14

(after Griffith et al, 1970)

I. EARLY TIME : INJECTION OF RICH GAS HAS FORMED "TERTIARY" OIL BANK WHICH, IN TURN, IS DISPLACING PRE-WATER BANK. PRE-WATER BANK IS DISSIPATING THROUGH BEING LEFT BEHIND AS "RESIDUAL" WATER SATURATION.

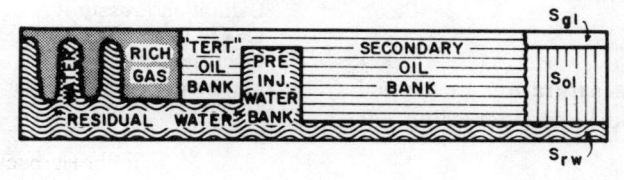

II. LATER TIME: PRE-WATER BANK HAS BECOME "SMEARED OUT" AS A "RESIDUAL" WATER SATURATION. PROCESS IS NOW SIMILAR TO THAT WITH NO PRE-INJECTED WATER.

Fig. 28. Schematic of alternate gas-water process.
(after Ballard and Smith, 1972)

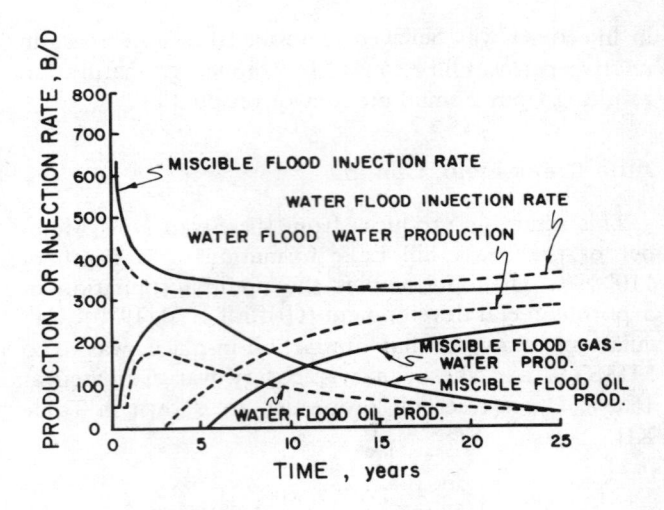

Fig. 29. Single five-spot performance prediction, waterflood and miscible flood (13 percent HCPV slug).
(after Ballard and Smith, 1972)

Studies showed that a rich gas of 65-70% methane content is miscible with the oil at reservoir conditions. Since 2-3 MMSCF of gas were being flared per day, it was decided to inject produced gas. However, reinjection of 90% of the separator gas as a solvent would result in reservoir voidage replacement of only 0.575 PV. Therefore, water had to be injected to replace remaining voidage and to improve the mobility ratio.

Since part of the volume of the injected rich gas is utilized in replacing reservoir voidage, the injected volume of rich gas must exceed the volume which is necessary to maintain miscibility with the oil. Miscibility is maintained as long as the rate of advance of the rich gas is equal to the rate of advance of the injected water. The ratio R of injected water to injected gas, at reservoir conditions, with rates of advance of water and gas the same, is given by the following equation:

$$R = (1 - S_{gf} - S_{wi})/S_{gt} \qquad (11)$$

where:

S_{gf} = average flowing gas saturation developed under drainage conditions (increasing gas saturation)

S_{gt} = gas saturation trapped under imbibition conditions (increasing water saturation)

S_{wi} = initial water saturation

Based on the values of $S_{gt} = 27.7\%$ and $S_{gf} = 30\%$ that were determined by core analyses, the above equation gives a maximum water-gas ratio (WGR) of about 1.6 reservoir volume per reservoir volume. Since injection of 90% of the separator gas replaces 57.5% of the voidage, a minimum of (1 - 0.575)/ 0.575 = 0.74 is required for a voidage replacement ratio of 1. Therefore, injection of WGR between 0.74 and 1.6 ensures both voidage replacement and satisfaction of miscibility requirements.

The injection schedule was formulated as follows: 5000-10000 Bbls of water in each injection well to reduce tendencies for fingering, injection of rich gas followed by water at a WGR of 0.80, injection of rich gas followed by water at a WGR of 1 or 1.2.

Through a numerical reservoir model and Higgins-Leighton calculations, an average recovery efficiency at breakthrough was calculated at 29%. At a GOR of 10, the recovery efficiency was estimated at 70%.

Block 31 Field, Texas (USA)

The field produces from three Devonian horizons. The Middle Devonian horizon is 350 ft thick and contains 85% of the oil-in-place and is the only portion which is being miscibly swept (Herbeck and Blanten, 1961). Its average depth is 8500 ft. Reservoir rock and fluid properties are summarized in Table XIII. The reservoir rock is crystalline limestone. The oil is trapped in a northeast-southwest trending anticline with the south end cut by a normal fault.

TABLE XIII
SUMMARY OF RESERVOIR ROCK AND FLUID CHARACTERISTICS FOR THE MIDDLE DEVONIAN HORIZON

Average Rock Characteristics	
Porosity	15%
Permeability	1 md
Connate Water	37%
Fluid Characteristics	
Reservoir Temperature	139° F
Initial Reservoir Pressure	4145 psig
Saturation Pressure	2764 psig
Gas Solubility @ P_s	1300 Scf/Bbl of oil
Viscosity @ P_s	0.30 cp
Formation Volume Factor	1.69 rb/STB
Oil Gravity	46° API

(after Herbeck and Blanton, 1961)

The miscible displacement project began in 1952 (Hardy and Robertson, 1975). It includes 24 gas injection wells and 86 oil producing wells. The field is developed on 80 acre spacing, but the gas injection is based on a 320 acre nine-spot pattern. Gas is injected at the rate of 134 MMCFPD at a pressure of 4200 psi. The miscible front displaces 95-100% of the contacted oil, but the sweep efficiency is low because of the gas - oil mobility ratio (10) and presence of permeability stratification and fractures.

The Devonian Reservoir has been carefully managed. When pressure in an injection block drops below miscibility the operator takes steps to reduce voidage or increase injection. Miscibility is regained by repressuring, since the gas manufactures its own miscibility by stripping the intermediate components from the crude.

Flue gas has been injected since 1966. This resulted in a saving of 30-50 MMCFPD, since one SCF of residue gas makes 9-11 SCF of flue gas. The flue gas is produced by a 54 MMCFPD plant. The flue gas is sweetened by a sulphur treating system, scrubbed in a quench tower, treated with ammonia and dehydrated by refrigeration and glycol injection. The ammonia neutralizes acid gases and prevents formation of carbonic and nitric acid. Nonetheless, injection well plugging occurred after flue gas injection was begun due to compressor lubricant and iron sulphide formed by mixing of flue gas and sour hydrocarbon gas in the injection well. A special lubricating oil eliminated the first problem and installation of cartridge type filters at the compressor discharge and wellhead reduced severity of the second problem.

The injected flue gas broke through into oil producing wells, which lowered the Btu quality of the produced hydrocarbon gas. Solution to this problem is still underway.

Swanson River Field, Alaska (USA)

The reservoir rock consists of interbedded fine to coarse grained sandstone, conglomerate, siltstone and coal. Reservoir and fluid properties are given in Table XIV (Young et al, 1977). Since 1966, lean gas at the rate of 300 MMCFPD has been injected at a pressure of 6000 psi. The current recovery at a GOR of 10 MSCF/STB is 38%.

TABLE XIV
SOLDOTNA CREEK UNIT FAULT BLOCK DATA
ROCK AND FLUID PROPERTIES

Datum, ft vss	10377
Reservoir temperature, °F	180
Bubble-point pressure, psi	1350
Original reservoir pressure, psi	5580
Solution GOR, SCF/STB	375
Oil gravity, °API	37
Formation volume factor — initial rb/STB	1.21
Formation volume factor — bubble point, rb/STB	1.26
Oil viscosity, cp	1.1
Oil compressibility (above bubble point)	10.8×10^{-6}
Porosity, %	21
Initial oil saturation, %	60
Permeability to air, md	197
Permeability to oil at reservoir conditions,* md	97

*Obtained from pressure buildup analyses.

(after Young et al, 1977)

Hassi-Messaoud Field, Algeria

This is the largest application of the lean gas process in the world. Since 1964, 175 MMCFPD of lean gas are being injected at a pressure of 6000 psi (Pottier et al, 1967).

The field is located in the northern Sahara some 500 miles from the Mediterranean sea coast. The reservoir consists of a domal structure about 30 miles in diameter. The reservoir rock is about 330 ft thick, at an average depth of 11000 ft. The rock is fine to coarse grained Cambrian sandstone and contains a large number of discontinuous intercalations of siltstone. The porosity is uniform, 6-10%, but the kh value differs considerably throughout the reservoir (Figure 30).

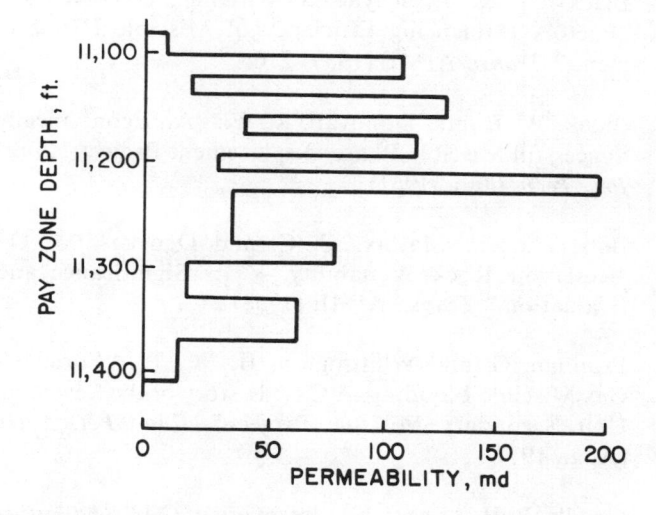

Fig. 30. High-pressure miscible gas flood, Hassi Messaoud reservoir. (after Holm, 1976)

The initial oil-in-place was estimated at 23 BSTB. The oil is undersaturated; the viscosity is 0.25 cp at reservoir conditions. The initial reservoir pressure is 7200 psi at 10550 ft, temperature is 245°F, GOR is 1235 SCF/STB, density at reservoir conditions 210 lbs/Bbl. Gas density is about half the density of oil; its viscosity is 0.035 cp.

Oil recovery at the bubble point (2700 psi) is estimated at 10%, and recovery at the abandonment pressure of 2250 psi is estimated at 16-17%.

Gas injection was preferred over waterflooding because of the availability of gas and because the reservoir is heterogeneous. Water fingering and uneven sweep could result in entrapping large amounts of oil. On the other hand, fingering and gravity override by the gas have no permanent detrimental effects. The gas, even if it breaks through prematurely, will be richer in recoverable distillates.

Extensive studies were made prior to commencement of gas injection. These studies illustrated that dispersion plays an important, favorable role. Because of dispersion, the adverse effects of gravity override and mobility ratios are minimized.

BIBLIOGRAPHY

Arnold, C. W., Stone, H. J. and Luffel, D. L.: "Displacement of Oil by Rich-Gas Banks," *Trans.*, AIME (1960) 219.

Ballard, J. R., Lanfranchi, E. E. and Vanags, P. A.: "Thermal Recovery in the Venezuelan Heavy Oil Belt," *Jour. Cdn. Pet. Tech.* (April-June, 1977).

Ballard, J. R. and Smith, L. R.: "Reservoir Engineering Design of a Low-Pressure Rich Gas Miscible Slug Flood," *Jour. Pet. Tech.* (May, 1972).

Blackwell, R. J., Rayne, J. R. and Terry, W. M.: "Factors Influencing Efficiency of Miscible Displacement," *Trans.*, AIME (1959) 216.

Block, W. E. and Donovan, R. W.: "An Economically Successful Miscible-Phase Displacement Project," *Jour. Pet. Tech.* (Jan., 1961).

Bobek, J. E., Mattax, C. C. and Denekas, M. O.: "Reservoir Rock Wettability — Its Significance and Evaluation," *Trans.*, AIME (1958) 213.

Brannan, G. and Whittington, H. M., Jr.: "Enriched-Gas Miscible Flooding: A Case History of the Levelland Unit Secondary Miscible Project," *Jour. Pet. Tech.* (Aug., 1977).

Caudle, B. H.: "Secondary Recovery of Oil," *The Future Supply of Nature Made Petroleum and Gas*, Pergamon Press, New York (1976).

Caudle, B. H. and Dyes, A. B.: "Improving Miscible Displacement by Gas-Water Injection," *Trans.*, AIME (1958) 213.

Craig, F. F., Jr.: "A Current Appraisal of Field Miscible Slug Projects," *Jour. Pet. Tech.* (May, 1970).

Craig, F. F., Jr. and Owens, W. W.: "Miscible Slug Flooding — A Review," *Jour. Pet. Tech.* (April, 1960).

Csaszar, A. K. and Holm, L. W.: "Oil Recovery from Watered-Out Stratified Porous Systems Using Water-Driven Solvent Slugs," *Jour. Pet. Tech.* (June, 1963).

Dean, M. R. and Brinkley, T. W.: "Measuring Solvent Production in Miscible Phase Oil Recovery Operations," *Jour. Pet. Tech.* (May, 1964).

Ferrer, J. and Farouq Ali, S. M.: "A Three-Phase, Two-Dimensional Compositional Thermal Simulator for Steam Injection Processes," *Jour. Cdn. Pet. Tech.* (Jan.-March, 1977).

Fitch, R. A. and Griffith, J. D.: "Experimental and Calculated Performance of Miscible Floods in Stratified Reservoirs," *Trans.*, AIME (1964) 231.

Garder, A. O., Peaceman, D. W. and Pozzi, A. L.: "Numerical Calculation of Multidimensional Miscible Displacement by Method of Characteristics," *Soc. Pet. Eng. Jour.* (March, 1964).

Griffith, J. D., Barton, N. and Steffensen, R. J.: "Ante Creek — A Miscible Flood Using Separator Gas and Water Injection," *Jour. Pet. Tech.* (Oct., 1970).

Habermann, B.: "The Efficiency of Miscible Displacement as a Function of Mobility Ratio," *Trans.*, AIME (1960) 219.

Handy, L. L.: "Oil Displacement Using Partially Miscible Gas-Solvent Systems," *Jour. Pet. Tech.* (Feb., 1963).

Hardy, J. H. and Robertson, N.: "Miscible Displacement by High Pressure Gas at Block 31," *Pet. Eng.* (Nov., 1975).

Herbeck, E. F. and Blanton, J. R.: "Ten Years of Miscible Displacement in Block 31 Field," *Jour. Pet. Tech.* (June, 1961).

Herbeck, E. F., Heintz, R. C. and Hastings, J. R.: "Fundamentals of Tertiary Oil Recovery — Part 2 — LPG Miscible Slug Process,' *Pet. Eng.* (Feb., 1976).

Herbeck, E. F., Heintz, R. C. and Hastings, J. R.: "Fundamentals of Tertiary Oil Recovery — Part 3 — Enriched Gas Miscible Process," *Pet. Eng.* (March, 1976).

Herbeck, E. F., Heintz, R. C. and Hastings, J. R.: "Fundamentals of Tertiary Oil Recovery — Part 4 — High Pressure Lean Gas Miscible Process," *Pet. Eng.* (April, 1976).

Holm, L. W.: "Status of CO_2 and Hydrocarbon Miscible Oil Recovery Methods," *Jour. Pet. Tech.* (Jan., 1976).

Jones, J.: "Cyclic Steam Reservoir Model for Viscous Oil, Pressure Depleted, Gravity Drainage Reservoirs," SPE 6544 presented at the SPE-AIME 47th Annual California Regional Meeting, Bakersfield, CA. (April, 1977).

Justen, J. J. et al: "The Pembina Miscible Displacement Pilot and Analysis of Its Performance," *Trans.*, AIME (1960) 219.

Kendall, G. H.: "Importance of Reservoir Description in Evaluating In-Situ Recovery Methods for Cold Lake Heavy Oil — Part 1 — Reservoir Description," *Jour. Cdn. Pet. Tech.* (Jan.-March, 1977).

Kieschnick, W. F., Jr.: "What Is Miscible Displacement?" *Pet. Eng.* (Aug., 1959).

Laue, L. C., Teubner, W. G. and Campbell, A. W.: "Gravity Segregation in a Propane Slug-Miscible Displacement Project — Baskinton Field, Louisiana," *Jour. Pet. Tech.* (June, 1965).

Marrs, D. G.: "Field Results of Miscible-Displacement Program Using Liquid Propane Driven by Gas, Parks Field Unit, Midland County, Texas," *Jour. Pet. Tech.* (April, 1961).

Moranville, M. B., Kessler, D. P. and Greenkorn, R. A.: "Directional Dispersion Coefficients in Anisotropic Porous Media," *Industrial and Engineering Chemistry Fundamentals,* Vol. 16. (1977).

Muskat, M.: *Physical Principals of Oil Production,* McGraw-Hill Book Co., Inc., New York (1949).

Peaceman, D. W. and Rachford, H. H., Jr.: "Numerical Calculation of Multidimensional Miscible Displacement," *Soc. Pet. Eng. Jour.* (Dec., 1962).

Pottier, J. et al: "Injection de Gaz Miscible A Haute Pression A Hassi-Messaoud," *New Methods in Secondary Recovery.* (1967).

Pozzi, A. L. and Blackwell, R. J.: "Design of Laboratory Models for Study of Miscible Displacement," *Soc. Pet. Eng. Jour.* (March, 1963).

Rose, D. A.: "Hydrodynamic Dispersion in Porous Materials," *Soil Science,* Vol. 123. (1977).

Settari, A., Price, H. S. and Dupont, T.: "Development and Application of Variational Methods for Simulation of Miscible Displacement in Porous Media," *Soc. Pet. Eng. Jour.* (June, 1977).

Shelton, J. L. and Schneider, F. N.: "The Effects of Water Injection on Miscible Flooding Methods Using Hydrocarbons and Carbon Dioxide," *Soc. Pet. Eng. Jour.* (June, 1975).

Thompson, J. L.: "A Laboratory Study of an Improved Water-Driven LPG Slug Process," *Soc. Pet. Eng. Jour.* (Sept., 1967).

Thompson, J. L. and Mungan, N.: "A Laboratory Study of Gravity Drainage in Fractured Systems Under Miscible Conditions," *Soc. Pet. Eng. Jour.* (June, 1969).

Todd, M. R. and Longstaff, W. J.: "The Development, Testing and Application of a Numerical Simulator for Predicting Miscible Performance," *Jour. Pet. Tech.* (July, 1972). *Trans.,* AIME (1972) 253.

Yarborough, L. and Smith, L. R.: "Solvent and Driving Gas Compositions for Miscible Displacement," *Soc. Pet. Eng. Jour.* (Sept., 1970).

Young, R. E., Fairfield, W. H. and Dykstra, H.: "Performance of a High-Pressure Gas Injection Project, Swanson River Field, Alaska," *Jour. Pet. Tech.* (Feb., 1977).

Chapter 8
CARBON DIOXIDE INJECTION

INTRODUCTION

The injection of carbon dioxide (CO_2) into oil reservoirs for enhanced oil recovery has received considerable attention by the industry for several years. This interest has intensified recently as the many waterflood fields approach the end of their productive lives.

Conventional gas or water-drive will usually leave 25-50% of the original oil in the reservoir. A considerable part of this oil can be recovered if the oil is contacted by a fluid with which it is miscible. A miscible fluid is formed when injected CO_2 mixes with reservoir oil under proper conditions. When miscibility occurs the forces of capillary pressure, which formerly held the oil immobile, disappear and the oil is then free to be carried forward to the producing well.

The concept of miscible displacement of oil by CO_2 has been known for many years and has been tried in a few full-scale field projects. Many unexpected problems have plagued these field projects and hampered economic success.

This does not mean that the CO_2 method does not have a promising future, only that the problems are many and the process cannot yet be termed a proven enhanced oil recovery technique. In order to improve chances of success, each project must be carefully planned and executed.

TECHNICAL DISCUSSION

Criteria for Application of Carbon Dioxide Injection

Each oil reservoir has a long list of characteristics and the sum total of these determines the personality of the reservoir and how it will behave when subjected to various methods of stimulation. The problem confronting the engineer is to determine the value of as many characteristics as possible and then predict the performance behavior.

In this context, each characteristic alone is not a determining factor. Therefore, the number assigned to a characteristic must not be considered a rigid boundary, but only an indication of an order of magnitude. For example, an oil gravity of less than 25° API is usually considered unfavorable for enhanced oil recovery by CO_2 injection. This does not automatically exclude all reservoirs having oil heavier than 25° API from consideration; there may be other favorable factors that will override an unfavorable one. The following criteria should be considered and put in proper perspective.

Residual oil saturation is of primary concern. If the field has been waterflooded the residual oil saturation may be insufficient for either technological or economic success. A saturation in the range of 25-30% is often quoted as the minimum.

Previous waterflooding does not automatically eliminate fields from consideration because simulation studies show that considerable oil can be recovered from waterflooded sands (Warner, 1977).

A large gas cap is usually an unfavorable factor. If reservoir pressure is considerably below miscibility pressure, large volumes of CO_2 will be needed to obtain miscibility. The density of CO_2 may be greater than that of the reservoir gas, thus promoting mixing. These problems are being circumvented at Weeks Island where CO_2 is being injected at the gas-oil contact and forced downward, aided by gravity factors (Perry, 1977).

A highly fractured reservoir is usually considered unfavorable because the fractures provide a conduit from injection to producing well. However, these fractures will also pose a serious problem for any other type of process being considered.

An adequate and reliable source of CO_2 at a reasonable cost is a primary prerequisite. The recent strong interest in nitrogen and flue gas as alternate gas injection methods is prompted largely by the lack of good CO_2 sources close to many oil fields of the world.

The horizontal permeability of the reservoir rock does not appear to be a critical factor; however, the ratio of vertical to horizontal permeability is critical. A reservoir simulator study on a waterflooded sandstone led to the conclusion that the k_v/k_h ratio is the most important reservoir parameter in the CO_2 process because this parameter controls the rate at which the CO_2 can segregate (Warner, 1977).

Relatively thin permeable zones in the reservoir (15-25 ft) are technically advantageous because they diminish the tendency of gravity override, but the thicker zones have an oil volume advantage.

Depth is important because the minimum miscibility pressure is usually above 1200 psi requiring a depth greater than 2500 ft in order not to exceed the fracture gradient. Temperature is not usually an important factor.

The lower limit of oil gravity is in the range of 25-30° API depending partly on whether the oil is aromatic, asphaltic, etc. The viscosities of reservoir oil in most CO_2 projects to date have been approximately 1 cp.

Pure CO_2 is best for injection but is rarely available. Contamination by methane increases the miscibility pressure; however, 5-10% methane can be tolerated. Hydrogen sulfide lowers the miscibility pressure, but causes serious problems due to corrosion, health hazard, effect on environment, and odor.

Field experience with CO_2 injection to date can be summarized by noting that it has been used to recover additional oil under the following wide range of conditions.

1. In sandstones, limestones, dolomites and cherts
2. To depths of 10800 ft with no known depth limitation
3. In formations with average permeabilities as low as 0.2 md
4. At bottom-hole temperatures up to 248°F with no known limitation
5. In formations varying from 8-600 ft in thickness with considerable variation in homogeneity
6. Where crudes vary in gravities from 16-45° API
7. Where crudes were displaced immiscibly
8. For crudes varying in viscosity from 0.15-188 cp
9. In reservoirs with oil saturations from 28-54%
10. With spacing up to 51 acres per well
11. When the injected mixture contains up to 29% hydrogen sulfide (McRee, 1977).

Phase Behavior and Miscibility

Carbon dioxide is a familiar substance in all three of it forms — gas, liquid and solid. As a gas, it provides the bubbles in champagne as the pressure is lowered below the saturation pressure. It is a liquid at a pressure above 300 psia if the temperature is 0°F or below. It is usually transported as a liquid in refrigerated trucks or tank cars. (See Figure 1.) It is a solid (dry ice) over a wide range of pressures if the temperature is sufficiently low.

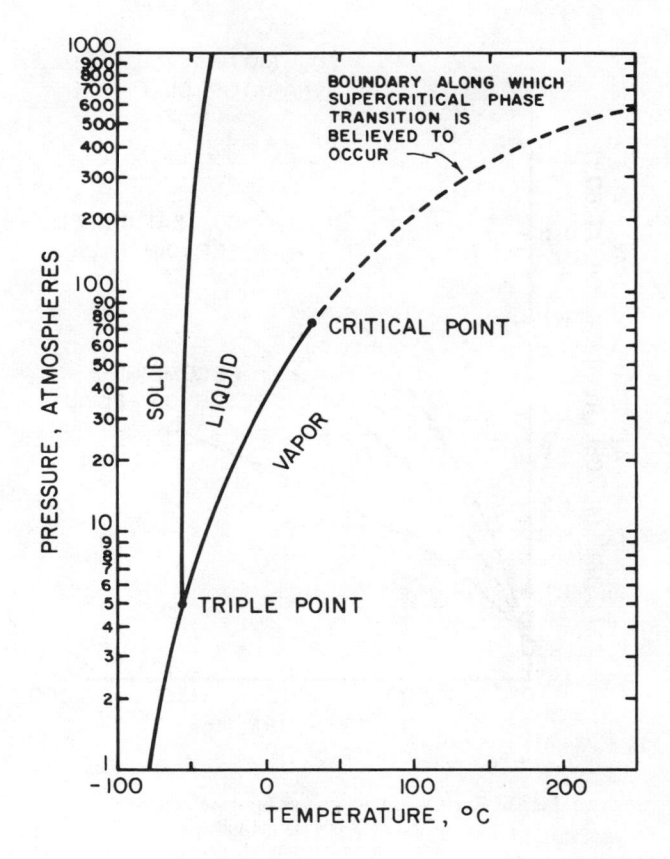

Fig. 1. Carbon dioxide phase diagram.

The critical temperature is 31.0C. Below this temperature pure CO_2 can be either a gas or a liquid over a rather wide range of pressures, but above this temperature CO_2 will be a gas regardless of the pressure applied. The pressure corresponding to the critical temperature (critical pressure) is 73.99 atm. Above the critical temperature and pressure pure CO_2 cannot be liquefied; however, at higher, supercritical pressures the vapor becomes more dense with increasing pressure and behaves more like a liquid. Most CO_2 pipelines operate in the supercritical region.

Some important properties of CO_2 are as follows:

Molecular Weight	44.01 g/mole
Critical Pressure	1073 psia
Critical Temperature	87.8°F
Critical Volume	0.0237 cu ft/lb
Density at 0°F, 300 psi	8.5 lb/gal
Specific Volume at 14.7 psia and 60°F	8.569 cu ft/lb
Specific Heat (Liquid) at 300 psi	0.5 Btu/lb-°F

One of the properties of CO_2 that makes it a useful enhanced oil recovery agent is the increase in volume of a crude oil when saturated with CO_2. Figure 2 shows how the volume of a west Texas reservoir fluid increases when saturated with CO_2 at various pressures (Holm and Josendal, 1974).

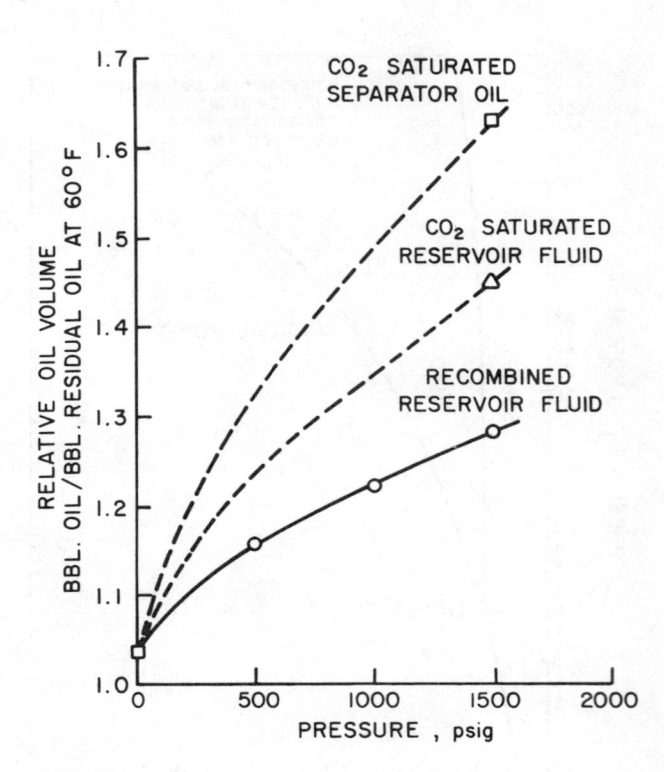

Fig. 2. Relative oil volume vs pressure at 144°F., west Texas reservoir fluid.
(after Holm and Josendal, 1974)

If CO_2 is injected into a reservoir and remains a gas, it will act as a gas drive and soon finger through to the producing well. Many field tests in the past have shown that CO_2 is not much better than methane if used as an immiscible oil recovery agent.

What is desired is for the CO_2 to mix with the crude oil in the reservoir and form a single-phase liquid that is much lighter than the original oil. This miscible bank of oil can be more easily displaced by the following gas or water-drive.

The important factors that determine whether CO_2 and reservoir oil are miscible or not are: the purity of the CO_2, characteristics of the reservoir oil, temperature, pressure and the degree of mixing of the fluids.

Several attempts have been made to develop a method to estimate miscibility. These correlations are based on many laboratory tests and theoretical concepts (Metcalfe, 1978; Shelton and Yarborough, 1976; Simon and Graue, 1965; and Yellig and Metcalfe, 1978).

The National Petroleum Council, in 1976, devised a rough formula for estimating whether or not miscibility pressure was attainable based on oil gravity, reservoir temperature and reservoir depth. The procedure for estimating pressure required for miscible displacement is given as follows:

Miscibility Pressure Versus Gravity

Gravity (°API)	Miscibility Pressure (PSI)
27	4000
27-30	3000
30	1200

Correction for Reservoir Temperature

Temperature (°F)	Additional Pressure Required (PSI)
120	None
120-150	+ 200
150-200	+ 350
200-250	+ 500

To estimate whether or not miscibility is attainable in each reservoir, the limiting fracturing pressure is estimated by multiplying the reservoir depth by an assumed fracturing gradient of 0.6 psi per foot of depth. A safety factor of 300 psi is subtracted from this limiting pressure to estimate a likely attainable pressure (NPC, 1976).

These correlation methods are useful for screening candidate reservoirs but are not sufficiently accurate for specific use. There is no alternative but to conduct laboratory tests to determine miscibility under a specific set of conditions.

Displacement Mechanisms

A critical step in a successful CO_2 injection project is to attain a zone in which the CO_2 and reservoir oil have mingled and formed a new fluid that is more easily displaced than the original reservoir oil. Almost as critical is the method used to push this expanded reservoir oil to the producing well.

At least four methods of CO_2 and water injection have been proposed as displacement mechanisms:

1. Continuous injection of CO_2 during the life of the project
2. Injection of a slug of CO_2 followed by water
3. Injection of alternate slugs of CO_2 and water
4. Simultaneous injection of CO_2 and water.

Sufficient field experience has not been accumulated to evaluate even one of these processes, let alone compare results among the four. However, a comparison was made using a four-component, miscible, mixing parameter reservoir simulator. This study simulated the recovery, by CO_2 injection, of oil left after waterflooding a sandstone reservoir.

The study led to the conclusions that:

1. The simultaneous injection of CO_2 and H_2O proved to be the best of the four oil recovery processes and recovered approximately 50% of the potential oil. Alternate slug injection of CO_2 and H_2O was second best. Straight CO_2 injection and CO_2 slug injection followed by water injection were equally poor and recovered only about 25% of the potential oil.
2. In all cases, gravity segregation between the CO_2 and water was complete before half of the reservoir rock had been swept by the mixture of the two fluids. The recovery success of each CO_2 process was controlled by the rate at which the gravity segregation of the CO_2 could occur (Warner, 1977).

An interesting application of the continuous CO_2 injection method is the gravity stable process. This process is applicable to steeply dipping beds where displacement is downward. The CO_2 builds up an expanding gas cap that pushes the oil downward to the producing well. The downward CO_2 displacement is designed to utilize gravity forces to stabilize the displacement and increase the sweep of the CO_2. This process is expected to be successful even if miscibility is not attained (Perry, 1977).

Project Design

The volume of CO_2 needed for the earliest floods was estimated by using past experience with gas injection projects. These early floods were immiscible displacement or carbonated waterfloods so the previous experience was of little use for later miscible displacement projects. Requirements for miscible projects were (and still are to a degree) estimated from results of laboratory tests made on physical models.

A physical model is intended to duplicate part of the reservoir as closely as possible. The porous media, temperature, pressure, fluid characteristics and initial fluid saturations are all closely duplicated in the model. The flow may be either horizontal, radial, part of a pattern or other configuration as desired. Provision is sometimes made for visual observation and analyses of the produced fluids.

Physical reservoir models have the advantage that several runs can be made under a variety of conditions so that the effects of pressure, sequence of injection fluids, well pattern, well spacing and other design parameters can be evaluated and quantified. The results of laboratory tests are quite useful and indicate, in a general way, the relative effect of each process variable.

Laboratory tests are necessary to determine the conditions necessary for miscibility between reservoir oil and injected CO_2. Although correlations have been developed to predict miscibility, such correlations are probably not sufficiently accurate to use in a specific reservoir.

A major difficulty with physical models is that they do not give reliable, quantitative answers. The model represents only itself in the laboratory and not the reservoir in the field. Many reservoir conditions (time, gravity, area, thickness, homogeneity, chemistry, etc.) cannot be properly scaled and duplicated in the laboratory. This does not mean that physical models should be abandoned, only that the results should be interpreted with caution and used in conjunction with newer, numerical reservoir simulators.

Several numerical models have been developed to predict the behavior and movement of fluids in the reservoir. Although some models are quite sophisticated, improvements will continue to be made. With added experience, the models will be much more reliable in the future.

The results of numerical calculations can only be as good as the reliabilty of the input data. As in the case of physical models, not all reservoir parameters can be taken into account. These deficiencies are largely overcome by history matching. If the numerical simulation does not match history, the magnitude of variables are adjusted to improve match. It is not always certain that the correct variables are adjusted to match real reservoir conditions.

Recent experience in a typical west Texas field illustrates the degree to which numerical simulation predicts field results. The North Cross (Devonian) Reservoir in the Crossett Field was discovered in 1944, produced under primary recovery until 1964, and then under partial pressure maintenance by gas injection until 1972. At that time CO_2 injection began. A source of CO_2 became available when the pipeline to the SACROC project was constructed nearby (Pontius and Tham, 1977).

In 1970 a leasing agreement was made between Shell and Canyon Reef Carriers (operator of the SACROC pipeline) whereby up to 20 MMSCFPD of that system's capacity would be made available for CO_2 transport to North Cross. Shell's miscible reservoir simulator was used in an attempt to define an optimum flood plan. The field's past production performance was history matched, and project predictions for numerous injection patterns and operating policies were generated and compared. An inverted nine-spot pattern was determined to be an optimum. Two wells were to be converted to CO_2 injection initially, with two more wells to be converted after four years of flood operations. Continuous CO_2 injection with recycling was planned. CO_2 injection began in 1972 and was continuing in 1977 when the results were reported.

Although the North Cross Unit CO_2 flood has not performed as dramatically as had been predicted by the original reservoir simulation model, the response to CO_2 injection has been definite and encouraging. Sustained production increases in responding wells and minimal breakthrough of CO_2 to date indicate that the CO_2 is efficiently displacing oil. Initial injectivity was considerably less than predicted by the simulation studies; it took six wells to inject the amount of CO_2 predicted for two wells (Pontius and Tham, 1977).

Sources of Carbon Dioxide

The planning for a CO_2 miscible project cannot proceed very far without considering a crucial factor — the source of supply of CO_2. The gas must be available in the volumes needed for a long period of time, up to 20 years or more in many cases. The gas must be relatively pure because some gases, such as methane, increase the pressure necessary for miscibility while others, such as hydrogen sulfide, are hazardous, odoriferous and increase environmental problems. The supply of gas must be continuous. If CO_2 injection is interrupted for a considerable period of time the miscible bank may deteriorate and threaten the success of the project.

There is always the possibility of interruption of supply if the source is the off-gas from a manufacturing process. Unforeseen conditions, such as fires, strikes or changing market conditions may cause temporary or permanent shutdowns.

On the other hand, care must be taken not to burden a project with more CO_2 under contract than is needed. The actual volumes and times of gas deliveries to the project are difficult to determine because of lack of experience with large-scale injection projects over a long period of time.

Another unknown factor is the volume of CO_2 that must be recycled. If CO_2 breaks through prematurely to producing wells, this gas must be processed and the CO_2 reinjected. Thus, early breakthrough might reduce the volume of contracted CO_2 needed, or if unusually severe, might lead to abandonment of the project. These and other factors suggest that it is good insurance for the field operator to have close control over the supply of CO_2.

A natural source is best, either from wells that produce relatively pure CO_2 or from plants that process hydrocarbon gas that contains considerable CO_2 as a contaminant. As future insurance, oil companies in the United States have leased a large part of the basins that are good prospects for CO_2 production and close to oil fields.

Most of the known natural deposits of CO_2 have been found while exploring for oil and gas. Relatively few wells have been drilled to explore for CO_2, although more such wells will be drilled in the future. Because of the accidental discoveries, most CO_2 wells are in the same geologic basins as those that also produce oil and gas. The oil-producing basins of Wyoming, Utah, Colorado and New Mexico also have the largest reserves of CO_2 presently known in the United States. Natural gas fields in the Delaware and Val Verde Basins of southwest Texas contain considerable CO_2 that is removed in gas processing plants. Most of this CO_2 is already committed to the pipeline supplying the SACROC flood.

Several sources other than natural deposits have been investigated to supply needed CO_2 for injection. Stack gas from large coal-fired power plants is often mentioned as a likely source. Present power plants produce large volumes of waste gas, and many more such plants will be built in the near future. Many of these plants will be close to coal deposits and also close to producing oil fields.

Unfortunately, stack gas from power plants is not as attractive as it first appears. The CO_2 content of the stack gas is not high, being usually in the range of 6-20% CO_2. The composition of a typical flue gas from a coal-fired power generating plant is as follows, with the gas and water given in mole percent: CO_2, 16.5; nitrogen, 64.6; oxygen, 5.6; and water 13.3 (Pullman Kellog, 1977). The cost of purifying, when added to other costs, will probably make CO_2 from flue gas prohibitively expensive for some time to come.

The gas vented from ammonia manufacturing plants is mostly CO_2 and this source seems the most promising of any of the non-natural sources. A recent study states that a future CO_2 injection project will use this source in south Louisiana because of these advantages:

1. There is a reasonable proximity between process vents and oil fields.
2. Multiple CO_2 vents tend to be clustered within a given industrial area.
3. The quantity of CO_2 from each source will be known.
4. There will not be any need for a purification scheme for CO_2 from the ammonia plant vents because of the 98% purity. Compression and piping is all that will be required (Pullman Kellog, 1977).

A study of the supply for CO_2 for enhanced oil recovery concludes: "Review . . . shows that more than sufficient CO_2 is available from above-ground sources to satisfy the projected future demand for EOR applications even without substantial new discoveries of high CO_2 content natural gases. Additionally, it appears that each state could be self-supporting in CO_2." This seems to be an encouraging prediction, but later in the same discussion the report states: "In the last analysis, it is impossible to predict whether any of the sources considered here will be developed for use in EOR by CO_2 miscible flooding. This will depend on a great many factors including the future cost of petroleum, the effectiveness of the process itself, and government policy insofar as it influences development of EOR programs." (Pullman Kellog, 1977).

CONCLUSIONS

The main advantages of miscible CO_2 injection are summarized as follows:

1. Swells oil and reduces viscosity.
2. Forms miscibility with oil by extraction, vaporization and chromatographic transport.
3. Acts as solution-gas drive, even if complete miscibility is not achieved.
4. The miscible front, if lost, will regenerate itself as it does with the lean gas process.
5. Unlike LPG, CO_2 will become miscible with oils that have been depleted of the C_2 - C_4 fractions.
6. Carbon dioxide, highly soluble in water, causes the water to swell and become slightly acidic.
7. Miscibility can be attained above pressures of 1500 psi in many reservoirs.
8. Carbon dioxide is a non-hazardous, non-explosive gas that causes no environmental concern if large quantities are lost to the atmosphere.
9. May be available as a waste gas (as from gas-processing plants or industrial plants) or from reservoirs containing CO_2.

Carbon dioxide provides an efficient low pressure miscible displacement for many reservoirs. The displacement efficiency is high, with the oil saturation being reduced to about 5% of pore volume in the contacted area.

Under some reservoir conditions, the density of CO_2 is close to that of crude oil and approaches that of water. This greatly minimizes the effects of gravity override.

Carbon dioxide is much more viscous than methane at higher pressures. For example, at 100°F and 1000 psi, the viscosity of CO_2 and methane are about the same, 0.015 cp. But at 2000 psi the viscosity of methane is still close to 0.015 cp, while that of CO_2 is 0.07 cp. At 5000 psi the viscosity of methane is about 0.027 cp and that of CO_2 is about 0.11 cp.

Some of the disadvantages of CO_2 injection as an enhanced oil recovery process are summarized as follows:

1. Solubility of CO_2 in water may increase volume needed for oil miscibility, but this disadvantage may be partly or wholly overcome by the increased volume of the CO_2-saturated water.
2. The low viscosity of any free CO_2 gas at low reservoir pressure will promote early breakthrough to the producing well, reducing sweep efficiency. Production of large volumes of diluted gas requires expensive cleanup and recycling facilities.
3. After miscibility is attained the oil is less viscous than reservoir oil, causing fingering and premature breakthrough.
4. Injection of slugs of water is often necessary to reduce fingering.
5. Carbon dioxide with water forms carbonic acid which is highly corrosive. Special metal alloys and coatings for facilities are needed. Corrosion mitigation can be a considerable part of the cost of the process.
6. The alternate injection of slugs of CO_2 and water requires a dual injection system, adding to the cost and complexity of the system.
7. Large volumes of CO_2 are needed. It may take 5-10 MCF of gas to produce 1 barrel of stock tank oil.
8. Carbon dioxide is usually not available locally, requiring long-distance pipelines. Experience has shown that CO_2 pipelines are more subject to breakdown than natural gas pipelines, thus causing expensive delays that may interfere with the technical success of the project.

Field experience has shown that initial rates of injection may be much less than those predicted. The cause of this is not entirely clear but may be related to the increased volume of interstitial water due to CO_2 absorption. More likely it is due to the increased viscosity of reservoir oil as the CO_2 front strips the oil of light ends and moves forward before miscibility is achieved and the front slows down.

Several projects have reported decreasing rates of injection as the miscible front advances. Usually this problem arises when the front is between 50-100 ft from the injection well. This loss of injectivity may require additional injection wells to maintain the desired or contracted rate of CO_2 injection. Although proponents of the process are reluctant to admit this disadvantage, it now appears likely that this increased resistance to flow is related to a resin or asphaltic precipitate in the reservoir. As the CO_2 strips the lighter hydrocarbons from the reservoir oil, the advancing front leaves ever-heavier oil behind. A point is reached where the asphaltic-like substances in the oil are no longer soluble in the stripped oil. These then precipitate and block small flow channels either by agglomeration or the "brush-pile" effect.

CASE HISTORIES

Carbon dioxide injection has been field tested on a small scale for many years, but most of these early, immiscible projects were abandoned because of discouraging results. The present high interest in CO_2 injection, which began in the early 1970's, is due to a combination of higher oil prices, the approaching end of many waterfloods and the need for good enhanced oil recovery processes. As shown in Table I, relatively few CO_2 projects have been started, and most of these have not been in operation long enough to give definitive engineering or economic results.

In addition to the list of CO_2 injection projects shown in the table, two projects have been reported in progress in Trinidad and two in Hungary, but little published information is available on these (O.&G. Jour., March 27, 1978).

TABLE I
MAJOR CO$_2$ PROJECTS

Field	Formation	Depth (ft)	BHT (°F)	Oil Gravity (°API)	Permeability (md)	Porosity (%)	Sand Thickness ft (Gross)	(Net)	Miscibility Pressure (psi)	Field Spacing (Ac/Well)	Start Date (year)	Field Productive (acres)
Arkansas												
Richie												
Union Co.	Baker - SS	2600	126	16	2750	31	8	8	None	20	1969	250
Lick Creek												
Bradley Co.	Meakin - SS	2250	118	17	1500	29	16	12	None	20	1976	1120
Louisiana												
*Weeks Island		12800	225	33	3500	27	100	70	5500		1977	679
Mississippi												
Little Creek												
Lincoln Co.	Tusc - SS	10700	248	39	65	23	360	29	5000	40	1974	6310
Texas												
North Cowden												
Ector Co.	Grayburg - Dol	4300	94	35	7	11	409	125	NA	40	1973	37000
Crossett												
Crane Co.	Devonion - Ch	5300	106	44	3	22	110	80	1650	40	1972	1120
South Gillock												
Galveston Co.	Frio - SS	9000	214	38	900	28	61	36	NA	40	1972	5900
Kelly-Snyder												
Scurry Co.	Canyon - LS	6700	132	42	19	8	213	139	1600	51	1972	50000
Mead-Strawn												
Jones Co.	Strawn - SS	4475	135	41	9	9	NA	9	850	NA	1964	NA
Slaughter												
Hockley Co.	San Andres - Dol	4950	105	28	8	10	150	89	1075	34	1976	87000
Twofreds												
Loving Co.	Delaware - SS	4800	104	36	33	20	42	25	1400	40	1974	3000
Wasson												
Yoakum Co.	San Andres - Dol	4890	107	32	2	11	450	111	1250	20	1972	63500
West Virginia												
*Granny's Creek												
Clay Co.	Big Injun - SS	2000	75	45	5	18	40	34	1000	10	1976	3000
*Griffithville												
Lincoln Co.	Berea - SS	2300	83	43	8	11	22	12	1000	10	1976	10000
*Rock Creek												
Roane Co.	Big Injun - SS	2000	73	45	20	22	40	35	1000	10	1976	11200

*Supported by U.S. Dept. of Energy.

(after McRee, 1977 and Perry, 1977)

Kelly-Snyder Field, Texas (USA)

A large CO$_2$ miscible flood is the SACROC Unit of the Kelly-Snyder Field, Scurry County, Texas. Five companies own the majority of the unit which is operated by Chevron Oil Company. There are three project areas of about 11000, 9000 and 13000 acres. These cover about 98% of the field which originally contained about 2.1 BSTB of oil in the Canyon Reef (Penn.) formation. Due to limited CO$_2$ supply, the pattern area was divided into three phase areas of about equal hydrocarbon pore volumes. Injection pattern is a 160 acre inverted 9-spot. CO$_2$ injection for Phase I, II, and III Project areas began in January 1972, March 1973 and November 1976, respectively. The initial design called for injection of an ultimate CO$_2$ slug of 20% hydrocarbon pore volume, but this volume has been reduced partly due to diminishing economic returns.

Most of the project area has not been waterflooded, but water was injected before CO$_2$ to raise reservoir pressure to miscibility pressure. Then, CO$_2$ and water were injected in alternating slugs until the total desired volume of CO$_2$ was injected. Since then water injection has been continuous. Water-CO$_2$ injection ratios have varied between one and three while CO$_2$ slug size has been in the range of 12-15% of hydrocarbon pore volume.

As of year-end 1977, 344 BCF CO$_2$ and 683 MMBbls of water have been injected. The CO$_2$ volume injected into Phase Areas I, II and III was 14.3%, 10.0% and 1.4% hydrocarbon pore volume, respectively.

Premature breakthrough of CO$_2$ forced installation of CO$_2$ removal and reinjection facilities. To maintain control of CO$_2$ production it was necessary to increase the water-CO$_2$ ratio and to conduct a costly zonal injection program.

Carbon dioxide is delivered to SACROC at 2400 psi (a supercritical pressure) through a 220-mile, 16-in pipeline system from gas fields in southwest Texas where it is removed as a contaminant from natural gas. The pipeline at first experienced rather severe operating difficulties, but these problems are now under control.

Injection rates have been in the range of 150-200 MMCFPD. Because this volume is less than that required for full, continuous injection in the 50000 acre unit, the CO_2 supply is alternatively switched from injection wells servicing the 9-spot patterns in one phase area to the injectors servicing like patterns in another phase area. This permits alternating slugs of CO_2 and water to be balanced to give maximum areal and vertical coverage in the reservoir.

Performance to date justifies the conclusion that substantial incremental oil will be recovered as compared to waterflooding in this carbonate reservoir. Incremental oil recovery will be in the range of 5.7-6.7% of OOIP. Vertical and areal reservoir heterogeneities have been the dominate factors in controlling oil recovery. The CO_2 process requires a high level of geologic and engineering sophistication. Large volumes of CO_2 can be transported long distances and injected successfully. Operational problems caused by CO_2 handling, scaling and corrosion are severe, but satisfactory solutions can be devised. The project is considered to be an economic success (Dicharry, 1973; Kane, 1978; and Stalkup, 1978).

BASIC RESERVOIR DATA
SACROC UNIT

Physical Reservoir Features

Formation	Canyon Limestone
Structure	Anticlinal
Approximate depth, ft	6700
Water-oil contact, ft	−4500
Average gross thickness, ft	268

Fluid and Rock Properties

Initial reservoir pressure @ −4300 ft, psig	3122
Reservoir temperature @ −4300 ft, °F	130
Oil gravity, °API	41
Bubble point, psia	1850
Solution GOR, SCF/STB	1000
Oil volume factor, flash @ P_b, rb/STB	1.5
Oil viscosity @ P_b, cp	0.35
Average porosity over gross thickness, %	3.93
Average permeability, md	19.4
Average initial water saturation, %	21.9
Water-oil mobility ratio	0.3
CO_2-oil mobility ratio	8.0

Drive Mechanism

Primary	Fluid Expansion & Solution Gas
Secondary	Waterflood
Tertiary	CO_2-WAG

Statistical Data

Surface area, acres	49900
Well status (12/31/77)	
Wells on artificial lift	648
Flowing wells	11
Shut-in wells	394
Injection wells	305
Original stock-tank oil-in-place, MMSTB	2113
Cumulative oil production (1/1/78), MMSTB	937
Cumulative injection (1/1/78)	
Water, MMB	1809
CO_2, BCF	346
Current production rates (12/77)	
Oil, BOPD	139886
Hydrocarbon gas, MCFPD	136847
CO_2, MCFPD	35864
Water, BPD	411974
Current injection rates (12/77)	
Water, BPD	596191
CO_2, MCFPD	146903

(Kane 1978)

Crossett Field, Texas (USA)

The Crossett Field in Crane and Upton Counties in west Texas was discovered in 1944, and the 28-well field was developed by the mid-1950's. When the field was unitized in 1964 it was renamed the North Cross (Devonian) Unit. The reservoir rock in this unit is composed of microscopic, tripolitic chert grains bound by limestone cement. The oil is in a stratigraphic trap bounded on three sides by decreasing porosity and on the other side by producible water. Net-to-gross pay ratios are 70-100%. Primary production by solution-gas drive would have recovered 12.9% of original oil.

A program of partial pressure maintenance by updip residue-gas injection was instituted, but a better secondary recovery process was needed. Waterflooding is not feasible because of extremely low permeability to water. Carbon dioxide became available when a pipeline from southwest Texas to the SACROC Unit farther north was planned to pass close to the North Cross Unit. A leasing agreement between the North Cross Unit operator (Shell Oil Company) and Canyon Reef Carriers, operator of the SACROC pipeline, was agreed upon whereby the pipeline would carry 20 BCFPD of gas (93% CO_2, 7% methane) to North Cross.

There are six injection wells in the total project area of 1700 acres. Produced gas is reinjected now into the gas cap, but future plans call for reinjection into the present CO_2 injectors. Carbon dioxide injection rates have been in the range of 15-20 MMCFPD with some curtailment of rates partly due to the inability of the pipeline to deliver larger volumes and partly to inability to sustain the targeted injection rate of 20 MMCFPD.

While the production increases at the North Cross Unit have not been as high as original predictions, response has been definite and encouraging. Evidence to date indicates that the displacement mechanism is very complex, probably more complex than predicted by numerical simulation. The results thus far are encouraging and the project is continuing.

The average rock and fluid properties in this unit are as follows:

Rock Properties

Porosity, %	22
Permeability, md	4
Connate-water saturation, %	35
Average thickness, gross, ft	110
Average pay thickness, net ft	80
Average depth to pay, ft	5300

Reservoir Fluid Properties

Crude Oil, °API	44
Initial Reservoir Pressure, psi	2500
Bubble-Point Pressure, psi	2328
Pressure @ Start of CO_2 Injection, psi	1600
Reservoir Temperature, °F	104
Formation Volume Factor @ P_b	1986
Solution Gas-Oil Ratio @ P_b, SCF/STB	1688
Molecular Weight of C_5+	214
Viscosity at 1650 psia, cp	0.42
Miscibility pressure, psia	1650

(Henderson, 1974; McRee, 1977; Pontius and Tham, 1977; and Stalkup, 1978).

Mead-Strawn Field, Texas (USA)

The Mead-Strawn Field pilot project was conducted from 1964 to 1968 by Union Oil Company in a primary-depleted reservoir in west Texas near Abilene. The two Strawn reservoirs, 4500 to 4900 ft, are sandstones of Pennsylvanian age in which porosity and permeability pinch out in all directions and there is no oil-water contact.

Primary production was by solution-gas expansion. The original reservoir pressure of 1807 psig dropped to 115 psig before the CO_2 injection test. After injecting water to raise the reservoir pressure, a 15% PV slug of CO_2 (5000 tons trucked 400 miles) was injected through four wells of a slightly irregular, 33.5 acre, normal five-spot pattern. This was followed by carbonated water and then brine. Water was also injected into 3 wells outside the test area to confine the higher pressure to one end of the field. The reservoir pressure in the CO_2 test area during the entire test period averaged about 2200 psi, but due to low permeability and low productivity of wells, there is doubt that miscibility pressure was maintained at all times at the flood front.

Although there was some water channeling during repressuring, less than 5% of the injected CO_2 was produced during the displacement phase. By comparing recovery from the three pilot test producers with recovery from other wells that had been waterflooded only, it was concluded that about 53-82% more oil was produced by the CO_2 flood. This amounted to an incremental recovery of about 10-12%, or 59% of the oil-in-place in conformal pore volume at the start of the CO_2 injection.

Cores taken with a pressurized core barrel showed 5% residual oil saturation in the CO_2 invaded zone with a lower saturation at a distance of 400 ft versus 60 ft from the input well. This indicated that the partial miscible displacement required more than 60 ft to develop. About 675 cu ft of CO_2 were required to produce a barrel of oil. There was no evidence of CO_2 channeling and less than 10% of the CO_2 injected was ever produced.

Low permeability, particularly around the producing wells, caused the flood life to be extended and the economics of the recovery process to be adversely affected. (Holm, 1971; McRee, 1977; and Stalkup, 1978).

The following are properties of the entire Upper Strawn Sand Reservoir in the Mead Field:

Porosity (average for 113.5 ft of core), %	9.4
Reservoir depth, Upper zone, ft	4475
Pay thickness (average), ft	9
Total acre-feet of net pay (in unit)	3900
Permeability to air (average), md	12
Connate water saturation, %	40
Original reservoir pressure, psig	1807
Reservoir pressure as of Jan. 1, 1964, psig	115
Reservoir temperature, °F	135
Original oil saturation pressure, psig	1500
Original formation volume factor, rb/STB	1.29
Present formation volume factor	1.12
Original solution GOR, SCF/STB	526
Stock-tank oil gravity, °API	41
Oil viscosity at reservoir temperature, cp	1.3

A study of core, electric log and primary production data yielded the following information for the pilot area:

Average thickness, ft	10.3
Average porosity, %	11
Average permeability to air, md	9
Reservoir oil-in-place, % PV	39
Water saturation, % PV	40
Oil recovered by primary, Bbl/acre-ft	
From area around Wells 1 through 5	105
From area around Wells 7, 9, 10	114

(Holm and O'Brien, 1971)

Wasson Field, Texas (USA)

The miscible carbon dioxide process was tested by Atlantic Richfield in the Willard Unit of the large Wasson Field in Yoakum and Gaines Counties of west Texas. The San Andreas, a tight, layered dolomite, produces from around 5000 ft. Secondary recovery in the Willard Unit was successful but left a considerable volume of unrecovered oil.

Two miscible carbon dioxide tests were conducted to investigate this method for field-wide application. A pilot test was begun in August 1972 and a larger test, Phase I, operated from November 1972 to February 1975 when it was discontinued following an accident. The CO_2 which came from a natural-gas processing plant was composed of 89% CO_2, 5% hydrogen sulfide and 6% hydrocarbon gas.

Four wells were used to evaluate the pilot injection test. These wells were located on a straight north-south line: 1) the CO_2 injector; 2) a logging and observation well located 100 ft north from the injector; 3) a fluid sampling and pressure monitoring well located 125 ft from the injector; and 4) a well drilled after CO_2 injection (and cored with a pressure core barrel) 35 ft north of the injector.

At the start of the test, August 1972, the reservoir in the vicinity of the three test wells was depleted by solution-gas-drive. Water was injected until December 1973 to waterflood the area, at which time CO_2 injection was begun with alternate slugs of water in 1:1 reservoir volume ratio. After February 1975 water was injected continuously and in May 1976 the formation was cored with a pressure core barrel.

Variable oil saturations were found in beds swept by CO_2, ranging from 3-30% PV. The operator used a miscible flood simulator to estimate that incremental oil recovery in a full-scale flood could be on the order of 12% of initial oil-in-place in a typical pattern for injection of a 20% hydrocarbon PV of CO_2.

The larger test, Phase I, on normal well spacing, was begun in November 1972 with injection into 8 wells. The test was stopped prematurely in Feburary 1975 after a fatal accident. Cumulative CO_2 injection was less than 5% hydrocarbon PV. This was insufficient to produce any significant improvement in oil recovery.

The results of these tests led to the following conclusions: injectivity performance was good and the reservoir was maintained above miscibility pressure. Profile surveys showed good vertical conformance. There was no evidence of gravity segregation or areal sweep problems. Oil displacement occurred even in watered-out areas. No significant gravity segregation of either initial free hydrocarbon gas or of injected CO_2 was observed. About 97% of the CO_2 injected in Phase I was retained in the reservoir. The 3% that was produced resulted from exceeding the fracture-extension pressure for short periods of time in some CO_2 injection wells. This CO_2 production was effectively controlled by lowering the CO_2 injection rates and pressures to balance with the alternate water injection. The test was terminated before profitability could be determined, but results were encouraging.

The following are average rock and fluid properties of the San Andreas reservoir in the Willard Unit:

Rock Properties
Porosity, %	10
Permeability, md	2
Average gross thickness, ft	180
Average depth to pay, ft	5100

Fluid Properties
Viscosity at bubble point, cp	0.97
Gravity of oil, °API	32
Reservoir temperature, °F	107

(Bilhartz, et al, 1978; Johnston, 1977; McRee, 1977; O.&G. Jour. August 7, 1972 and March 27, 1978; and Stalkup, 1978).

BIBLIOGRAPHY

Beeson, S. M. and Ortloff, G. D.: "Laboratory Investigations of the Water-Driven Carbon Dioxide Process for Oil Recovery," *Trans.*, AIME (1959) 216.

Benham, A. L., Dowden, W. F. and Kunzman, W. J.: "Miscible Fluid Displacement Prediction of Miscibility," *Trans.*, AIME (1960) 219.

Bilhartz, H. L., Jr., Charlson, G. S., Stalkup, F. I. and Miller, C. C.: "A Method for Projecting Full-Scale Performance of CO_2 Flooding in the Willard Unit," SPE 7051 presented at the SPE-AIME Symposium on Enhanced Oil Recovery, Tulsa, OK. (April, 1978).

Blackwell, R. J., Rayne, J. R. and Terry, W. M.: "Factors Influencing the Efficiency of Miscible Displacement," *Trans.*, AIME (1959) 216.

Blackwell, et al: "Recovery of Oil by Displacements With Water-Solvent Mixtures," *Trans.*, AIME (1960) 219.

Blanton, J. R., McCaskill, N. and Herbeck, E. F.: "Performance of a Propane Slug Pilot in a Watered-Out Sand — South Ward Field," *Jour. Pet. Tech.* (Oct., 1970).

Bouck, L. S., Hearn, C. L. and Dohy, G.: "Performance of a Miscible Flood in the Bear Lake Cardium Unit, Pembina Field, Alberta, Canada," *Jour. Pet. Tech.* (June, 1975).

Brummet, W. M., Jr., Emanuel, A. S. and Ronquille, J. D.: "Reservoir Description by Simulation at SACROC — A Case History," *Jour. Pet. Tech.* (Oct., 1976).

Buxton, T. S. and Campbell, J.: "Compressibility Factors for Lean Natural Gas Carbon Dioxide Mixtures at High Presure," *Soc. Pet. Eng. Jour.* (March, 1967).

Byars, C.: "Carbon Dioxide Grows as a Recovery Tool," *O. & G. Jour.* (Aug., 1972).

Camy, J. P. and Emanuel, A. S.: "Effect of Grid Size in the Compositional Simulation of CO_2 Injection," SPE 6894 prepared for the SPE-AIME 52nd Annual Fall Meeting, Denver, CO. (Oct., 1977).

Caudle, B. H. and Dyes, A. B.: "Improving Miscible Displacement by Gas-Water Injection," *Trans.*, AIME (1958) 213.

Craig, F. F., Jr.: "A Current Appraisal of Field Miscible Slug Projects," SPE 2418 presented at the SPE-AIME Mid Continent Section Improved Oil Recovery Symposium, Tulsa, OK. (April, 1969).

Craig, F. F., Jr., Sanderlin, J. L., Moore, D. W. and Geffen, T. M.: "A Laboratory Study of Gravity Segregation in Frontal Drives," *Trans.*, AIME (1957) 210.

Cramer, T. D. and Massey, J. A.: "Carbon Dioxide Injection May Give High SACROC Oil Recovery," *O. & G. Jour.* (April 10, 1972).

Crawford, H. R., Neill, G. H., Bucy, B. J. and Crawford, P. B.: "Carbon Dioxide — A Multipurpose Additive for Effective Well Stimulation," *Jour. Pet. Tech.* (March, 1963).

Crockett, D. H.: "CO_2 Profile Control at SACROC," *Pet. Eng.* (Nov., 1975).

Deming, W. E. and Deming, L. S.: "The Fugacity of CO_2," *Phys. Review.* (July, 1939).

de Nevers, H.: "A Calculation Method for Carbonated Water Flooding," *Soc. Pet. Eng. Jour.* (March, 1964). *Trans.*, AIME (1964) 231.

Dicharry, R. M.: "Landmark CO_2 Injection Project Paying Off at SACROC," *Pet. Eng.* (Dec., 1974).

Dicharry, R. M., Perryman, T. L. and Ronquille, J. D.: "Evaluation and Design of a CO_2 Miscible Flood Project — SACROC Unit, Kelly-Snyder Field," *Jour. Pet. Tech.* (Nov., 1973). *Trans.*, AIME (1973) 255.

Dietz, D. N.: "A Theoretical Approach to the Problem of Encroaching and Bypassing Edge Water," *Proc.*, Royal Academy of Science, The Hague, The Netherlands. (1953).

Donohoe, C. W. and Buchanan, R. D., Jr.: "Economic Evaluation of Cycling Gas-Condensate Reservoirs With Nitrogen," SPE 7494 prepared for the SPE-AIME 43rd Annual Fall Meeting, Houston, TX. (Oct., 1978).

Doscher, T. M.: "CO_2 Flooding Best for Enhanced Recovery in Gulf of Mexico Area," *O. & G. Jour.* (May 2, 1977).

Doscher, T. M. and Wise, F. A.: "Enhanced Crude Oil Recovery Potential — An Estimate," *Jour. Pet. Tech.* (May, 1976).

Dumore, J. M.: "Stability Considerations in Downward Miscible Displacement," *Soc. Pet. Eng. Jour.* (Dec., 1964).

"Enhanced Oil Recovery — An Analysis of the Potential for Enhanced Oil Recovery from Known Fields in the United States — 1976 to 2000," *National Petroleum Council.* (Dec., 1976).

Farouq, Ali, S. M. and Colmont, G. J.: "Appraisal of Micellar Flooding, Carbon Dioxide and Surfactant Flooding Projects," SPE 6624 presented at the SPE-AIME Eastern Regional Meeting, Pittsburgh, PA. (Oct., 1977).

Frey, R. P.: "Operating Practices in the North Cross CO_2 Pilot Project," SPE 7090 presented at the SPE-Course Seminar, Lubbock, TX. (April, 1975).

Frey, R. P.: "West Texas Unit Responds to CO_2 Flood," *O. & G. Jour.* (July 28, 1975).

Gill, D.: "EOR-Carbon Dioxide," *The Western Oil Reporter.* (1978).

George, C. J. and Stiles, L. H.: "Improved Techniques for Evaluating Carbonate Waterfloods in West Texas," SPE 6739 presented at the SPE-AIME 52nd Annual Fall Meeting, Denver, CO. (Oct., 1977).

Graue, D. J. and Blevins, T. R.: "SACROC Tertiary CO_2 Pilot Project," SPE 7090 presented at the SPE-AIME Symposium on Improved Oil Recovery, Tulsa, OK. (April, 1978).

Graue, D. J. and Zana, E.: "Study of Possible CO_2 Flood in the Rangely Field, Colorado," SPE 7060 presented at the SPE-AIME Symposium on Enhanced Oil Recovery, Tulsa, OK. (April, 1978).

Griffith, J. D. and Horne, A. L.: "South Swan Hills Solvent Flood," *Proc.*, Ninth World Petroleum Congress, Tokyo. (1975).

Hansen, P. W.: "A CO₂ Tertiary Recovery Pilot, Little Creek Field, Mississippi," SPE 6747 presented at the SPE-AIME 52nd Annual Fall Meeting, Denver, CO. (Oct., 1977).

Henderson, L. E.: "Carbon Dioxide Miscible Displacement in the North Cross (Devonian) Unit, Project Design and Performance," SPE 4737 presented at the SPE-AIME Symposium on Improved Oil Recovery, Tulsa, OK. (April, 1974).

Henderson, L. E.: "The Use of Numerical Simulations to Design a Carbon Dioxide Miscible Displacement Project," *Jour. Pet. Tech.* (Dec., 1974).

Herbeck, E. F.: "Fundamentals of Tertiary Oil Recovery — Carbon Dioxide Miscible Process," *Pet. Eng.* (May, 1976).

Hing, N. C., Luks, K. D. and Kohn, J. P.: "Phase Equilibria Behavior of the Systems Carbon Dioxide — Normal Eicosane and Carbon Dioxide — Normal Decane — Normal Eicosane," *Jour. Chem. & Eng. Data.* (July, 1973).

Holm, L. W.: "Carbon Dioxide Solvent Flooding for Increased Oil Recovery," *Trans.*, AIME (1959) 216.

Holm, L. W.: "A Comparison of Propane and CO₂ Solvent Flooding Processes," *AIChE Jour.* (1961).

Holm, L. W.: "Status of CO₂ and Hydrocarbon Miscible Oil Recovery Methods," *Jour. Pet. Tech.* (Jan., 1976).

Holm, L. W. and Josendal, V. A.: "Mechanism of Oil Displacement by Carbon Dioxide," *Jour. Pet. Tech.* (Dec., 1974). *Trans.*, AIME, (1974) 257.

Holm, L. W. and O'Brien, L. J.: "Carbon Dioxide Test at the Mead-Strawn Field," *Jour. Pet. Tech.* (April, 1971).

Hutchinson, C. A. and Braun, P. H.: "Phase Relationships of Miscible Displacement in Oil Recovery," *AIChE Jour.* (1961).

Jacobson, H. A.: "Acid Gases and Their Contribution to Miscibility," *Jour. Cdn. Pet. Tech.* (April-June, 1972).

Jacoby, R. H. and Rzasa, M. J.: "Equilibrium Vaporization Ratios for Nitrogen, Methane, Carbon Dioxide, Ethane and Hydrogen Sulphide in Absorber Oil-Natural Gas and Crude Oil-Natural Gas Systems," *Trans.*, AIME (1952) 195.

Johnston, J. W.: "A Review of the Willard (San Andres) Unit CO₂ Injection Project," SPE 6388 presented at the SPE-AIME Permian Basin Conference, Midland, TX. (March, 1977).

Kane, A. V.: "Performance Review of a Large Scale CO₂-WAG Project SACROC Unit — Kelly Snyder Field," SPE 7091 presented at the SPE-AIME Symposium On Improved Oil Recovery, Tulsa, OK. (April, 1978).

Kennedy, J. T. and Thodos, G.: "The Transport Properties of Carbon Dioxide," *AIChE Journal.* (Dec., 1961).

Koch, H. A., Jr., and Slobod, R. L.: "Miscible Slug Process," *Trans.*, AIME (1957) 210.

Koval, E. J.: "A Method for Predicting the Performance of Unstable Miscible Displacement in Heterogeneous Media," *Soc. Pet. Eng. Jour.* (June, 1963). *Trans.*, AIME (1963) 228.

Lewin and Associates, Inc.: "The Potential and Economics of Enhanced Oil Recovery," prepared for Federal Energy Administration, (April, 1976).

Lewin and Associates, Inc.: "Research and Development In Enhanced Oil Recovery — Final Report," Vol. 1, 2, 3 prepared for ERDA. (Dec., 1976).

McRee, B. C.: "CO₂: How It Works, Where It Works," *Pet. Eng.* (Nov., 1977).

Menzie, D. E. and Nielsen, R. F.: "A Study of the Vaporization of Crude Oil by Carbon Dioxide Repressuring," *Jour. Pet. Tech.* (Nov., 1963). *Trans.*, AIME (1963) 228.

Metcalfe, R. S. and Yarborough, L.: "Discussion 'Mechanisms of Oil Displacement By Carbon Dioxide'," *Jour. Pet. Tech.* (Dec., 1974).

Metcalfe, R. S. and Yarborough, L.: "Effect of Phase Equilibria on the CO₂ Displacement Mechanism," SPE 7061 presented at the SPE-AIME Symposium on Improved Oil Recovery, Tulsa, OK. (April, 1978).

Michels, A., Botzen, A. and Schurman, W.: "Viscosity of CO₂ Between 0° and 75° at Pressures Up to 2,000 Atmospheres," *Physica.* (1957).

Miller, M. C.: "Gravity Effects in Miscible Displacement," SPE 1531 presented at the SPE-AIME 41st Annual Fall Meeting, New Orleans, LA. (Oct., 1966).

Moses, P. L. and Wilson, K.: "Phase Equilibrium Considerations in Utilizing Nitrogen for Improved Recovery from Retrograde Condensate Reservoirs," SPE 7493 prepared for the SPE-AIME 53rd Annual Fall Meeting, Houston, TX. (Oct., 1978).

Newton, L. E., Jr. and McClay, R. A.: "Corrosion and Operational Problems, CO₂ Project, SACROC Unit," SPE 6391 prepared for the SPE-AIME Permian Basin Meeting, Midland, TX. (March, 1977).

Noran, D.: "Growth Marks Enhanced Oil Recovery," *O. & G. Jour.* (March 27, 1978).

O. & G. Jour. (Aug. 7, 1972).

O. & G. Jour. (April 5, 1976).

O. & G. Jour. (March 27, 1978).

Pappano, W. W., Sears, J. T. and Parr, W. R.: "Availability and Economics of CO_2 for Enhanced Oil Recovery in Appalachia," ERDA Grant #G0155014. (Aug., 1976).

Pease, R. W., Lohse, E. A. and Lane, R. N.: "Selection and Evaluation of West Virginia Oil Reservoirs as Candidates for Enhanced Oil Recovery Technology," SPE 7636 presented at the SPE-AIME Regional Meeting, Franklin Station, Washington, D.C. (Nov., 1978).

Perkins, T. K., Johnston, O. C. and Hoffman, R. N.: "Mechanics of Viscous Fingering in Miscible Systems," *Trans.,* AIME (1965) 234.

Perry, G. E.: "Weeks Island "S" Sand Reservoir B Gravity Stable Miscible CO_2 Displacement, Iberia Parish," *ERDA Enhanced Oil, Gas Recovery and Improved Drilling Methods,* Vol. 1 - Oil, Tulsa, OK. (Sept., 1977).

Peterson, A. V.: "Optimal Recovery Experiments with N_2 and CO_2," *Pet. Eng.* (Nov., 1978).

Poettmann, F. H.: "Vaporization Characteristics of CO_2 in a Natural Gas-Crude Oil System," *Trans.,* AIME (1951) 192.

Poettmann, F. H. and Katz, D. L.: "CO_2 in a Natural Gas-Condensate System," *I. & E. C.* (1946).

Pontius, S. B. and Tham, M. J.: "North Cross (Devonian) Unit CO_2 Flood — Review of Flood Performance and Numerical Simulation Model," SPE 6390 presented at the SPE-AIME Permian Basin Meeting, Midland, TX. (March, 1977).

Pullman Kellog Co.: "Enhanced Oil Recovery By Carbon Dioxide Injection," *ERDA Enhanced Oil and Gas Recovery and Improved Drilling Methods,* No. 12. (Sept. 30, 1977).

Ramsay, H. J., Jr. and Small, F. R.: "Use of Carbon Dioxide for Water Injectivity Improvement," *Jour. Pet. Tech.* (Jan., 1964).

Rathmell, J. J., Stalkup, F. I. and Hassinger, R. C.: "A Laboratory Investigation of Miscible Displacement by Carbon Dioxide," SPE 3483 presented at the SPE-AIME 46th Annual Fall Meeting, New Orleans, LA. (Oct., 1971).

Reamer, H. H. and Sage, B. H.: "Volumetric and Phase Behavior of the n-Decane-CO_2 Systems," *Jour. Chem. Eng. Data.* (Oct., 1963).

Rosman, A. and Zana, E.: "Experimental Studies of Low Lift Displacement by CO_2 Injection," SPE 6723 prepared for the SPE-AIME 52nd Annual Fall Meeting, Denver, CO. (Oct., 1977).

Rushing, M. D., Thomasson, B. C., Reynolds, B. and Crawford, P. B.: "Miscible Displacement with Nitrogen," *Pet. Eng.* (Nov., 1977).

Sage, B. H. and Lacey, W. N.: "Some Properties of the Lighter Hydrocarbons, Hydrogen Sulfide and Carbon Dioxide," *Monograph on API Research Project 37.* (1955).

Schremp, F. W. and Roberson, G. R.: "Effect of Supercritical CO_2 on Materials of Construction," SPE 4667 presented at the SPE-AIME 48th Annual Fall Meeting, Las Vegas, NV. (Sept. 1973).

Secondary Recovery and Pressure Maintenance Operations in Louisiana — 1973 Report, Department of Conservation, Baton Rouge, Louisiana, Section B.

"Shell Starts Miscible CO_2 Pilot in Mississippi's Little Creek Field," *O. & G. Jour.* (Sept. 2, 1974).

Shelton, J. L. and Schneider, F. N.: "The Effects of Water Injection on Miscible Flooding Methods Using Hydrocarbons and Carbon Dioxide," *Soc. Pet. Eng. Jour.* (June, 1975).

Shelton, J. L. and Yarborough, L.: "Multiple Phase Behavior in Porous Media During CO_2 or Rich-Gas Flooding," *Jour. Pet. Tech.* (Sept., 1977).

Simon, R. and Graue, D. J.: "Generalized Correlations for Predicting Solubility, Swelling and Viscosity Behavior of CO_2-Crude Oil Systems," *Jour. Pet. Tech.* (Jan., 1965). *Trans.,* AIME (1965) 234.

Simon, R., Rosman, A. and Zana, E., "Phase-Behavior Properties of CO_2-Reservoir Oil Systems," *Soc. Pet. Eng. Jour.* (Feb., 1978).

Smith, C. R.: "Mechanics of Secondary Oil Recovery," Reinhold Publishing Corp., New York (1966).

Smith, R. L.: "SACROC Initiates Landmark CO_2 Injection Project," *Pet. Eng.* (Dec., 1971).

Snyder, R. E.: "Shell Starts CO_2 Injection Project in West Texas," *World Oil.* (Sept., 1972).

Stalkup, F. I.: "Carbon Dioxide Miscible Flooding: Past, Present, and Outlook for the Future," SPE 7042 presented at the SPE-AIME Symposium On Improved Oil Recovery, Tulsa, OK. (April, 1978).

Stewart, F. M., Garthwaite, D. L. and Krebill, F. K.: "Pressure Maintenance by Inert Gas Injection in the High Relief Elk Basin Field," *Jour. Pet. Tech.* (March, 1955). *Trans.,* AIME (1955) 204.

Thompson, J. L.: "A Laboratory Study of an Improved Water-Driven LPG Slug Process," *Soc. Pet. Eng. Jour.* (Sept., 1967).

Todd, M. R., and Longstaff, W. J.: "The Development, Testing and Application of a Numerical Simulator for Predicting Miscible Flood Performance," *Jour. Pet. Tech.* (July, 1972). *Trans.,* AIME (1972) 253.

U.S. Bureau of Mines Bulletins 486, 576 and 617.

U.S. Bureau of Mines Information Circulars 8221, 8239, 8241, 8304, 8316, 8356, 8395, 8443, 8475, 8518, 8554, 8607, 8658 and 8684.

Vukalovich, M. P. and Altunin, V. V.: "Thermophysical Properties of Carbon Dioxide," Collet's Publishers, London (1968).

Warner, H. R.: "An Evaluation of Miscible CO_2 Flooding in Waterflooded Sandstone Reservoirs," *Jour. Pet. Tech.* (Oct., 1977).

Welker, J. R. and Dunlop, D. D.: "Physical Properties of Carbonated Oils," *Jour. Pet. Tech.* (Aug., 1963).

West, J. M. and Kroh, C. G., "Carbon Dioxide Compressed to Supercritical Values," *Pet. Eng.* (Dec., 1971).

White, R. W.: "Oil Recovery by CO_2: Past and Future," *Pet. Eng.* (Dec., 1971).

Whorton, L. P., Brownscombe, F. R. and Dyes, A. B.: "Method for Producing Oil by Means of Carbon Dioxide," U.S. Patent No. 2,623,596. (Dec., 1952).

Wilson, K.: "Enhanced-Recovery Inert Gas Processes Compared," *O. & G. Jour.* (July 31, 1978).

Yellig, W. F. and Metcalfe, R. S.: "Determination and Prediction of CO_2 Minimum Miscibility Pressures," SPE 7477 presented at the SPE-AIME Annual Fall Meeting, Houston, TX. (Oct., 1978).

Zane, E. T. and Thomas, G. W.: "Some Effects of Contaminants on Real Gas Flow," *Jour. Pet. Tech.* (Sept., 1970). *Trans.* (1970) 249.

Chapter 9
INERT GAS INJECTION

INTRODUCTION

Natural gas has been injected successfully into oil reservoirs throughout the world for many years. In most of these reservoirs the injected gas maintained pressure and stored gas for future use. In relatively few, the goal was to achieve miscibility between the injected gas and the reservoir oil. When miscibility is achieved, capillary effects disappear and displacement efficiency approaches 100% in the swept zone.

In more recent years, the ever-rising cost and limited supply of natural gas, especially in the United States, has prompted operators to search for a substitute. "Inert gas" (either pure N_2 or a mixture that is predominately N_2) is the substitute receiving the most attention.

Three processes are available to generate inert gas for injection: 1) boiler flue gas; 2) gas engine exhaust; and 3) nitrogen from cryogenic air separation. Many factors will dictate the first choice of a gas generation method, but cost, reliability of supply and control of corrosion will always be primary concerns.

Of course, the cheapest and most readily available gas is air. Air injection was the earliest secondary recovery method. It is still used and experimented with in a few projects. Air injection usually increases production for a short time but soon leads to severe operational problems.

Most of the problems arising from air injection are connected with the oxygen in the air. Oxygen, being highly reactive, causes intolerable problems throughout the surface system and in the reservoir. Some of the major problems are: spontaneous ignition of oil near the injection well, corrosion, formation of explosive mixtures, alteration of reservoir oil and formation of emulsions.

These and other problems have prevented the use of air injection as an accepted enhanced oil recovery method. The method has been seriously studied in the past (Fettke, 1928 and Mills, 1928) and is being used in an interesting experiment in the Willow Draw Field, Wyoming (Wilson, 1977).

TECHNICAL DISCUSSION

Criteria for Application

The phenomena of miscible displacement of reservoir fluids by inert gas occurs only in a narrow range of fluid composition, pressure, volume and temperature. The number of reservoirs that are candidates for this enhanced oil recovery method is therefore limited.

In addition to miscibility, the following factors must be considered:

1. Prevention of pressure drop in the reservoir that would cause fluid loss by retrograde condensation.
2. Reservoir permeability which may be too low for other enhanced oil recovery methods using liquids.
3. Prevention of fluid migration into an original gas cap with attendant loss of crude.
4. Replacement of natural gas in the gas cap and oil zone by inert gas that will remain in the reservoir at abandonment.
5. Enhancement of gravity drainage. N_2 is a light gas and will stay in the gas cap, while CO_2 at high pressure may be heavier than the gas cap fluid and thus have a tendency to migrate down-dip.
6. Crudes with oil gravities of 35° API and higher present the best prospects.
7. Reservoir depth must be great enough to assure that the miscibility pressure can be attained without fracturing.
8. Reservoir heterogeneity and/or fracturing may detract from this method; however, these factors may also be highly unfavorable to other candidate enhanced oil recovery methods.

Phase Behavior and Miscible Displacement

Results of some laboratory studies seem to indicate that N_2 is not a suitable agent for miscible displacement because of the high pressure needed to obtain miscibility. For example, the bubble point pressure of the reservoir oil in the Painter Field, Wyoming was about 4000 psi. The addition of 10% N_2 increased the bubble point pressure to approximately 5100 psi, 20% to more than 9000 psi and 30% to more than 10000 psi, which excluded total single-contact miscibility below 10000 psi (Peterson, 1978).

Although such data appear to make a strong case against an N_2 injection program, these experiments only measured first contact miscibility. Under actual reservoir conditions, additional inert gas will repeatedly contact reservoir fluids. Vaporization of the condensed liquids will occur as they are contacted with fresh N_2.

Experiments were performed on the Painter reservoir oil to determine miscibility pressures by multiple contacts of reservoir oil with injected inert gas. Reservoir oil was displaced by N_2 in a 56-ft long tube with injection pressure of 4280 psi at reservoir temperature. Results showed that miscibility was obtained after multiple contacts, and approximately 90% of the oil was recovered after injection of about 90% PV of N_2 (Peterson, 1978).

The results of such laboratory experiments, confirmed by field experience, can be generalized to state that the composition of the injected gas is relatively unimportant in establishing the miscibility pressure for a given reservoir fluid. This generalization is correct for the compositional range of gases typically encountered in the field (Koch and Hutchinson, 1958).

The process whereby miscibility between N_2 and reservoir oil is obtained by multiple contacts is outlined on the ternary diagram in Figure 1. The corners of the diagram represent 100% N_2, 100% C_7+ and 100% intermediates, C_1-C_6. The point midway on the bottom line is pure N_2 contacting a crude composed of 50% C_1-C_6 and 50% heavy oil, or C_7+. The crude and N_2 reach equilibrium at some temperature and pressure. This equilibrium point, M_1, lies within the 2-phase region of the phase envelope and has some liquid and vapor constituents L_1 and G_1. The gas (G_1) being more mobile than the Liquid (L_1) moves ahead to contact fresh oil. Gas G_1 and crude will then come to equilibrium. For convenience, the equilibrium point of the mixture is on the tie line at M_2 in Figure 1, resulting in the gas G_2 and the liquid L_2.

Gas G_1 has about 35% light hydrocarbons, gas G_2 has about 40% and gas G_3 about 50%. As enriched gas flows ahead in the oil sand, the process repeats until gas moving ahead is miscible with the oil-in-place. No phase boundary exists between oil and gas at the critical point because gas and liquid compositions are equal. Reservoir crude displacement then approaches 100% at the leading edge of the miscible zone.

McNiese (1963) briefly described the process as follows:

> In summary, the high pressure gas process generated a transition zone that establishes miscibility between the flowing gas and oil phases. There is a constant, small attrition of liquid from the transition zone during the displacement. Part of this liquid will remain as a residual phase in the reservoir. The pressure required to establish miscibility between an oil and a gas depends on the pressure and temperature of the system and the properties of the reservoir fluid. The true miscibility pressure is essentially independent of the composition of the displacing gas. Some finite displacement length is required before miscibility can be achieved. This path length will often be substantially longer when the displacing gas contains large amounts of nitrogen. Certain volatile reservoir fluids, however, will achieve miscibility nearly as easily with nitrogen or flue gas as they will with a hydrocarbon gas.

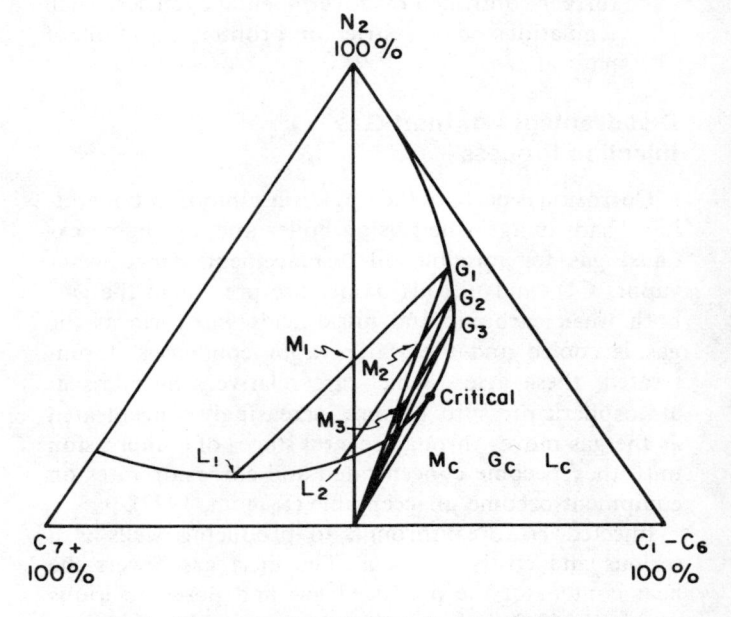

Fig. 1. Nitrogen mixes with oil to form a miscible phase. (after Rushing et al, 1977).

CONCLUSIONS

Advantages of Inert Gas Injection Process

A primary advantage of inert gas over natural gas is that combustion increases the volume of inert gas by a factor of 5-10. Quite often this larger volume of natural gas would not be available for injection, regardless of cost.

Sale of released natural gas may be economically advantageous. For example, installation of a flue-gas generation plant in the Block 31 Field in Texas resulted in: not having to purchase 30-40 MMCFPD of natural gas, permitted the sale of 20-30 MMCFPD of produced gas and assured a good supply of displacement gas for the reservoir (Caraway and Lowrey, 1975).

Inert gas is almost as good a displacement agent as is lean natural gas. If the displacement is conducted at a sufficiently high pressure, essentially all the oil is recovered by miscible displacement. This occurs even though the injected gas is not miscible with the oil at first contact at the injection pressure (McNeese, 1963 and Peterson, 1978).

Other advantages of inert gas are:
1. If a gas cap is present, gas injection will prevent oil encroachment into the gas cap. Residual gas at abandonment will be inert gas rather than saleable natural gas.
2. Gas injection will give higher recoveries as compared to water drive in many reservoirs with very low permeability.
3. The reliability of the supply may be unstable since the future supply and price of natural gas may be further controlled by governmental agencies. Such regulations may restrict or prohibit injection of natural gas.

Disadvantages of Inert Gas Injection Process

Corrosion is perhaps the single most important operating disadvantage when using boiler and/or engine exhaust gas for miscible oil displacement. Since water vapor, CO_2 and nitrous oxides are present in the gas, both weak carbonic and nitric acids will form as the gas is cooled and the water vapor condenses. If untreated, these acids, which are relatively harmless at atmospheric pressure, become increasingly concentrated as the gas moves through several stages of compression until they become concentrated and corrosion rates on equipment become unacceptable (Kuehm, 1977).

Injected N_2 breakthrough to producing wells is a serious and costly problem. The inert gas lowers the heat content of the produced gas and poses a serious problem whether the produced gas is sold or used for fuel in the field. Various schemes have been devised to solve this problem such as well re-completion and closing in high GOR wells. Regardless of the method used, the economic burden increases as the project matures.

A large front-end investment is required for inert gas projects. All these disadvantages and costs must be carefully weighed against the advantages and the increased production expected.

FIELD CASE HISTORIES

Block 31 Field, Texas (USA)

The world's first large-scale miscible displacement project by high-pressure gas injection has produced 130 MMBO as of 1975, almost double the original estimated primary recovery of 69 MMBO. This has occurred at the University Block 31 Field in Crane County, Texas (Hardy and Robertson, 1975).

The field was discovered by Atlantic Richfield in 1945. Partial pressure maintenance by gas injection began in 1949 and full-scale gas injection for miscible displacement, using lean hydrocarbon gas, in 1952. The flue-gas operation began in 1966. Waterflooding was impractical because of the low permeability of the reservoir.

The Devonian formation is crystalline limestone with average porosity of 15% and average permeability of 1 md. Gross thickness is 1000 ft. There are three reservoirs in the Devonian column of which the Middle is the major producer. Reservoir temperature, not considered a critical factor, is 140° F. Block 31 crude has a saturation pressure of 2764 psi and is 48° API gravity.

Miscibility requires high injection pressures and volumes, which mean large horsepower. Block 31 has 53500 bhp developed by 29 compressors which inject 134 MMCFPD of gas to maintain 3500 psi pressure at the miscible bank. To accomplish this, compressors discharge at 4200 psi. In some less permeable zones the pressure is boosted to 5000 psi. The formation depth of 8500 ft permits high pressure injection without fracturing.

The flue gas operation started in 1966. This operation has eliminated the need to purchase 30-40 MMCFPD of natural gas, permitted the sale of 20-30 MMCFPD of produced gas and assured a good supply of displacement gas for the reservoir. Feedstock for the flue-gas plant is adsorber residue gas and air. By burning a portion of the residue gas in a controlled atmosphere, the volume can be increased 9 to 11 times (Caraway and Lowrey, 1975).

The operation has not been without problems. Injected flue gas has broken through into oil producing wells. Because it contains CO_2 and N_2, flue gas lowers the quality of the produced hydrocarbon gas, some of which is used for fuel. Four different gathering systems, operating at pressures from 40-1200 psi, have been installed because of gas breakthrough. Corrosion due to oxygen, carbon dioxide and nitrous oxide would be severe if not controlled. Some injection well plugging has occurred, especially during periods of excessive compressor maintenance. Plugging was attributed to several factors. One was related to compressor lubricants entrained in the gas stream, another associated with solid compounds which formed when injection was switched from flue gas to sour hydrocarbon gas (Caraway and Lowrey, 1975 and Hardy and Robertson, 1975).

Block 31 is considered to be a classic example of miscible displacement beginning with lean hydrocarbon gas and later changing to flue gas injection.

Hawkins Field, Texas (USA)

The Hawkins Field, Wood County, Texas, produces 24° API oil from the Woodbine sandstone from a depth of approximately 4350 ft. The oil in the high porosity, high permeability (28% and 3400 md) reservoir is trapped between a strong water drive below and a gas cap above. The field covers 10590 acres and produces 85000 BOPD from 351 wells, of which 28000 BOPD is attributed to enhanced oil recovery (O & G Jour., 1978).

Extensive reservoir studies prior to field unitization indicated a need to use gas drive rather than the natural water drive to maximize field recovery and to minimize or to prevent residual oil loss. This residual oil loss results from oil movement into the gas cap because of the gradual decline in the gas cap pressure. Sufficient natural gas was not available either in the Hawkins Field or elsewhere on an economical basis, so a plan was devised to use steam boiler exhaust gas for pressure maintenance injection needs. Steam produced in the inert gas generators is used to drive the steam turbines that in turn drive the compressors which compress the boiler exhaust gas (Kuehm, 1977).

The inert plant is made up of three identical trains operating in parallel. Each train has two inert gas generators (steam boilers), two catalytic reactors, a direct contact cooler, a 20000 bhp compressor train driven by a steam turbine and a gas dry bed dehydration unit. Each train is designed to produce 44 MMCFPD of inert gas. Considering normal downtime, annual average injection rates were targeted for 120 MMCFPD from the plant (Kuehm, 1977).

The expected cost of the project will be in excess of $70 million. The plant was expected to be operational in 1978 and it will be a few years before the economic and technical success of the project is known.

BIBLIOGRAPHY

Baker, L. E.: "Effects of Dispersion and Dead-End Pore Volume in Miscible Flooding," *Soc. Pet. Eng. Jour.* (June, 1977).

Barstow, W. F.: "Engine Exhaust Boosts Oil Recovery," *O. & G. Jour.* (March 26, 1973).

Barstow, W. F.: "Inert Gas Systems for Secondary Recovery," 76-PET-86 prepared for *Am. Soc. Mechanical Eng.* (Sept., 1976).

Barstow, W. F. and Hendricks, H.: "Engine Exhaust Inert Gas System for Secondary Oil Recovery," 75-DPG-3 prepared for *Am. Soc. Mechanical Eng.* (April, 1975).

Barstow, W. F. and Watt, G. W.: "Fifteen Years of Progress in Catalytic Treating of Exhaust Gas," SPE 5347 prepared for the SPE-AIME Rocky Mountain Regional Meeting, (April, 1975).

Beecher, C. E.: "Repressing During Early Stages of Development," *SPE*. Tulsa, OK. (Oct., 1928).

Bell, A. M.: "Summary of Repressuring Experiments in California Fields," *SPE*. New York. (Feb., 1928).

Bowman, R. W., Dunlop, A. K. and Tralmer, J. P.: "CO/CO_2 Cracking in Inert Gas-Miscible Flooding," *Materials Performance*. (April, 1977).

Breston, J. N. et al: "Conditioning Engine Exhaust Gas for Injection into Oil Reservoirs," *Producers Monthly*. (April, 1953).

Brown, A., Harrison, J. T. and Wilkins, R.: "Transgranular Stress Corrosion Cracking of Ferric Steels," *Corrosion Sci.*, No. 10. (1970).

Camy, J. P. and Emanuel, A. S.: "Effect of Grid Size in the Compositional Stimulation of CO_2 Injection," SPE 6894 prepared for the SPE-AIME 52nd Annual Fall Meeting, Denver, CO. (Oct., 1977).

Caraway, G. E., Jr. and Lowrey, L. L.: "Flue Gas Generation Problems, Solutions and Costs — Block 31 Field," SPE 2631 prepared for the SPE-AIME 44th Annual Fall Meeting, Denver, CO. (Sept., 1969).

Caraway, G. E., Jr. and Lowrey, L. L.: "Generating Flue Gas for Injection Releases Sales Gas," *O. & G. Jour.* (July 28, 1975).

Donohoe, C. W. and Buchanan, R. D., Jr.: "Economic Evaluation of Cycling Gas-Condensate Reservoirs with Nitrogen," SPE 7494 prepared for the SPE-AIME 43rd Annual Fall Meeting, Houston, TX. (Oct., 1978).

Engineers Data Book, Natural Gas Processors Suppliers Association, 8th ed. Tulsa, OK. (1966).

Espach, R. M. and Fry, J.: "Variable Characteristics of the Oil in the Tensleep Sandstone Reservoir, Elk Basin Field," *Trans.,* AIME (1951) 192.

Farouq Ali, S. M. and Colmont, G. J.: "Appraisal of Micellar, CO_2 and Surfactant Flooding Projects," SPE 6624 prepared for the SPE-AIME Eastern Regional Meeting, Pittsburgh, PA. (Oct., 1977).

Fettke, C. R.: "Ten Years' Application of Compressed Air at Hamilton Corners, Pa., with Core Samples of the Producing Sand," *SPE*. New York. (Feb., 1928).

Foran, E. V.: "Effect of Repressuring Producing Sands During the Flush Stage of Production," *SPE*. New York. (Feb., 1928).

George, C. J. and Stiles, L. H.: "Improved Techniques for Evaluating Carbonate Waterfloods in West Texas," SPE 6739 prepared for the SPE-AIME 52nd Annual Fall Meeting, Denver, CO. (Oct., 1977).

Hansen, P. W.: "A CO_2 Tertiary Recovery Pilot, Little Creek Field, Mississippi," SPE 6747 prepared for the SPE-AIME 52nd Annual Fall Meeting, Denver, CO. (Oct., 1977).

Hardy, J. H. and Robertson, N.: "Miscible Displacement by High-Pressure Gas at Block 31," *Pet. Eng.* (Nov., 1975).

Hutchinson, C. A., Jr. and Braun, P. H.: "Phase Relations of Miscible Displacement in Oil Recovery," AIChE 28 presented at the Joint Symposium on Fundamental Concepts of Miscible Displacement: Part 1, 52nd Annual Meeting of AIChE, San Francisco, CA. (Dec., 1959).

Joplin, J. L. and Joplin, J. F.: "Cryogenic Nitrogen Plants Ready for Offshore," *Pet. Eng.* (May, 1978).

Journal of Petroleum Technology Forum: "Nitrogen May Be Used for Miscible Displacement in Oil Reservoirs," *Jour. Pet. Tech.* (Dec., 1978).

Kaira, M. et al: "The Equilibrium Phase Properties of the Nitrogen-n-Pentane System," *Jour. of Chem. and Eng. Data*, Vol. 22, No. 2. (1977).

Kastrop, J. E.: "Converted Engine Exhaust Supplies Make-Up Gas," *Pet. Eng.* (May, 1961).

Koch, M. A., Jr. and Hutchinson, C. A., Jr.: "Miscible Displacement of Reservoir Oil Using Flue Gas," *Trans.*, AIME (1958) 312.

Kowaka, M. and Nagata, S.: "Transgranular Stress Corrosion Cracking of Mild Steels and Low Alloy Steels in the $H_2O/CO/CO_3$ Systems," *Corrosion-NACE*. (Dec., 1968).

Kuehm, H. G.: "Hawkins Inert Gas Plant: Design and Early Operation," SPE 6793 prepared for the SPE-AIME 52nd Annual Fall Meeting, Denver, CO. (Oct., 1977).

McNeese, C. R.: "The High Pressure Gas Process and the Use of Flue Gas," American Chemical Society Symposium, Los Angeles, CA. (March-April, 1963).

McRee, B. C.: "CO_2: How It Works, Where It Works," *Pet. Eng.* (Nov., 1977).

Mills, R. Van: "Oil Recovery Investigation of the Petroleum Experiment Station of the U.S. Bureau of Mines," Tulsa, OK. (Oct., 1928).

Moses, P. L. and Wilson, K.: "Phase Equilibrium Considerations in Utilizing Nitrogen for Improved Recovery From Retrograde Condensate Reservoirs," SPE 7493 prepared for the SPE-AIME 53rd Annual Fall Meeting, Houston, TX. (Oct., 1978).

Oil and Gas Journal: "Growth Marks Enhanced Oil Recovery," *O. & G. Jour.* (March 27, 1978).

Perry, G. E.: "Weeks Island "S" Sand Reservoir B Gravity Stable Miscible CO_2 Displacement, Iberia Parish, Louisiana," *ERDA Enhanced Oil, Gas Recovery and Improved Drilling Methods*, Vol. 1 - Oil, Tulsa, OK. (Sept., 1977).

Peterson, A. V.: "Optimal Recovery Experiments with N_2 and CO_2," *Pet. Eng.* (Nov., 1978).

Porter, R. E. and Cover, A. E.: "The Carbon Dioxide Supply Situation for Miscible Flooding Operations," *ERDA Enhanced Oil, Gas Recovery and Improved Drilling Methods*, Vol. 1 - Oil, Tulsa, OK. (Sept., 1977).

Rosman, A. and Zana, E.: "Experimental Studies of Low IFT Displacement by CO_2 Injection," SPE 6723 prepared for the SPE-AIME 52nd Annual Fall Meeting, Denver, CO. (Oct., 1977).

Rushing, M. D., Thomasson, B. C., Reynolds, B. and Crawford, P. B.: "High Pressure Air Injection," *Pet. Eng.* (Nov., 1976).

Rushing, M. D., Thomasson, B. C., Reynolds, B. and Crawford, P. B.: "High Pressure Nitrogen or Air May Be Used for Miscible Displacement in Deep, Hot Oil Reservoirs," SPE 6445 prepared for the SPE-AIME Deep Drilling and Production Symposium, Amarillo, TX. (April, 1977).

Rushing, M. D., Thomasson, B. C., Reynolds, B. and Crawford, P. B.: "Miscible Displacement by High Pressure Nitrogen Injection," Southwestern Petroleum Course, Texas University. (1977).

Rushing, M. D., Thomasson, B. C., Reynolds, B. and Crawford, P. B.: "Miscible Displacement with Nitrogen," *Pet. Eng.* (Nov., 1977).

SanFilippo, G. P. and Guckert, L. G.: "Development of a Pilot Carbon Dioxide Flood in the Rock Creek-Big Injun Field, Roane County, Texas," SPE 6626 prepared for the SPE-AIME Eastern Regional Meeting, Pittsburgh, PA. (Oct., 1977).

Slobod, R. L. and Koch, M. A., Jr.: "High Pressure Gas Injection — Mechanism of Recovery Increase," API Drilling and Production Practice. (1953).

Stewart, F. M., Garthwaite, D. L. and Krebill, F. K.: "Pressure Maintenance by Inert Gas Injection in the High Relief Elk Basin Field," *Trans.,* AIME (1955) 204.

Whorton, L. P. and Kieschnick, W. F., Jr.: "A Preliminary Report on Oil Recovery by High Pressure Gas Injection," *API Drilling and Production Practice.* (1950).

Wilson, Quentin T.: "Willow Draw Field Attic Air Injection Project, Park County, Wyoming," *ERDA Enhanced Oil, Gas Recovery and Improved Drilling Methods,* Vol. 1 - Oil, Tulsa, OK. (Sept., 1977).

Part IV

CONCLUSIONS

Chapter 10
STATE OF THE ART

The present status of enhanced oil recovery in the United States can best be illustrated by the following estimates of daily production attributed to enhanced processes:

Methods	Estimated Production BOPD
Steam flood (and stimulation)	250000
In Situ Combustion	10000
Chemical	3000
Miscible Hydrocarbon	110000
	373000

Estimates have also been made for the present world total enhanced oil recovery production as follows:

Method	Estimated Production BOPD
Steam flood (and stimulation)	405000
In situ combustion	55000
Others	221000
	681000

Thermal Processes

Although the potential for enhanced oil recovery is immense, the application of the various techniques is in its infancy. Until the technologies are well understood and the economic return and risks are acceptable, the growth of enhanced oil recovery application will remain slow. Notable exceptions are the steam drive and steam stimulation thermal processes which are now considered proven enhanced recovery techniques. Thermal methods account for about 70% of the world's enhanced oil recovery production. Their application to reservoirs having low gravity, high viscosity and high porosity have become almost routine. There is every indication that this segment of enhanced oil technology will continue to grow.

Steam drive and steam stimulation are not technically confined to heavy oils, but only a limited amount of field testing has been conducted in lower viscosity crude reservoirs.

In situ combustion offers attractive economics when applied to reservoirs containing high oil saturation ($\approx$50%), and a fuel content that will support combustion at a relatively low air-oil ratio (<15-20 MCF/bbl). Although laboratory testing will provide some basic parameters, a field application is the primary evaluation tool. There is probably very little laboratory research that can be done to improve this process. Its growth in use will be related directly to operators that are willing to risk the required capital for a field project based on the profit potential estimated from a thorough engineering and geological evaluation.

Chemical Processes

The chemical processes for recovering additional oil account for less than 1% of the enhanced oil recovered in the United States. Although these processes have the best chance for recovering oil from reservoirs that have been successfully waterflooded (but still contain considerable oil), development has been slow because of associated high costs, high risk and complicated technology. Although the chemical cost for polymers and caustic solutions are relatively low compared to surfactant-polymer processes, the incremental oil recovery is also low. In the United States active projects from 1970 to 1977 have increased from 5 to 22 for surfactant-polymer, 0 to 3 for caustic and 14 to 21 for polymer. Projects in the planning stage will add to the active projects in the near future.

Chemical flooding research is taking place in many laboratories. The chemical processes will become more widely accepted as more is learned about the mechanism of oil displacement and processes are improved and proven by field testing.

All of these chemical processes are complicated and not well understood. The chemicals react with the in-

CONCLUSIONS

place fluids and rock and their effectiveness decreases. Also, the heterogeneities of the reservoir itself add to the complexity.

The state of the art of the caustic and surfactant-polymer processes is still in the research stage. No economic field projects have been reported. The polymer flooding process has been field tested extensively and can be classified as proven. Improvements in polymer chemicals will continue to take place by new developments. These improvements will not only benefit the polymer flooding process but also the caustic and surfactant-polymer processes because of their use as mobility control agents.

Miscible Displacement Processes

Although CO_2 miscible projects have been encouraging, sufficient information is not available to determine economic potential. It has been estimated that about 4-10 MCF of CO_2 will be required per incremental barrel of oil recovered. However, until further testing is carried out, this requirement will be just an estimate. In the United States the number of active CO_2 projects has risen from one in 1970 to 14 in 1978.

Projections from a number of reports have indicated that CO_2 miscible flooding could potentially recover as much as 40% of the total oil to be recovered by all enhanced oil recovery methods. In order to produce these billions of barrels of oil a tremendous volume of CO_2 must be available at reasonable costs.

The number of active miscible hydrocarbon displacement projects in the Western Hemisphere peaked at 21 in 1957 and has declined to 15 in 1977. The main problem associated with all miscible hydrocarbon processes is the high cost of the injected hydrocarbons. Because of high costs and consumption demands for natural gas, propane and other hydrocarbons used in the process, miscible hydrocarbon displacement will probably not be utilized to any great degree in the future.

Nitrogen, flue gas and air are other gases that have been used for miscible displacement. Active inert and flue gas projects increased from one project in 1970 to 6 projects in 1977. It is believed that future use of this technique will be limited because the method has application over such a very narrow range of reservoir conditions.

Application of any enhanced oil recovery technique will require a detailed reservoir description incorporating both engineering and geological considerations. Field production data must be analyzed to determine the oil remaining as the target for enhanced oil recovery application. A thorough understanding of the well conditions is a necessity to insure proper control of injected and produced fluids. The various enhanced oil recovery techniques should be screened for possible field application. Appropriate laboratory programs should be designed to focus on the most profitable processes. This data might also be incorporated in a mathematical reservoir model to estimate field performance. Case histories of similar projects should also be researched to provide response estimates.

On the basis of all this work a field pilot project should be designed to provide results that can be extrapolated to a field-wide application.

In the final analysis a great deal of research, professional effort and creative thought must precede the actual application of any enhanced oil recovery process to assure the best possible chances for economic success.